DATE DUE

DEMCO, INC. 38-2931

that had proven so vulnerable during the catastrophic events of the 1930s and 1940s. The civil rights struggles and urban uprisings in the United States (the murders of Malcolm X, Martin Luther King and the frontal attack upon the Black Panthers which culminated in the state assassination of Fred Hampton in Chicago) also called for serious re-evaluations in thought and political practice.

It seemed important to engage with Marx for two compelling reasons: first, to understand why it was that a doctrine so denigrated and despised within official circles in the English-speaking world could have such widespread appeal to those actively struggling for emancipation everywhere else; secondly, to see if a reading of Marx could help ground a critical theory of society to embrace and interpret the social conflicts that culminated in high political drama (bordering on cultural and political revolution) in the climacteric years of 1967–73.

My own work on these topics originated as part of a general effort to come to terms with these questions during the early 1970s. It was, of course, helpful to discover that the embers of Marxist scholarship were still glowing strongly in certain quarters (the work of Paul Baran and Paul Sweezy shone out in the United States and of Maurice Dobb, E. P. Thompson and Raymond Williams in Britain) and that various currents of Marxist thought remained strong in Europe. At first attention had to be paid to recuperating these achievements while developing fresh insights from the classical Marxian texts appropriate to the times. Marx's writings subsequently became more widely studied and commonly accepted, but later still were seen increasingly as repressive dogma or as anachronistic and reactionary: it was then important to show that there was life in his ideas when they were adapted and extended to deal with unfamiliar circumstances.

The specific angle of my work was, however, somewhat unusual since it was almost as uncommon for those working in the Marxist tradition to pay any mind to questions of geography (or of urbanization, except as a historical phenomena) as it was for geographers to consider Marxian theory as a possible foundation for their thinking. If anything, the radical tradition of geography (which was never very strong) harked back to the anarchists, particularly those at the end of the nineteenth century when geographer-anarchists like Peter Kropotkin and Elisée Reclus were prominent thinkers and activists. There is much of value in that tradition. It was, for example, much more sensitive to issues of environment and urban organization (albeit critically) than has generally been the case within Marxism. But the influence of such thinkers was either strictly circumscribed or was transformed, through the influence of town planners like Patrick Geddes, into a communitarianism framed in gentle and

Spaces of Capital

Towards a Critical Geography

David Harvey

Routledge

New York

Published in 2001 by
Routledge
29 West 35th Street
New York, NY 10001-2299

Routledge is an imprint of the Taylor & Francis Group.

© David Harvey, 2001

Published by arrangement with
Edinburgh University Press, Edinburgh, in 2001.

Typeset in Ehrhardt by Pioneer Associates, Perthshire
Printed and bound in Great Britain by
The Cromwell Press, Trowbridge, Wiltshire

Library of Congress Cataloging-in-Publication Data

Harvey, David, 1935–
    Spaces of capital : towards a critical geography / David Harvey.
        p. cm.
    Includes bibliographical references and index.
    ISBN 0-415-93240-8 – ISBN 0-415-93241-6 (pbk.)
    1. Communism and geography. 2. Urban geography. 3. Urban eco-
nomics. 4. Marxian economics. 5. Space in economics. 6. Capitalism.
I. Title.

HX550.G45 H35 2001
330.12′2–dc21   2001031921

Contents

Part 2
THE CAPITALIST PRODUCTION OF SPACE

Preface

No one who aspires to change the way we think about and understand the world can do so under circumstances of their own choosing. Everyone has to take advantage of the raw materials of the intellect at hand. Each must also try to combat the presumptions, prejudices and political predilections that at any time constrain thinking in ways which may at best be understood as repressive tolerance and at worst as merely repressive. The essays collected here, written over some thirty years, record my attempts to change ways of thought in the discipline of geography (until recently my institutional home within the increasingly dysfunctional disciplinary division of knowledge characteristic of the academy), in cognate areas (such as urban studies) and among the public at large. They also reflect the changing circumstances of knowledge production within the English-speaking world during those years.

The onset of the Cold War and the devastations wrought on freedom of thought by McCarthyism during the 1950s, aided and abetted by disturbing revelations about the excesses of Stalinism in the Soviet Union, made it extremely difficult during the 1950s and early 1960s to treat Marx's writings as serious raw materials for shaping new understandings and modes of political action. Indeed, as the case of Owen Lattimore (see Chapter 5) so clearly shows, it was dangerous in the United States to voice any dissident opinion (no matter whether grounded in Marxism or not) which did not fit exactly into the mould demanded by US foreign policy. This policy was dominated by the doctrine of containment of Soviet influence and the co-option or outright suppression of all political movements that sought a socialist rather than a capitalist path to economic betterment. Yet by the mid-1960s it was clear to many that prevailing systems of knowledge were failing badly when it came to understanding the numerous revolutionary thrusts and struggles over decolonization (often inspired by Marxist thought) occurring throughout much of Africa, Latin America and Asia. As the Vietnam War evolved, so the US was increasingly seen as not defending freedom and liberty but working to establish a new kind of imperialism in support of the US-based capitalist system that had proven so vulnerable during the catastrophic events of the 1930s and 1940s. The civil rights struggles and urban uprisings in the United States (the murders of Malcolm X, Martin Luther King and the frontal attack upon the Black Panthers which culminated in the state assassination of Fred Hampton in Chicago) also called for serious re-evaluations in thought and political practice.

It seemed important to engage with Marx for two compelling reasons: first, to understand why it was that a doctrine so denigrated and despised within official circles in the English-speaking world could have such widespread appeal to those actively struggling for emancipation everywhere else; secondly, to see if a reading of Marx could help ground a critical theory of society to embrace and interpret the social conflicts that culminated in high political drama (bordering on cultural and political revolution) in the climacteric years of 1967–73.

My own work on these topics originated as part of a general effort to come to terms with these questions during the early 1970s. It was, of course, helpful to discover that the embers of Marxist scholarship were still glowing strongly in certain quarters (the work of Paul Baran and Paul Sweezy shone out in the United States and of Maurice Dobb, E. P. Thompson and Raymond Williams in Britain) and that various currents of Marxist thought remained strong in Europe. At first attention had to be paid to recuperating these achievements while developing fresh insights from the classical Marxian texts appropriate to the times. Marx's writings subsequently became more widely studied and commonly accepted, but later still were seen increasingly as repressive dogma or as anachronistic and reactionary: it was then important to show that there was life in his ideas when they were adapted and extended to deal with unfamiliar circumstances.

The specific angle of my work was, however, somewhat unusual since it was almost as uncommon for those working in the Marxist tradition to pay any mind to questions of geography (or of urbanization, except as a historical phenomena) as it was for geographers to consider Marxian theory as a possible foundation for their thinking. If anything, the radical tradition of geography (which was never very strong) harked back to the anarchists, particularly those at the end of the nineteenth century when geographer-anarchists like Peter Kropotkin and Elisée Reclus were prominent thinkers and activists. There is much of value in that tradition. It was, for example, much more sensitive to issues of environment and urban organization (albeit critically) than has generally been the case within Marxism. But the influence of such thinkers was either strictly circumscribed or was transformed, through the influence of town planners like Patrick Geddes, into a communitarianism framed in gentle and acceptable opposition to what Lewis Mumford, for example, considered the dystopian trajectory of technological change under capitalism. Part of the radical geography movement in the late 1960s was dedicated to revitalizing the anarchist tradition, while geographers with strong sympathies with, say, national liberation and anti-imperialist revolutionary movements wrote in a more directly historical-materialist and experiential mode and eschewed Marxian abstractions. Geographers of this sort (Lattimore and Keith Buchanan come to mind) were marginalized, often treated like pariahs, within their discipline. Radical geographers sought nevertheless both to uphold this tradition (in the face of fierce opposition) but also, as in the radical geography journal *Antipode* (founded in 1968) to underpin it by appeal to the texts of Marx and Engels, Lenin, Luxemburg, Lukacs, and the like.

The initial essays in Part Two of this collection, all published in *Antipode*, were part of that collective effort. There was very little written on the geography of capital accumulation, the production of space and of uneven geographical development from a Marxist perspective. Marx, though he promised a volume of *Capital* dedicated to the formation of the state and the world market, never completed his project. I therefore set out to do a comprehensive reading of all of his texts to see what he might have said on these matters had he lived to complete his argument. There are two ways to conduct such a reading. One is to treat Marx as the 'master thinker' whose statements bear the imprimatur of absolute truth no matter what. The second, which I much prefer, is to treat his statements as tentative suggestions and rough ideas that need to be consolidated into a more consistent theoretical form of argument that respects the dialectical spirit rather than the verbal niceties of his largely unpublished studies, notes and letters. Read in this second mode, I found in Marx a fertile basis for a whole range of subsequent studies (some of which appear in this volume) as well as later books such as *The Limits to Capital* (1982), *The Condition of Postmodernity* (1989), and *Spaces of Hope* (2000).

But the learning of Marx's method also opened up all sorts of other avenues for intellectual work and political commentary on matters as diverse as the politically-contested nature of geographical knowledges, environmental issues, local political-economic developments, and the general relation between geographical knowledge and social and political theory. A whole field of endeavor emerged to understand the uses of geographical knowledges (however defined) by political power. In parallel this indicated a pressing need to define a critical geography (and a critical urban theory) that could 'deconstruct' (to use the current jargon) how certain kinds of knowledge, seemingly 'neutral,' or 'natural' or even 'obvious' could in fact be an instrumental means to preserve political power.

The essays assembled in Part One hover around this question. Enough partial evidence is here assembled to make such a connection more than merely plausible even though a satisfactory systematic presentation of the idea has yet to see the light of day. I consider these essays as studies preparatory to a broader project, deserving the deepest consideration, on the role of geographical knowledges in the perpetuation of political-economic power structures and in transforming by opposition the political-economic order.

Over the thirty years of writing on these topics I have had the good fortune to be engaged with many scholars and activists who have risked a great deal to develop alternative views to the standard technocratic evasions – bordering on capitalist apologetics – that dominate geography and the social sciences more generally. I owe an immense debt to these many others who are simply too numerous to mention (I trust they know who they are). But the untimely death of one long-standing comrade, Jim Blaut, leads me to dedicate this book to his memory. His recently published *Eight Eurocentric Historians* is a courageous example of the kind of salutary critical work I have in mind. It is my fervent hope that the embers which glow brightly in Jim's work as well, I hope, as in my own may be used by a younger generation to light a fire in critical geography that will remain burning until we have constructed a more just, equitable, ecologically sane, and open society than we have experienced heretofore.

David Harvey
New York, April 2001

Sources

Chapter 10 was first delivered to the Conference on Model Cities in Singapore in April 1999 and published by the Urban Redevelopment Authority of Singapore as conference proceedings, *Model Cities: Urban Best Practices in 2000*; the revised version, published here, was presented to the Man and City: Towards a more Human and Sustainable Development Conference in Naples, September 2000. Chapter 11 was first presented to the conference on Social Sciences at the Millennium sponsored by Hong Kong Baptist University in June 2000; the revised version, published here, was presented to the twenty-ninth International Geographical Congress in Seoul in August 2000. Chapter 18 was prepared for the Conference on Global and Local, held at the Tate Modern in London, February, 2001.

The sources for the remaining chapters are as follows:

Chapter 1:   an interview published in *New Left Review*, August 2000
Chapter 2:   *Transactions of the Institute of British Geographers*, 1974
Chapter 3:   *Economic Geography*, 1974
Chapter 4:   *Comparative Urban Research*, 1978
Chapter 5:   *Antipode*, 1983
Chapter 6:   *The Professional Geographer*, 1984
Chapter 7:   *New Perspectives Quarterly*, 1992
Chapter 8:   *The Baltimore Book: New Views on Urban History*, 1992
Chapter 9:   *Social Text*, 1995
Chapter 12: *Antipode*, 1975
Chapter 13: *Antipode*, 1976
Chapter 14: *Antipode*, 1981
Chapter 15: *Social Relations and Spatial Structures*, edited by
                 Derek Gregory and John Urry, 1985
Chapter 16: *Geografiska Annaler*, 1989
Chapter 17: *Socialist Register*, 1998.

The references have been collated in one bibliography. Otherwise reprinted articles appear unchanged in all substantive respects.

PROLOGUE

CHAPTER 1

Reinventing geography:
an interview with the editors of
New Left Review

First published in New Left Review *in August 2000.*

Since the war, the typical field for Marxist research has been history. Your path was more original. How did you become a geographer?

There's a trivial answer to this, which actually has profundity. When I was a kid, I often wanted to run away from home but every time I tried, I found it very uncomfortable, so I came back. So I decided to run away in my imagination, and there at least the world was a very open place, since I had a stamp collection, which showed all these countries with a British monarch on their stamps, and it seemed to me that they all belonged to us, to *me*. My father worked as a foreman in the shipyards at Chatham, with its very strong naval traditions. We lived in Gillingham. Once every year during the War, we would be taken for tea in the dockyards, on a destroyer; the romance of the high seas and of empire left a strong impression. My earliest ambition was to join the Navy. So that even in the very gloomy days of 1946–7, just after the war, there was still an imaginary that encompassed this whole imperial world. Reading about it, drawing maps of it, became a childhood passion. Later, when I was in my teens, I cycled all over north Kent, getting to know a great deal about the geology, agriculture and landscape of our local area. I greatly enjoyed this form of knowledge. So I've always been drawn to geography. At school I was also strongly attracted to literature. When I got into Cambridge, which was still a bit unusual for a boy from my background, I took Geography rather than Literature partly because I had a teacher who had been trained in Cambridge, who made it clear to me that if you studied English there, you didn't so much read literature as deal with F. R. Leavis. I felt I could read literature on my own, and didn't need Leavis to tell me how to do it. So I preferred to follow the track of geography, though of course I never ceased to be interested in history and literature.

Geography was quite a big, well-established school at Cambridge, which gave a basic grounding in the discipline as it was practised in Britain at the time. I went on to do a PhD there, on the historical geography of Kent in the nineteenth century, focusing on the cultivation of hops. My first publication was actually in the house journal of Whitbread, the brewing concern – as a graduate student I earned a tenner for a piece published side by side with an article by John Arlott.

Your first book, Explanation in Geography, *published in 1969, is a very confident intervention, of ambitious scope, in the discipline. But it seems to come out of a very specific positivist setting – a horizon of reference that is exclusively Anglo-Saxon, without any sense of the powerful alternative traditions in geography in France or Germany?*

Explanation in Geography was looking for an answer to what I regarded as a central problem of the discipline. Traditionally, geographical knowledge had been extremely fragmented, leading to a strong emphasis on what was called its 'exceptionalism'. The established doctrine was that the knowledge yielded by geographical enquiry is different from any other kind. You can't generalize about it, you can't be systematic about it. There are no geographical laws; there are no general principles to which you can appeal – all you can do is go off and study, say, the dry zone in Sri Lanka, and spend your life understanding that. I wanted to do battle with this conception of geography by insisting on the need to understand geographical knowledge in some more systematic way. At the time, it seemed to me that the obvious resource here was the philosophical tradition of positivism – which, in the 1960s, still had a very strong sense of the unity of science embedded in it, coming from Carnap. That was why I took Hempel or Popper so seriously; I thought there should be some way of using their philosophy of science to support the construction of a more unitary geographical knowledge. This was a moment when, inside the discipline, there was a strong movement to introduce statistical techniques of enquiry, and new quantitative methods. You could say my project was to develop the philosophical side of this quantitative revolution.

What about the external role of the discipline, as these internal changes took hold? Historically, geography seems to have had a much more salient position in the general intellectual culture of France or Germany than Britain – it's been more closely linked to major public issues. The line of Vidal de la Blache's geography, descending into the Annales *School, is clearly concerned with a problematic of national unity; von Thünen's, in Germany, with industrialization; Haushofer's, with geopolitical strategies of imperial expansion – there was an Edwardian version of this in Mackinder, but more peripheral. How should postwar British geography be situated?*

By the 1960s, it was connected here far more than anywhere else to planning – regional planning and urban planning. By that time there was a certain embarrassment about the whole history of empire, and a turning away from the idea that geography could or should have any global role, let alone shape geopolitical strategies. The result was a strongly pragmatic focus, an attempt to reconstruct geographical knowledge as an instrument of administrative planning in Britain. In this sense, the discipline became quite functionalist. To give you an indication of the trend, I think there are hardly any areas where, if you put the word 'urban' in front of research, you would say this is the center of the field. Urban history is essentially a rather marginal form; urban economics is an equally marginal thing; so, too, is urban politics. Whereas urban geography was really the center of a lot of things going on in the discipline. Then, too, on the physical side, environmental management is often about the handling of local resources in particular kinds of ways. So that in Britain, the public presence of geography – and I think it was quite strong – operated in these three particular areas; it wasn't projected outwards in any grander intellectual formulation of the sort we might find in Braudel or the French tradition. You need to remember that for many of us who had some political ambitions for the discipline, rational planning was not a bad word in the sixties. It was the time of Harold Wilson's rhetoric about the 'white heat of technology', when the efficiency of regional and urban planning was going to be a lever of social betterment for the whole population.

Yet a striking feature of Explanation *is the absence of any political note in it. It reads as a purely scientific treatise, without any mention of concerns of this kind. One would never guess from it that the author might become a committed radical.*

Well, my politics at that time were closer to a Fabian progressivism, which is why I was very taken with the ideas of planning, efficiency and rationality. I would read economists like Oskar Lange, who were thinking along these lines. So in my mind, there was no real conflict between a rational scientific approach to geographical issues, and an efficient application of planning to political issues. But I was so absorbed in writing the book that I didn't notice how much was collapsing around me. I turned in my *magnum opus* to the publishers in May 1968, only to find myself acutely embarrassed by the change of political temperature at large. By then, I was thoroughly disillusioned with Harold Wilson's socialism. Just at that moment, I got a job in the United States, arriving in Baltimore a year after much of the city had burnt down in the wake of the assassination of Martin Luther King. In the US, the anti-war movement and the civil rights movement were really fired up; and here was I, having written this neutral tome that seemed somehow or other just not to fit. I realized I had to rethink a lot of things I had taken for granted in the 1960s.

What took you to the United States?

At that time, American universities were expanding their geography departments. Training in the discipline was much stronger in Britain than in the US, so there was quite an inflow of British geographers to fill the new positions. I had taught in the US on visiting appointments at various times, and when I was offered a job at Johns Hopkins, felt it was an attractive opportunity. The department there was interdisciplinary, combining Geography and Environmental Engineering. The idea was to put together a whole group of people from the social sciences and the natural sciences, to attack issues of environment in a multidisciplinary way. I was one of the first to come into the new program. For me, this was a tremendous situation, particularly in the early years. I learnt a great deal about how engineers think, about political processes, about economic problems: I didn't feel constrained by the discipline of geography.

What was the political atmosphere?

Hopkins is an extremely conservative campus, but it has a long history, of harboring certain maverick figures. For instance, someone who interested me a great deal when I first arrived there – his *Inner Frontiers of Asia* is a great book – was Owen Lattimore, who had been at Hopkins for many years, before he was targeted by McCarthyism. I spent a lot of time talking to people who were there about what had happened to him, and went to see Lattimore himself. Eventually I tried to get Wittfogel, who had been his accuser, to explain why he had attacked Lattimore so violently. So I was always fascinated by the political history of the university, as well as of the city. It's a small campus, which has always remained very conservative. But for that reason, even a small number of determined radicals could prove quite effective – at the turn of the 1970s, there was quite a significant anti-war movement, as well as civil rights activism around the university. Baltimore itself intrigued me from the start. In fact, it was a terrific place to do empirical work. I quickly became involved in studies of discrimination in housing projects, and ever since the city has formed a backdrop to much of my thinking.

What is the particular profile of Baltimore as an American city?

In many ways, it is emblematic of the processes that have moulded cities under US capitalism, offering a laboratory sample of contemporary urbanism. But, of course, it has its own distinctive character as well. Few North American cities have as simple a power structure as Baltimore. After 1900, big industry largely moved out of the city, leaving control in the hands of a rich elite whose wealth was in real estate and banking. There are no corporate headquarters in Baltimore today, and the city is often referred to as the biggest plantation in the South, since it is run much like a plantation by a few major financial institutions. Actually, in social structure, the city is half Northern and half Southern. Two-thirds of the population are African-American, but there is nowhere near the level of black militancy you find in Philadelphia, New York or Chicago. Race relations are more Southern in pattern. Mayors may be African-American, but they are largely dependent on the financial nexus, and are surrounded by white suburbs who don't want anything to do with the city. Culturally, it is one of the great centers of American bad taste. John Waters's movies are classic Baltimore – you can't imagine them anywhere else. Architecturally, whatever the city tries to do it gets a little bit wrong, like an architect who builds a house with miscalculated angles, and then, many years later, people say, 'Isn't that a very interesting structure?' One ends up with a lot of affection for it. At one time, I thought I might write a book called *Baltimore: City of Quirks*.

Your second book, Social Justice and the City, *which came out in 1973, is divided into three sections: Liberal Formulations, Marxist Formulations, Syntheses. Did you write these as a deliberate sequence from the start, to trace an evolution of your own, or did they just emerge* en cours de route?

The sequence was more fortuitous than planned. When I started the book, I would still have called myself a Fabian socialist, but that was a label which didn't make much sense in the US context. Nobody would understand what it meant. In America, I would then have been termed a card-carrying liberal. So I set out along these lines. Then I found they weren't working. So I turned to Marxist formulations to see if they yielded better results. The shift from one approach to the other wasn't premeditated – I stumbled on it.

But you were engaged in a reading group studying Marx's Capital *from 1971 onwards, not long after you got to Baltimore – an experience you have recently described as a decisive moment in your development. Were you the main animator of this group?*

No, the initiative came from graduate students who wanted to read *Capital* – Dick Walker was one of them – and I was the faculty member who helped to organize it. I wasn't a Marxist at the time, and knew very little of Marx. This was anyway still a period when not much Marxist literature was available in English. There was Dobb, and Sweezy and Baran, but little else. Later, you people brought out French and German texts, and the Penguin Marx Library. The publication of the *Grundrisse* in that series was a step in our progression. The reading group was a wonderful experience, but I was in no position to instruct anybody. As a group, we were the blind leading the blind. That made it all the more rewarding.

At the conclusion of Social Justice and the City, *you explain that you encountered the work of Henri Lefebvre on urbanism after you'd written the rest of the book, and go on to make some striking observations about it. How far were you aware of French thinking about space at this stage? Looking back, one would say there were two distinct lines of thought within French Marxism that would have been relevant to you: the historical geography of Yves Lacoste and his colleagues at* Herodote, *and the contemporary urban theory of Lefebvre, which came out of the fascination of surrealism with the city as a landscape of the unexpected in everyday life.*

Actually there was another line in France, which was institutionally more important than either of these, connected to the Communist Party, whose most famous representative was Pierre Georges. This group was very powerful in the university system, with a lot of control over appointments. Their kind of geography was not overtly political at all: it focused essentially on the terrestrial basis on which human societies are built, and its transformations as productive forces are mobilized on the land. Lefebvre was not regarded as a geographer. Georges was a central reference point in the discipline.

Your response to Lefebvre's ideas strikes quite a distinctive note, one that recurs in your later work. On the one hand, you warmed to Lefebvre's radicalism, with a generous appreciation of the critical utopian charge in his writing; on the other hand, you point to the need for a balancing realism. This two-handed response becomes a kind of pattern in your work – one thinks of the way you both imaginatively take up, and empirically limit, the notion of 'flexible accumulation' in The Condition of Postmodernity, *or your reaction to ecological apocalyptics in your more recent writing: an unusual combination of passionate engagement and cool level-headedness.*

One of the lessons I learnt in writing *Social Justice and the City* has always remained important for me. I can put it best with a phrase Marx used, when he spoke of the way we can rub different conceptual blocks together to make an intellectual fire. Theoretical innovation so often comes out of the collision between different lines of force. In a friction of this kind, one should never altogether give up one's starting-point – ideas will only catch fire if the original elements are not completely absorbed in the new ones. The liberal formulations in *Social Justice and the City* don't entirely disappear, by any means – they remain part of the agenda that follows. When I read Marx, I'm very aware that this is a critique of political economy. Marx never suggests that Smith or Ricardo are full of nonsense, he's profoundly respectful of what they had to say. But he's also setting their concepts against others, from Hegel or Fourier, in a transformative process. So this has been a principle of my own work: Lefebvre may have some great ideas, the Regulationists have developed some very interesting notions, which should be respected in their own right, but you don't give up on everything you've got on your side – you try to rub the blocks together and ask: is there something that can come out of this which is a new form of knowing?

What was the reception of Social Justice *in the discipline? The early 1970s were a time of widespread intellectual shift to the left – did it get a sympathetic hearing?*

In the US there was already a radical movement within geography, built around the journal *Antipode* produced at Clark University in Worcester, Massachusetts – traditionally one of the major schools of geography in the country. Its founders were strongly anti-imperialist, hating the history of geography's entanglement with Western colonialism. The journal spawned strong interventions at national meetings in the US, and the formation of a group called Socialist Geographers. In Britain, Doreen Massey and others represented a similar sort of movement. So I'd say, at the beginning of the 1970s, there was a very widespread kind of movement amongst younger people in geography, to explore this particular dimension. *Social Justice and the City* was one of the texts which recorded that moment, becoming a reference point, as time went on. It was also read outside the discipline, particularly by urban sociologists, and some political scientists. Radical economists, of course, were interested in urban questions, too – they had become central political issues in the US. So the setting was quite favorable for the reception of the book.

The Limits to Capital appeared some nine years later, in 1982. It is a major work of economic theory – a startling leap from your previous writing. What is the history of this mutation?

I had some background in neoclassical economics and planning theory, from Cambridge. For any geographer, von Thünen's location theory was a very important point of reference, from the start. Then, of course, in writing *Explanation in Geography* I had steeped myself in positivist discussions of mathematical reason, so that when I came across works by Marxist economists like Morishima or Desai, I had no major difficulties in understanding what was going on. Morishima's work and, naturally, Sweezy's *Theory of Capitalist Development* were very helpful to me. But to be honest, in writing *The Limits to Capital* I stuck with Marx's own texts most of the way. What I realized after *Social Justice and the City* was that I didn't understand Marx, and needed to straighten this out, which I tried to do without too much assistance from elsewhere. My aim was to get to the point where the theory could help me understand urban issues – and that I couldn't do without addressing questions of fixed capital, which no one had written much about at the time. There was the problem of finance capital, fundamental in housing markets, as I knew from Baltimore. If I had just stopped with the first part of the book, it would have been very similar to many other accounts of Marx's theory that were appearing at the time. It was the later part, where I looked at the temporality of fixed-capital formation, and how that relates to money flows and finance capital, and the spatial dimensions of these, that made the book more unusual. That was hard to do. Writing *Limits to Capital* nearly drove me nuts; I had a very difficult time finishing it, also struggling to make it readable – it took me the best part of a decade. The book grounded everything that I've done since. It is my favorite text, but ironically it's probably the one that's least read.

What was the response to it at the time? NLR certainly paid no attention, but what about other sectors of the Left?

I can't really recall anyone who would call themselves a Marxist economist taking it seriously. I always found that guild spirit odd, because it is so unlike Marx's own way of proceeding. Of course, there were some circumstantial reasons for the blank reaction. The controversy over Sraffa and Marx's concept of value was still going on, which I think put off many people from any attempt to consider Marx's theories of capitalist development. There were other versions of crisis theory available – Jim O'Connor's or John Weeks's. The ending of the book could be made to seem like a prediction of inter-imperialist wars, which was easy to dismiss. The only real debate about the book occurred when Michael Lebowitz attacked it in *Monthly Review*, and I replied, some time after it appeared. Overall, the book didn't seem to go anywhere.

Well, you were in good company. After all, Marx was so short of responses to Capital *he was reduced to writing a review of it under a pseudonym himself. In retrospect, what is striking is the extent to which your theory of crisis anticipates later work by two Marxists, who also came from outside the ranks of economists: Robert Brenner, from history, and Giovanni Arrighi, from sociology. In both, space becomes a central category of explanation in a way nowhere to be found in the Marxist tradition, prior to your book. The register is more empirical – detailed tracking of postwar national economies in one case, long-run cycles of global expansion in the other – but the framework, and many of the key conclusions, are basically similar. Your account offers the pure model of this family of explanations, its tripartite analysis of the ways in which capital defers or resolves tendencies to crisis – the structural fix, the spatial fix and the temporal fix – laid out with unexampled clarity.*

Looking back, you can say it was prophetic in that way. But what I hoped to be producing was a text that could be built on, and I was surprised that it wasn't taken in that spirit, but just lay there, rather flat. Of course, it had some currency among radical geographers, and maybe a few sociologists, but no one really used it as I'd have liked it to be. So today, for example, I might take this account of crisis and rub it against, say, world systems theory – in fact, that's probably what I will try to do in a course next year.

The deeper obstacle to a ready acceptance of what you were doing must lie in the difficulty Marxists have always had in confronting geography as a domain of natural contingency – the arbitrary shifts and accidents of the terrestrial crust, with their differential consequences for material life. The main propositions of historical materialism have a deductive structure independent of any spatial location, which never figures in them. The curious thing is that your theory of crisis in The Limits to Capital, *in one sense, respects this tradition – it develops a beautifully clear deductive structure. But it builds space into the structure as an ineliminable element of it. That was quite new. The geographically undifferentiated categories of* Capital *are put to work on natural-historical terrain – still represented abstractly, of course, in keeping with the demands of a deductive argument. That combination was calculated to throw conventional expectations.*

My own intention was, originally, to bounce some historical enquiries into urbanization off *The Limits to Capital*, but this became too massive a project, and I eventually decanted this stuff into the two volumes of essays that appeared in 1985, *Consciousness and the Urban Experience* and *The Urbanization of Capital*. Some of the material in them predates *Limits* itself. In 1976–7 I spent a year in Paris, with the aim of learning from French Marxist discussions, when I was still struggling with *Limits* – but it didn't work out that way. To tell the truth, I found Parisian intellectuals a bit arrogant, quite unable to handle anyone from North America – I felt a touch of sympathy when Edward Thompson launched his famous attack on Althusser, a couple of years later. On the other hand, Castells – who was not part of the big-name circus – was very warm and helpful, along with other urban sociologists, so my time was not lost. But what happened, instead, is that I became more and more intrigued by Paris as a city. It was much more fun exploring that than wrestling with reproduction schemes, and out of this fascination came the piece on Sacré-Coeur and the Commune, which appeared in 1978. Then I backed into the Paris of the Second Empire, a wonderful subject, which became the topic of the longest essay in the two volumes. My interest was: how far might the sort of theoretical apparatus in *The Limits to Capital* play out in tangible situations?

A notable departure in the Second Empire essay – which could have been published as a short book – is the sudden appearance of so many literary sources, quite absent in your writing up till then. Now they cascade across the pages: Balzac, Dickens, Flaubert, Hardy, Zola, James. Had you been holding back a side of yourself, or was this in a sense a new horizon?

I'd always been reading this literature, but I never thought of using it in my work. Once I started to do so, I discovered how many historical ideas poetry or fiction can set alight. And once I made that turn, everything came flooding out. This had something to do with my position in academia: by then I was fairly secure; I didn't feel I had to stay within any narrow professional channels – not that I'd done that too much anyway. But I certainly felt a liberation in deliberately breaking out of them, not to speak of the pleasure of the texts themselves, after the hard grind of *Limits*.

It looks as if the change also prepared the way for the panoramic style of The Condition of Postmodernity. *Presumably by the mid-1980s your antennae were starting to twitch a bit, as talk of the postmodern took off. But what prompted the idea of a comprehensive book on the subject?*

My first impulse was one of impatience. Suddenly, there was all this talk of postmodernism as a category for understanding the world, displacing or submerging capitalism. So I thought: I've written *The Limits to Capital*; I've done all this research on Second Empire Paris; I know a certain amount about the origins of modernism, and a lot about urbanization, which features strongly in this new dispensation; so why not sit down and produce my own take on it? The result was one of the easiest books I've ever written. It took me about a year to write, flowing out without problems or anxieties. And once I embarked on it, of course, my response became more considered. I had no wish to deny the validity of some idea of postmodernity. On the contrary, I found the notion pointed to many developments to which we should be paying the closest attention. On the other hand, this shouldn't mean surrendering to the hype and exaggeration which was then surrounding it.

The book brings together your interdisciplinary interests in a remarkable way, starting, logically enough, from the urban in its strictest sense, with a discussion of redevelopment in Baltimore that makes two fundamental points against the uncritical celebrations of postmodernism as an 'overcoming' of the blights of architectural modernism. The standard argument of the time – blend of Jacobs and Jencks – went: modernism ruined our cities by its inhuman belief in rational planning, and its relentless monolithism of formal design; postmodernism, by contrast, respects the values of urban spontaneity and chaos, and engenders a liberating diversity of architectural styles. You displace both claims, pointing out that it was not so much devotion to principles of planning that produced so many ugly developments, but the subjection of planners to market imperatives, which have continued to zone cities as rigidly under postmodern as modern conditions; while greater diversity of formal styles has been as much a function of technological innovations, allowing use of new materials and shapes, as any aesthetic emancipation.

Yes, I thought it was important to show the new kinds of serial monotony that the supposed flowering of architectural fantasy could bring, and the naïveté of a good many postmodernist staging effects – the simulacra of community you often find them striving for. But I also wanted to make it clear that to understand why these styles had taken such powerful hold, one needed to look at the underlying shifts in the real economy. That brought me to the whole area most famously theorized by the Regulation School in France. What had changed in the system of relations between capital and labor, and capital and capital, since the recession of the early 1970s? For example, how far could we now speak of a new regime of 'flexible accumulation', based on temporary labor markets? Was that the material basis of the alterations in urban fabric we could see around us? The Regulationists struck me as quite right to focus on shifts in the wage contract, and reorganizations of the labor process; one could go quite a way with them there, but not to the notion that capitalism itself was somehow being fundamentally transformed. They were suggesting that one historical regime – Fordism – had given way to another – Flexible Accumulation – which had effectively replaced the first. But empirically, there is no evidence of such a wholesale change – 'flexible accumulation' may be locally or temporarily predominant here or there, but we can't speak of systemic transformation. Fordism plainly persists over wide areas of industry, although of course it has not remained static, either. In Baltimore, where Bethlehem Steel used to employ 30,000 workers, it now produces the same quantity of steel with less than 5,000, so the employment structure in the Fordist sector itself is no longer the same. The extent of this kind of downsizing, and the spread of temporary contracts in the non-Fordist sector, have created some of the social conditions for the fluidity and insecurity of identities that typify what can be called postmodernity. But that's only one side of the story. There are many different ways of making a profit – of gaining surplus value: whichever way works, you are likely to find increasing experiments with it, so there might be a trend towards flexible accumulation; but there are some key limits to the process. Imagine what it would mean for social cohesion if everyone was on temporary labor – what the consequences would be for urban life or civic security. We can already see the damaging effects of even partial moves in this direction. A universal transformation would pose acute dilemmas and dangers for the stability of capitalism as a social order.

That goes for capital–labor; what about capital–capital relations?

What we see there is a dramatic asymmetry in the power of the state. The nation-state remains the absolutely fundamental regulator of labor. The idea that it is dwindling or disappearing as a centre of authority in the age of globalization is a silly notion. In fact, it distracts attention from the fact that the nation state is now more dedicated than ever to creating a good business climate for investment, which means precisely controlling and repressing labor movements in all kinds of purposively new ways: cutting back the social wage, fine-tuning migrant flows, and so on. The state is tremendously active in the domain of capital–labor relations. But when we turn to relations between capitals, the picture is quite different. There the state has truly lost power to regulate the mechanisms of allocation or competition, as global financial flows have outrun the reach of any strictly national regulation. One of the main arguments in *The Condition of Postmodernity* is, that the truly novel feature of the capitalism that emerged out of the watershed of the 1970s is not so much an overall flexibility of labor markets, as an unprecedented autonomy of money capital from the circuits of material production – a hypertrophy of finance, which is the other underlying basis of postmodern experience and representation. The ubiquity and volatility of money as the impalpable ground of contemporary existence is a key theme of the book.

Yes, adapting Céline's title, Vie à Crédit. *Procedurally,* The Condition of Postmodernity *actually follows Sartre's prescription for a revitalized Marxism very closely. He defined its task as the necessity to fuse the analysis of objective structures with the restitution of subjective experience, and representations of it, in a single totalizing enterprise. That's a pretty good description of what you were doing. What do you regard as the most important upshot of the book?*

The Condition of Postmodernity is the most successful work I've published – it won a larger audience than all the others put together. When a book hits a public nerve like that, different kinds of readers take different things away from it. For myself, the most innovative part of the book is its conclusion – the section where I explore what a postmodern experience means for people in terms of the way they live, and imagine, time and space. It is the theme of 'time-space compression', which I look at in various ways through the last chapters, that is the experimental punchline of the book.

The Condition of Postmodernity *came out in 1989. Two years earlier, you had moved from Baltimore to Oxford. What prompted the return to England?*

I felt I was spinning my wheels a bit in Baltimore at the time, so when I was asked if I would be interested in the Mackinder Chair at Oxford I threw my hat into the ring, for a different experience. I was curious to see what it would be like. I stayed at Oxford for six years, but I kept on teaching at Hopkins right the way through. My career has, in that sense, been rather conservative compared with most academics – I've been intentionally loyal to the places I've been. In Oxford, people kept treating me as if I'd just arrived from Cambridge, which I'd left in 1960 – as if the intervening twenty-seven years had just been some waiting-room in the colonies, before I came back to my natural roosting-place at Oxbridge, which drove me nuts. I do have strong roots in English culture, which I feel very powerfully to this day. When I go back to the Kentish country-side that I cycled around, I still know all its lanes like the back of my hand. So in that sense, I've got a couple of toes firmly stuck in the native mud. These are origins I would never want to deny. But they were ones that also encouraged me to explore other spaces.

What about the university or city, themselves?

Professionally, for the first time for many years I found myself in a con-ventional geography department, which was very useful for me. It renewed my sense of the discipline, and reminded me what geographers think about how they think. Oxford doesn't change very fast, to put it mildly. Working there had its pleasurable sides, as well as the more negative ones. By and large, I liked the physical environment, but found the social environment – particularly college life – pretty terrible. Of course, you quickly become aware of the worldly advantages afforded by a position at Oxford. From being seen as a kind of maverick intellectual sitting in some weird transatlantic department, I was transformed into a respectable figure, for whom various unexpected doors subsequently opened. I first really discovered class when I went to Cambridge, in the 1950s. At Oxford I was reminded of what it still means in Britain. Oxford as a city, of course, is another matter. Throughout my years in Baltimore, I always tried to maintain some relationship to local politics: we bought up an old library, and turned it into a community action center, took part in campaigns for rent control, and generally tried to spark radical initiatives; it always seemed to me very important to connect my theoret-ical work with practical activity, in the locality. So when I got to Oxford, the local campaign to defend the Rover plant in Cowley offered a natural extension of this kind of engagement. For personal reasons, I couldn't become quite as active as in Baltimore, but it provided the same kind of connection to a tangible social conflict. It also led to some very interesting political discussions – recorded in the book, *The Factory and the City*, which Teresa Hayter and I produced around it – a fascinating experience. Soon afterwards I read Raymond Williams's novel, *Second Generation*, which is exactly about this, and was astonished by how well he captured so much of the reality at Cowley. So one of the first essays in *Justice, Nature and the Geography of Difference* became a reflection on his fiction.

Isn't there a range of affinities between the two of you? Williams's tone was always calm, but it was uncompromising. His stance was consistently radical, but it was also steadily realistic. His writing ignored disciplinary frontiers, crossing many intellectual boundaries and inventing new kinds of study, without any showiness. In these respects, your own work has a likeness. How would you define your relationship to him?

I never met Williams, though of course I knew of his writing from quite early on. *The Country and the City* was a fundamental text for me in teaching Urban Studies. At Hopkins I always felt an intense admiration for him, in a milieu where so many high-flying French intellectuals were overvalued. Williams never received this kind of academic validation, although what he had to say about language and discourse was just as interesting as any Parisian theorist, and often much more sensible. Of course, when I got to Oxford, I re-engaged with his work much more strongly. The account Williams gives of how he felt on arriving as a student in Cambridge matched almost exactly my own experience there. Then there was this powerful novel, set in Oxford, where I was now working, with its extraordinary interweaving of social and spatial themes. So I did feel a strong connection with him.

There seems to be an alteration of references in Justice, Nature and the Geography of Difference *in other ways, too. Heidegger and Whitehead become much more important than Hempel or Carnap. It is a very wide-ranging collection of texts. What is its main intention?*

It must be the least coherent book I've written. There may even be some virtue in its lack of cohesion, since the effect is to leave things open, for different possibilities. What I really wanted to do was to take some very basic geographical concepts – space, place, time, environment – and show that they are central to any kind of historical-materialist understanding of the world. In other words, that we have to think of a historical-*geographical* materialism, and that we need some conception of dialectics for that. The last three chapters offer examples of what might result. Geographical issues are always present – they have to be – in any materialist approach to history, but they have never been tackled systematically. I wanted to ground the need to do so. I probably didn't succeed, but at least I tried.

One of the strands of the work is a critical engagement with radical ecology, which strikes a characteristic balance. You warn against environmental cata-strophism on the Left. Should we regard this as the latter-day equivalent of economic Zusammenbruch *theories of an older Marxism?*

There was quite a good debate about this with John Bellamy Foster in *Monthly Review*, which laid the issues out very plainly on the table. I'm extremely sympathetic to many environmental arguments, but my experience of working in an engineering department, with its sense for pragmatic solutions, has made me chary of doomsday prophesies, even when these come from scientists themselves, as they sometimes do. I've spent a lot of time trying to persuade engineers that they should take the idea that knowledge, including their own technical ingenuity, is still socially constructed. But when I argue with people from the humanities, I find myself having to point out to them that when a sewage system doesn't work, you don't ring up the postmodernists, you call in the engineers – as it happens, my department has been incredibly creative in sewage disposal. So I am on the boundary between the two cultures. The chapter on dialectics in *Justice, Nature and the Geography of Difference* was designed to try to explain to engineers and scientists what this mystery might be about. That's why it is cast more in terms of natural process than philosophical category. If I had been teaching dialectics in a Humanities program, I would, of course, have had to talk of Hegel; but addressing engineers, it made more sense to refer to Whitehead or Bohm or Lewontin – scientists, familiar with the activities of science. This gives a rather different take on dialectical argumentation, compared to the more familiar, literary-philosophical one.

Another major strand in the book – it's there in the title – is an idea of justice. This is not a concept well-received in the Marxist tradition. Historically, it is certainly true that a sense of injustice has been a powerful, if culturally variable, lever of social revolt, as Barrington Moore and others have shown. This hasn't seemed to require, however, any articulated theory of rights or justice. In modern times, there have been many attempts to found these, without much success. Marx, following Bentham, was withering about their philosophical basis. Why do you think these objections should be overridden?

Marx reacted against the idea of social justice, because he saw it as an attempt at a purely distributive solution to problems that lay in the mode of production. Redistribution of income within capitalism could only be a palliative – the solution was a transformation of the mode of production. There is a great deal of force in that resistance. But in thinking about it, I was increasingly struck by something else Marx wrote – his famous assertion in the introduction to the *Grundrisse*, that production, exchange, distribution and consumption are all moments of one organic totality, each totalizing the others. It seemed to me that it's very hard to talk about those different moments without implying some notion of justice – if you like, of the distributive effects of a transformation in the mode of production. I have no wish to give up on the idea that the fundamental aim is just this transformation, but if you confine it to that, without paying careful attention to what this would mean in the world of consumption, distribution and exchange, you are missing a political driving-force. So I think there's a case for reintroducing the idea of justice, but not at the expense of the fundamental aim of changing the mode of production. There's also, of course, the fact that some of the achievements of social democracy – often called distributive socialism in Scandinavia – are not to be sneered at. They are limited, but real gains. Finally, there is a sound tactical reason for the Left to reclaim ideas of justice and rights, which I touch on in my latest book, *Spaces of Hope*. If there is a central contradiction in the bourgeoisie's own ideology throughout the world today, it lies in its rhetoric of rights. I was very impressed, looking back at the UN Declaration of Rights of 1948, with its Articles 21–4, on the rights of labor. You ask yourself: what kind of world would we be living in today if these had been taken seriously, instead of being flagrantly violated in virtually every capitalist country on the globe? If Marxists give up the idea of rights, they lose the power to put a crowbar into that contradiction.

Wouldn't a traditional Marxist reply be: but precisely, the proof of the pudding is in the eating. You can have all these fine lists of social rights, they've been sitting there, solemnly proclaimed for fifty years, but have they made a blind bit of difference? Rights are constitutionally malleable as a notion – anyone can invent them, to their own satisfaction. What they actually represent are interests, and it is the relative power of these interests that determines which – equally artificial – construction of them predominates. After all, what is the most universally acknowledged human right, after the freedom of expression, today? The right to private property. Everyone should have the freedom to benefit from their talents, to transmit the fruits of their labours to the next generation, without interference from others – these are inalienable rights. Why should we imagine rights to health or employment would trump them? In this sense, isn't the discourse of rights, though teeming with contrary platitudes, structurally empty?

No, it's not empty, it's full. But what is it full of? Mainly, those bourgeois notions of rights that Marx was objecting to. My suggestion is that we could fill it with something else, a socialist conception of rights. A political project needs a set of goals to unite around, capable of defeating its opponents, and a dynamic sense of the potential of rights offers this chance – just because the enemy can't vacate this terrain, on which it has always relied so much. If an organization like Amnesty International, which has done great work for political and civil rights, had pursued economic rights with the same persistence, the earth would be a different place today. So I think it's important that the Marxist tradition engage in dialogue in the language of rights, where central political arguments are to be won. Around the world today, social rebellions nearly always spontaneously appeal to some conception of rights.

In the first essay of your new book, Spaces of Hope, *'The difference a generation makes', you contrast the situation of a reading group on* Capital *in the early 1970s with a comparable one today. Then, you remark, it required a major effort to connect the abstract categories of a theory of the mode of production with the daily realities of the world outside where, as you put it, the concerns of Lenin rather than those of Marx held the stage, as anti-imperialist struggles and revolutionary movements battled across the world. By the 1990s, on the other hand, there was little or no revolutionary ferment left, but the headlines of every morning's paper, as corporate acquisitions or stock prices relentlessly dominated the news, read like direct quotations from* Theories of Surplus Value. *Reviewing the contemporary scene at the end of the essay, you criticize the over-use of Gramsci's adage – taken from Romain Rolland – 'optimism of the will, pessimism of the intellect', arguing for the validity of a robust optimism of the intellect, too. The conclusion is quite unforced, it comes as entirely natural. But it casts an interesting light on your development. For what it suggests is that the whole Communist experience, unfolding across a third of the earth's land-mass, scarcely registered in your line of sight at all – as if you were neither anti-communist, nor pro-communist, but developed your own very energetic and creative Marxism, while bypassing this huge drama altogether. If the collapse of the USSR, and the hopes once invested in it, has been the principal background to pessimism of the intellect on the Left, it is logical that you would be rather unaffected. But it still raises the question, how you could mentally avoid such a large object on the horizon?*

Part of the answer is circumstance. I had no background in Soviet geography, and though I was interested in China, I was never involved in anything to do with it. But if that was in a sense fortuitous, there was a temperamental preference as well. Marx was my anchor, and what Marx wrote was a critique of capitalism. The alternative comes out of that critique, and nowhere else. So I was always more interested in trying to apply the critique and see the alternative where I actually was, in Baltimore, or Oxford, or wherever I happened to be. That may be my own form of localism. On the one hand, I develop a general theory, but on the other, I need to feel this rootedness in something going on in my own backyard. Marxism was so often supposed to be mainly about the Soviet Union or China, and I wanted to say it was about capitalism, which is rampant in the US, and that must have priority for us. So one effect of this was to insulate me a bit from the fall-out of the collapse of Communism. But I should also concede that this is a real limitation of my own work. For all my geographical interests, it has remained Eurocentric, focused on metropolitan zones. I have not been exposed much to other parts of the world.

In your most recent writing, you turn a number of times to the theme of evolution, engaging with E. O. Wilson's work in a sympathetic if critical spirit, very unlike most responses to his writing on the Left. His notion of the 'consilience' of the sciences might well appeal to anyone once attracted to Carnap, though you make clear your own reservations. But it is Wilson's emphasis on the genetic dispositions of every species that offers the occasion for a remarkable set of reflections on human evolution, which you suggest has left the species a 'repertoire' of capacities and powers – competition, adaptation, cooperation, environmental transformation, spatial and temporal ordering – out of which every society articulates a particular combination. Capitalism, you argue, requires all of these – not least its own forms of cooperation – yet gives primacy to a particular mode *of competition. But if competition itself could never be eliminated, as an innate propensity of humanity, its relations with the other powers are in no way unalterable. Socialism is thus best conceived as a reconfiguration of the basic human repertoire, in which its constituent elements find another and better balance. This is a striking response to the claims of sociobiology on its own terrain. But a committed champion of the existing system would reply: yes, but just as in nature the survival of the fittest is the rule whatever the ecological niche, so in society the reason why capitalism has won out is its competitive superiority. It is competition that is the absolute center of the system, lending it an innovative dynamic that no alternative which relativized or demoted the competitive drive into another combination could hope to withstand. You might try to mobilize competition for socialism, but you would want to subordinate it as a principle within a more complex framework, whereas we don't subordinate it – that is our unbeatable strength. What would be your reply to this kind of objection?*

My answer is – oh, but you do: you do subordinate competition in all kinds of areas. Actually, the whole history of capitalism is unthinkable without the setting-up of a regulatory framework to control, direct and limit competition. Without state power to enforce property and contract law, not to speak of transport and communications, modern markets could not begin to function. Next time you're flying into London or New York, imagine all those pilots suddenly operating on the competitive principle: they all try to hit the ground first, and get the best gate. Would any capitalist relish that idea? Absolutely not. When you look closely at the way a modern economy works, the areas in which competition genuinely rules turn out be quite circumscribed. If you think of all the talk of flexible accumulation, a lot of it revolves around diversification of lines and niche markets. What would the history of capitalism be without diversification? But actually the dynamic behind diversification is a flight from competition – the quest for specialized markets is, much of the time, a way of evading its pressures. In fact, it would be very interesting to write a history of capitalism exploring its utilization of each of the six elements of the basic repertoire I outline, tracing the changing ways it has brought them together and put them to work, in different epochs. Kneejerk hostility to Wilson isn't confined to the Left, but it is not productive. Advances in biology are teaching us a great deal about our make-up, including the physical wiring of our minds, and will tell us much more in the future. I don't see how one can be a materialist and not take all this very seriously. So in the case of sociobiology, I go back to my belief in the value of rubbing different conceptual blocks together – putting E. O. Wilson in dialogue with Marx. There are obviously major differences, but also some surprising commonalities, so let's collide the two thinkers against each other. I'm not going to claim I've done it right, but this is a discussion we need. The section of *Spaces of Hope* which starts to talk about this is called 'Conversations on the Plurality of Alternatives', and that's the spirit in which we should approach this. I have questions, not solutions.

What is your view of the present prospect for the system of capital? Limits *set out a general theory of its mechanisms of crisis – over-accumulation, tied to the rigidity of blocs of fixed capital, and of its typical solutions – devalorization, credit expansion, spatial reorganization.* Postmodernity *looked at the way these surfaced in the 1970s and 1980s. Where are we now? There seem to be two possible readings of the present conjuncture, of opposite sign, allowed by your framework, with a third perhaps just over the horizon. The first would take as its starting-point your observation in* The Condition of Postmodernity *that the devalorization necessary to purge excess capital is most effective when it occurs, not in the classic form of a crash, but rather slowly and gradually, cleansing the system without provoking dangerous turmoil within it. On one view, isn't this what has been imperceptibly happening, through successive waves of downsizing and line-shifting, since the start of the long down-turn of the 1970s – the kind of cumulative transformation you cited at Bethlehem Steel; finally unleashing a new dynamic in the mid-1990s, with a recovery of profits, stable prices, surge of high-tech investment and increase in productivity growth, giving the system a new lease of life? On another view, equally compatible with your framework, this is not the underlying story. Rather, what we have mainly been seeing is an explosion of the credit system, releasing a tremendous wave of asset inflation – in other words, a runaway growth of fictitious capital – one that is bound to lead to a sharp correction when the stock bubble bursts, returning us to the realities of continued and unresolved over-accumulation. There is also a third alternative, which would give principal weight to the fall of Soviet Communism in Eastern Europe and Russia, and the Open Door to foreign trade and investment in China. These developments pose the question: isn't capitalism in the process of acquiring, in your terms, a gigantic 'spatial fix' with this sudden, huge expansion in its potential field of operations? This would still be in its early phase – as yet the US has a large negative trade balance with China – but aren't we witnessing the construction of a WTO order that promises to be the equivalent of a Bretton Woods system for the new century, in which for the first time the frontiers of capitalism reach to the ends of the earth? These are three different scenarios, all of which could be grounded in your work. Do you have a provisional judgement of their relative plausibilities?*

I don't think there's any simple choice between these explanations. Both a process of steady, ongoing devalorization – downsizing, reorganizing and outsourcing – and of spatial transformation, along lines traditionally associated with imperialism, are very much part of the real story. But these massive restructurings wouldn't have been possible without the incredible power of fictitious capital today. Every major episode of devalorization or geographical expansion has been imprinted by the role of financial institutions, in what amounts to a quite new dynamic of fictitious capital. Such capital is, of course, no mere figment of the imagination. To the extent that it brings about profitable transformations of the productive apparatus, running through the whole cycle of money being transformed into commodities and back into the original money plus profits, it ceases to be fictitious and becomes realized. But to do so it always depends on a basis in expectations, which must be socially constructed. People have to believe that wealth – mutual funds, pensions, hedge funds – will continue to increase indefinitely. To secure these expectations is a work of hegemony that falls to the state, and its relays in the media. This is something the two great theorists of the last world crisis understood very well – it is instructive to read Gramsci and Keynes side by side. There may be objective processes that block devalorization, or resist geographical incorporation; but the system is also peculiarly vulnerable to the subjective uncertainties of a runaway growth in fictitious capital. Keynes was haunted by the question: how are the animal spirits of investors to be sustained? A tremendous ideological battle is necessary to maintain confidence in the system, in which the activity of the state – we need only think of the role of the US Federal Reserve in the 1990s – is all-important. Someone who has written well about this, in a non-economic way, is Žižek. So the three explanations are not mutually exclusive: they need to be put together, under the sign of a new drive for hegemony. This is a system that has withstood the shocks from the East Asian financial crisis of 1998–9 and the collapse of a major New York-based hedge fund, Long-Term Capital Management, owing billions of dollars. But, each time it was a near-run thing. How long it will last no one can say.

But while the adaptability of capitalism is one of its prime weapons in class struggle, we should not underestimate the vast swathe of opposition it continues to generate. That opposition is fragmented, often highly localized, and endlessly diverse in terms of aims and methods. We have to think of ways to help mobilize and organize this opposition, both actual and latent, so that it becomes a global force and has a global presence. The signs of coming together are there: think only of Seattle. At the level of theory, we need to find a way to identify commonalities within the differences, and so develop a politics that is genuinely collective in its concerns, yet sensitive to what remains irreducibly distinctive in the world today, particularly geographical distinctions. That would be one of my key hopes.

PART 1

GEOGRAPHICAL KNOWLEDGES/
POLITICAL POWER

CHAPTER 2

What kind of geography for what kind of public policy?

First published in Transactions of the Institute of
British Geographers, *1974.*

Can geographers contribute successfully, meaningfully and effectively to the formation of public policy?

General Pinochet is a geographer by training, and by all accounts he is successfully putting geography into public policy. As President of the military Junta that overthrew the elected government of Salvador Allende in Chile on 11 September 1973, General Pinochet does not approve of 'subversive' academic disciplines such as sociology, politics and even philosophy. He has asked that 'lessons in patriotism' be taught in all Chilean schools and universities and he is known to look with great favor upon the teaching of geography – such a subject is, he says, ideally suited to instruct the Chilean people in the virtues of patriotism and to convey to the people a sense of their true historic destiny. Since the military have taken full command of the universities and frequently supervise instruction in the schools, it appears that geography will become a very significant discipline in the Chilean educational system.

General Pinochet is also actively changing the human geography of Chile. An example is here in order. The healthcare system of Chile has, for some time, comprised three distinct components: the rich paid for services on a 'free-market' basis; the middle classes made use of hospital-based medicine financed by private insurance schemes; while the lower classes and poor (some 60 per cent of the population) received free medical care in community-based health centers paid for out of a National Health Service (Navarro 1974). Under Allende, resources were switched from the first two sectors into the community health services which had previously been poorly financed and largely ignored. The geography of the healthcare system began to be transformed from a centralized, provider-controlled, hospital–centred system catering exclusively to the middle and upper classes, to a decentralized, community-controlled, free healthcare system primarily catering to the needs of the lower classes and the poor. This transformation did not occur without resistance – the

27

providers of hospital-based medicine organized strikes to preserve the old social geography of healthcare against the emergence of the new. But during the Allende years the community health centers grew and flourished. Also, community control through the creation of community health councils had a profound political impact and many aspects of life began to be organized around the community health centers. The emphasis also shifted from curative medicine (with all of its glamour and expensive paraphernalia) to preventive medicine which sought to treat medical care as something integral to a wide range of environmental issues (water supply, sewage disposal, and the like). The human geography of social contact, political power and distribution changed as hitherto never before, as the lower classes and poor people began to realize the potential for controlling social conditions of their own existence.

But military power and General Pinochet have changed all that. The community health councils have been disbanded and many of those who participated in them have been imprisoned or executed. The community health centers have been severely curtailed in their operation. The administration of the healthcare system has been given back to the providers of medicine; and the system is reverting to a centralized, hospital-based system catering to the upper and middle classes. Curative medicine is once more the order of the day and open-heart surgery for the few replaces sanitation for the many as the primary goal of medical care. The old geography has been reasserted and the new has been effectively dismantled. Thus has the intervention of the geographer, General Pinochet, become a determining force in the human geography of the healthcare system of Chile.

Chile may seem a long way from Britain. My purpose in quoting this example is not, however, to seek parallels with Britain (although it is disconcerting to note that the government of a country which so actively resisted the advance of fascism from 1939–45 has so hastily extended the hand of friendship to General Pinochet, and that the reorganization of the British National Health Service in the summer of 1973 eliminated all trace of community control and placed the provision of healthcare firmly in the hands of the providers who favor a centralized, hospital-based, healthcare delivery system). I am concerned, rather, to use this example of the successful injection of geography into public policy to pose two very basic questions that must be asked prior to any kind of commitment of geography to public policy: 'What kind of geography ?' and 'Into what kind of public policy?'

These are profoundly difficult questions to answer. It is perhaps useful to begin by asking why we might feel the urge to put any kind of geography into any kind of public policy in the first place. If we reflect upon our motivations for a moment, it seems that this urge arises out of an odd blend of personal ambition, disciplinary imperialism, social necessity and moral obligation. Some of us may be governed (or think we are governed) more by one factor than another, but none of us, surely, can claim total immunity from any of these motivations.

Personal ambition is very significant for us all since we are raised in an economic and social system that is inherently both individualistic and competitive. Since much of the power in society (both economic and political) resides in the public domain, it is natural for academics to be drawn to the locus of that power. Vaunting personal ambition is probably the most significant of all motivating factors in explaining individual behavior. But it does not explain too well the behavior of the geographer as distinct from any other academic and it is to be doubted if an academic possessed of enormous personal ambition would choose to start from what, in Britain at least, must surely be a disadvantageous base in the pecking order of academic disciplines.

The reputation and status of the discipline is, in a way, personal ambition mediated by group consciousness. Disciplines inevitably serve to socialize individuals to the point where they come to locate their identity in terms of 'geography', 'economics', 'biology', etc. In reply to the question 'who are you ?' we frequently reply, 'I am a geographer (economist, biologist, etc)'. Disciplines are important for they help us to understand our role and to feel secure. But geography is one amongst many disciplines which compete for status and prestige in the public eye. Disciplines also compete for public funds. The security of these who identify themselves as 'geographers' is, as a consequence, wrapped up in the position of geography with respect to other disciplines. And so we come to think, 'what is good for geography is good for me' and to recognize that 'a threat to geography is a threat to me'. By promoting geography we promote ourselves and we defend ourselves by defending geography.

Personal ambition and disciplinary imperialism explain a great deal when it comes to understanding individual and professional behaviors. But as explanations they are, I believe, far too simplistic. In what follows, therefore, I shall largely ignore the question of personal ambition and concentrate on the deeper problems of social necessity (mediated by disciplinary imperialism) and moral obligation.

Geography and social necessity

The evolution of geography as a discipline has to be understood against a background of changing social necessities. Since these necessities vary somewhat from society to society I shall confine attention, for the most part, to the recent history of geography in Britain.

In Britain an 'epistemological break' occurred in geographical thinking and activity somewhere around 1945. This break was perhaps best symbolized, first by Professor Wooldridge's influential invocation of the slogan 'the eyes of the fool are on the ends of the earth', and, second, by the foundation of the Institute of British Geographers as a breakaway organization from the Royal Geographical Society. Prior to the Second World War, geography had been more of a non-academic, practical activity than a strong academic discipline. It was oriented, primarily through the activities of the Royal Geographical Society, to what can best be called 'the technics and mechanics of the management of Empire'. The university-based component of geography was relatively weak, while much of what there was (the tie to the Colonial Survey being a good example) related to the concern for Empire. This situation has now changed quite remarkably. Professional university-based geography, strongly aspiring to the status of a distinctive intellectual discipline, is now in the ascendant. Geographers now seek, by and large, to contribute to what can best be called 'the technics and mechanics of urban, regional and environmental management'. Like all such epistemological shifts, elements of the new can be discerned in the old (Dudley Stamp's Land Use Survey of the 1930s surely being the most outstanding example) and residuals from the old are still with us today. But there is no doubt that a major shift in style and in focus has occurred.

How and why did this shift occur? We certainly cannot attribute it to an inner struggle within the intellectual tradition of geography itself (in the fashion, say, of certain shifts in the paradigms of mathematics). It has to be viewed, rather, as an adaptation within geography to external conditions. The end of Empire is in itself sufficient to explain the demise of the old-style geography of the Royal Geographical Society (and it was the end of this era that Wooldridge was heralding). But how are we to explain the transformation to the new style of geography? What were the social necessities that pushed us into concern for the technics and mechanics of urban, regional and environmental management? And why did we move to a professional stance and a university base? To answer these questions we need to say something about our own contemporary history.

If we could return to earth in some future century and if the inhabitants at that time still care (or are able) to write history, then what will the textbooks say of the period 1930–70? I suspect that the relevant chapter will be headed: 'The birth-pangs of the corporate state'. The prototype for the corporate state began to be designed by Bismarck. Mussolini's Italy (particularly in the early years) developed the model while the appalling excesses of Hitler's Germany tend to conceal from us the real meaning of the Fascist Form. Today we sit quietly by and observe Spain, Uruguay, Greece, Brazil, Guatemala, Chile... and at home we accept a growing state interventionism in the name of economic stability (Lord Keynes) and distributive justice (Lord Beveridge). It should be clear to us that Western capitalism is undergoing some sort of radical transformation. Each of the advanced capitalist nations has been fumbling its way to some version of the corporate state (Miliband 1969). Exactly how this is manifest in any particular nation depends upon its existing institutional framework, political traditions, ruling ideology, and opportunities for economic growth and development.

How can we characterize the general form of the corporate state as a mode of sociopolitical organization? It appears as a relatively tightknit, hierarchically ordered structure of interlocking institutions – political, administrative, legal, financial, military, and the like – which transmits information downwards and 'instructs' individuals and groups down the hierarchy as to what behaviors are appropriate for the survival of society as a whole. The slogan for such an operation is 'the national interest'. The corporate state is dominated by the ethics of 'rationality' and 'efficiency' (the two concepts being regarded as interchangeable). Since neither efficiency nor rationality can be defined without a goal, the national interest – the survival of the corporate state – becomes the *de facto* 'purpose'. Within the corporate state a ruling class emerges which, in the advanced capitalist nations, is almost exclusively drawn from the ranks of the industrial and financial interests. In the communist nations, many of which have assumed the corporate state form, the ruling elite is drawn from the party.

In Britain, much of the infrastructure for the corporate state was laid by the Labour Party in the name of distributive justice. But it soon became apparent that 'the social good' could not be achieved without subsuming it under 'the national interest'. It has taken the bureaucratic and technocratic conservativism of Edward Heath to demonstrate how far we have come since 1945 and how easily an infrastructure created in the name of distributive justice can be converted into an instrument for class war. There is, of course, resistance. The free-market capitalism promoted by Enoch Powell coincides with deep misgivings on both the left and right as the law, education, research, the social services, all became subservient to the needs of the corporate state. Even the *Financial Times* (14 January 1974) argues that:

We are now only a decade away from the kind of modern state, with its technological and bureaucratic capacities, that can create and sustain an Orwellian control of the citizen's life. If we are to avoid the totalitarian systems, so chillingly depicted in Huxley's *Brave New World* and Orwell's *1984*, the law as declared from the courts will need to be deployed ever-increasingly to protect the individual's rights.

The legal decisions, which were the focus of the *Financial Times'* concern, went in favor of the government and against the individual's rights.

Consistent with this trend towards a corporate-state form of social and political organization, education has increasingly come to be regarded purely as investment in manpower. Concern for individual health, welfare and sanity has been notably lacking in our calculations. We have been forced, as a consequence, to market the graduate in geography as a commodity. The corporate state requires a technically proficient bureaucracy if it is to function. The commodity we now produce is in part tailored to fit the needs of this market in addition to the market for teachers. We also had to ensure appropriate mechanisms for quality control over the production of this commodity – hence the growth of professional standards within the discipline. Research has likewise become a commodity. National priorities and needs (the pervasive national interest once more) condition the market, and we are progressively pushed to sell research to a client who has a specific need – and the client is, increasingly, the government itself.

And what are these 'national needs and priorities'? Within the overarching concern for the survival of the corporate state itself, we can distinguish the need for designing and implementing a variety of techniques of manipulation, control and co-optation, such that: (1) economic growth, the rate of accumulation of capital, and the competitive position of the state in world markets, are preserved and enhanced; (2) cyclical crises in the economy can be managed; and (3) discontent can be contained and defused. Geographers have sought to respond to these needs by contributing, in both research and education, to the discovery and diffusion of such techniques in the sphere of urban, regional and environmental management. The tightening structure of the corporate state during the 1960s put more and more pressure on us to move in these directions. We are, by now, more subservient to the state in Britain than ever before. We have, in short, been co-opted. Yet there has been virtually no sign of any resistance on our part. Indeed, it looks as if we have been eager to participate in such a process. We certainly have spent little time worrying about the possible consequences.

The reasons why we have not worried are complex. In the first place, the co-optation of the academic into the corporate-state structure provided certain channels through which the academic could approach the locus of power in society. Whether or not the geographer, *qua* academic, could exercise real power or not is beside the point – the illusion was enough to gain the acquiescence of that part of us that responds to vaunting personal ambition. More crucial, however, is the mediating power of disciplinary imperialism. Geographers had to demonstrate that geography did indeed have something to contribute to the fulfillment of national needs and priorities. Much of the debate over the nature of geography in the 1960s was, in fact, a debate over how best to fulfill that tacit commitment. This was a question of survival, for universities were by no means persuaded of the necessity to invest in geography. We had to compete with other disciplines and in the process we were forced, if we were to survive as a collectivity, to carve out a niche, to establish a 'turf' which it was distinctly ours to command.

And it was, of course, the job of the profession (and the Institute of British Geographers in particular) to establish such a niche. There were plenty of fights and some interminable arguments over where that niche should be. In order to demonstrate that geography was an academic discipline occupying a certain turf of academic knowledge, we had to be seen to know what geography was and to present a united front on the matter. The consequences of this were legion. Strong constraints had to be placed on what could or could not be done within the discipline. The Kantian conception of 'synthesis in space' was far too broad and unspecific and so the tortuous search was begun for an analytical methodology which we would call our own. The tendency for geographers to spin off in all directions had to be controlled and the profession sought means to suppress its own dissidents. A corporate structure arose within the discipline – a mini-corporate state within geography that faithfully replicated the corporate structure of the state. We equipped ourselves with powerbrokers within the discipline, self-appointed arbiters of good taste and ultimately with the loosely hegemonic power of the Institute itself.

By such adaptations we have come to define a niche for ourselves to facilitate our own survival in a world of changing social necessities. In the process we have learned to be good citizens, to prostrate ourselves and to prostitute our discipline before 'national priorities' and 'the national interest'. We have survived, in short, by adopting an Eichmann mentality. The only solace to be gained, apart from our survival, is that this mentality is on a clear collision course with our sense of moral obligation.

Geography and moral obligation

Most geographers seem to go about their work with an easy conscience. The self-image of the geographer at work appears to be one of doing good. Tune into any discussion among geographers and as likely as not the discussion unfolds from the standpoint of the benevolent bureaucrat, a person who knows better than other people and who will therefore make better decisions for others than they will be able to make for themselves. The self-image of benevolence appears to contradict the actual behavior of the geographer battling the social necessities laid out in the preceding section. How can we interpret this self-image?

To some degree, it has its source in the broad tradition of humanistic creative scholarship that has permeated Western thought since the Renaissance. The dynamism of the capitalist economic order required technological and social innovation to sustain it. The tradition of creative individualism which grew with the evolution of capitalism (hindered here and artificially fostered there) was functional to the sustenance of the capitalist order and it applied as much to scholarship as it did to practical invention. And this tradition was regarded as an essential ingredient to the progress of mankind (which is some times regarded as a euphemistic phrase for the accumulation of capital). We have undoubtedly been affected by this tradition; the more so as we have created a base within the universities. Western humanism as an intellectual tradition is still quite strong. It has its negative features of course; it is strongly elitist and therefore paternalistic. But it is in this tradition that a good deal of unalienated truly creative scholarship lies.

The source of humanism within the geographic tradition is more problematical. While it is possible to point to some writings in the humanistic vein, the more traditional geographical literature is dominated by racism, ethnocentrism and, at best, a strong paternalism. Even someone as lauded as Humboldt had a quite appalling perspective on 'the natives' which Malthus gleefully quoted in later editions of his celebrated *Essay on Population*. The geography textbooks of today continue in this vein and they are something of which we cannot be proud. Attitudes gleaned from many years devoted to the technics and mechanics of the management of Empire have yet to be expunged from our school texts. Although there is more of which to be ashamed than proud in the geographic tradition, there is a thread to geographic thinking which, at its best, produces an acute sensitivity to place and community, to the symbiotic relations between individuals, communities and environments. This sensitivity to locale and interaction produces a kind of parochial humanism – a humanism that is, in certain senses deep and penetrating, but which is locked into the absolute spaces generated by the regional concept.

But our move away from concern for Empire and into the technics and mechanics of urban, regional and environmental management has brought us into contact with another tradition which has strong humanistic roots. The tradition of Edwin Chadwick and Ebenezer Howard is a strong one in Britain. It is permeated by benevolence and reformism. Contact with this literature has had its effect so that there is an emergent reformist tradition within geography itself which bears the standard of Chadwick and Howard into the contemporary arena.

If anything, humanism and its associated sense of moral obligation have increased in geographic work and thought since 1945. This appears contradictory to the growing power and influence of the corporate state. Since scholarship (of the Western sort) abhors a contradiction much as nature abhors a vacuum, I shall endeavor to resolve it. The humanism of contemporary geography is parochial and elitist (by and large) and in this form it poses less of a threat to the operations of the corporate state. Indeed, it can be argued that such a form of humanism is a positive advantage, for it is functional to have those working to devise and implement techniques of manipulation, control and co-optation to perform their task in the self-image of benevolence. When it is important to co-opt and defuse discontent in a community, for example, it is useful to have people do it with a smile.

But it would be unfair to the strengths of the humanistic tradition and to the potential for individualistic but creative scholarship, to shrug them off as easily co-opted and rendered subservient to the needs of the corporate state. There is no question that the tradition of creative, active, intellectual humanism is in a very fundamental way alien to the functioning of the corporate state. This intellectual tradition labors under certain disadvantages, of course. The professionalization of geography and the ability of the profession to repress its own dissidents is a barrier to be overcome, but this barrier is not unbreakable and indeed there is a peculiar kudos attached to breaking it. And the corporate state cannot afford to seal up all such barriers for it is itself caught in a contradictory position: on the one hand it needs a flexible educational system and an adaptable labor force to meet changing social necessities, while on the other it cannot abide free creative individualism (Gorz 1973).

These tensions have in part been resolved by a simple stratagem. If we can accept that 'facts' and 'values' are separate and distinct from each other and that the former are the subject of scientific investigation whereas the latter are mere personal opinion (subjective), then we can snap the tension by a neat methodological device. If geography is a 'science' and therefore concerned with facts and models and abstract theories, we can then relegate our humanism to personal opinion to be expressed outside geography but not within. The move towards a more 'scientific' geography in the 1960s was consistent with professionalization and the need to produce a commodity with specific skills and abilities. But it also had the deeper effect of resolving the growing tension between the

Eichmann-mentality necessary for our successful social adaptation and the humanism to which we became increasingly prone. Such a solution appeared stable in the 1960s, but it appears less so today. For critical scholarship exposes the artificiality of the separation between fact and value and shows that the claim of science to be ideology-free is itself an ideological claim. The debate over relevance in geography was not really about relevance (whoever heard of irrelevant human activity?), but about whom our research was relevant to and how it was that research done in the name of science (which was supposed to be ideology-free) was having effects that appeared somewhat biased in favor of the status quo and in favor of the ruling class of the corporate state. In other words, reflection on our practice has been leading us to the questions which I started out by posing: 'What kind of geography?' 'Into what kind of public policy?'

The moral obligation of geographers

In order to change the world, we have first to understand it. In order to change the world, we have to create new human practices with respect to the realities around us. So where do we go from here? We live in a corporate state that is tightening its organization. It operates in the name of the national interest. But if we accept that the only meaning to be attached to an individual's life and existence is that which derives from this national interest then we are close to embracing the ideology of fascism. The corporate state is proto-fascist. Perhaps this explains why 'democratic' governments are so friendly with regimes that are overtly repressive and authoritarian and why General Pinochet was made so welcome.

Marx took the view that there were two possible future states for mankind: communism or barbarism. We need urgently to clarify what we mean by these terms. By the former we certainly do not mean what is going on in Russia, Poland or even Cuba and China. By the latter we do not mean a return to a neolithic age. I believe the choice down the road (and perhaps not so far down it either) is between an 'incorporated state' which reflects the creative needs of people struggling to control the social conditions of their own existence in an essentially human way (which is what Marx meant by the phrase 'dictatorship of the proletariat') and a corporate state which instructs downwards in the interests of finance capitalism (the advanced capitalist nations) or the party bureaucracy (Russia and Eastern Europe). The corporate state appears to be the transitional form towards the barbarism of Orwell's *1984* and the incorporated state the transitional form towards communism. The corporate state is in the ascendant and its supersession by the incorporated state requires quite massive organization with worker-control and community-control but two small steps down a long and tortuous path.

It is, of course, the task of critical and reflective thought to understand our condition and to reveal the potentiality for the future imminent in the present. As geographers, we have certain limited contributions to make, as academics and intellectuals ranging synthetically over a wide range of issues we have much more to contribute. We live in a corporate state, we need jobs and we have to conform to a certain extent merely to survive. But we are intelligent and we can live by our wits; we can attempt to subvert the ethos of the corporate state from within. In fact, universities provide quite strong bases for resistance – the essentially backward-looking tradition of free creative scholarship pits the academic against the corporate state as does the forward-looking tradition that seeks for means to transcend its power. And within geography there are some small, but quite significant tasks. To start with there is that tradition of racism, ethnocentrism and condescending paternalism – a residual from the imperial era – that has to be expunged from our textbooks. There is the task of building a genuinely humanistic literature which collapses the artificial (almost schizophrenic) dualisms between fact and value, subject and object, man and nature, science and human interests. The moral obligation of the geographer, *qua* geographer, is to confront the tension between the humanistic tradition and the pervasive needs of the corporate state directly, to raise our consciousness of the contradiction and thereby to learn how to exploit the contradiction within the corporate state structure itself.

The moral obligation of the geographer becomes a social necessity when placed against a broader background. None of us, after all, can afford to think of ourselves merely as geographers, as academics, or even as British. We are human beings struggling, like all other human beings, to control and enhance the social conditions of our own existence. Struggles conducted from a parochial perspective – no matter whether that parochialism emanates from territory (the community, the nation) or from the para-technical division of labor in society – are self-defeating and doomed from the start. Only struggles which overcome the parochialisms inherent in the geography of our situation and in the situation of geography hold out any prospect for success.

CHAPTER 3

Population, resources, and the ideology of science

First published in Economic Geography, *1974.*

It would be convenient indeed if such a contentious issue as the relationship between population and resources could be discussed in some ethically neutral manner. In recent years scientific investigations into this relationship have multiplied greatly in number and sophistication. But the plethora of scientific investigation has not reduced contentiousness; rather, it has increased it. We can venture three possible explanations for this state of affairs:

1. science is not ethically neutral;
2. there are serious defects in the scientific methods used to consider the population-resources problem; or
3. some people are irrational and fail to understand and accept scientifically established results.

All of these explanations may turn out to be true, but we can afford to proffer none of them without substantial qualification. The last explanation would require, for example, a careful analysis of the concept of *rationality* before it could be sustained (Godelier 1972). The second explanation would require a careful investigation of the capacities and limitations of a whole battery of scientific methods, techniques, and tools, together with careful evaluation of available data, before it could be judged correct or incorrect. In this paper, however, I shall focus on the first explanation and seek to show that the lack of ethical neutrality in science affects each and every attempt at 'rational' scientific discussion of the population-resources relationship. I shall further endeavor to show how the adoption of certain kinds of scientific methods inevitably leads to certain kinds of substantive conclusions which, in turn, can have profound political implications.

The ethical neutrality assumption

Scientists frequently appear to claim that scientific conclusions are immune from ideological assault. Scientific method, it is often argued, guarantees the objectivity and ethical neutrality of 'factual' statements as well as the conclusions drawn therefrom. This view is common in the so-called natural sciences; it is also widespread in disciplines such as economics and sociology. The peculiarity of this view is that the claim to be ethically neutral and ideology-free is itself an ideological claim. The principles of scientific method (whatever they may be) are normative and not factual statements. The principles cannot, therefore, be justified and validated by appeal to science's own methods. The principles have to be validated by appeal to something external to science itself. Presumably this 'something' lies in the realms of metaphysics, religion, morality, ethics, convention, or human practice. Whatever its source, it lies in realms that even scientists agree are freely penetrated by ideological considerations. I am not arguing that facts and conclusions reached by means of a particular scientific method are false irrelevant, immoral, unjustifiable, purely subjective, or non-replicable. But I am arguing that the use of a particular scientific method is *of necessity* founded in ideology, and that any claim to be ideology free is *of necessity* an ideological claim. The results of any enquiry based on a particular version of scientific method cannot consequently claim to be immune from ideological assault, nor can they automatically be regarded as inherently different from or superior to results arrived at by other methods.

The ideological foundation of the ethical neutrality assumption can be demonstrated by a careful examination of the paradigmatic basis of enquiry throughout the history of science (both natural and social) (Harvey 1973; Kuhn 1962; Mesjaros 1972), as well as by examining the history of the ethical neutrality assumption itself (Mesjaros 1972; Tarascio 1966). The ideological foundation can also be revealed by a consideration of those theories of meaning in which it is accepted that there cannot be an ethically neutral language because meaning in language cannot be divorced from the human practices through which specific meanings are learned and communicated (Hudson 1970; Wittgenstein 1958). It is not, however, the purpose of this paper to document the problems and defects of the ethical neutrality assumption, critical though these are. I shall, rather, start from the position that scientific enquiry cannot proceed in an ethically neutral manner, and seek to show how the inability to sustain a position of ethical neutrality inevitably implies some sort of an ideological position in any attempt to examine something as complex as a population-resources system.

Lack of ethical neutrality does not, in itself, prove very much. It does serve, of course, to get us beyond the rather trivial view that there is one version of some problem that is scientific and a variety of versions which are purely ideological. For example, the Malthusian terms 'overpopulation' and 'pressure of population on the means of subsistence' are inherently no more or less scientific than Marx's terms 'industrial reserve army' and 'relative surplus population,' even though there is a predilection among unsophisticated analysts to regard the former phrases as adequately scientific and the latter as purely ideological. Unfortunately, it is not very informative to aver also that *all* versions of a problem are ideological, and it is downright misleading to suggest that our views on the population-resources problem depend merely upon whether we are optimists or pessimists, socialists or conservatives, determinists or possibilists, and the like. To contend the latter is not to give sufficient credit to that spirit of scientific endeavor that seeks to establish 'truth' without invoking subjective personal preferences; to say that there is no such thing as ethical neutrality is not to say that we are reduced to mere personal opinion.

We are, however, forced to concede that 'scientific' enquiry takes place in a social setting, expresses social ideas, and conveys social meanings. If we care to probe more deeply into these social meanings, we may observe that particular kinds of scientific method express certain kinds of ethical or ideological positions. In something as controversial as the population-resources debate an understanding of this issue is crucial; yet it is all too frequently ignored. If, as I subsequently hope to show, the dominant method of logical empiricism inevitably produces Malthusian or neo-Malthusian results, then we can more easily understand how it is that scientists raised in the tradition of logical empiricism have, when they have turned to the population-resources question, inevitably attributed a certain veracity to the Malthusian and neo-Malthusian view. When they have found such a view distasteful such scientists have rarely challenged it on 'scientific' grounds; they have, rather, resorted to some version of subjective optimism as a basis for refutation. This kind of refutation has not been helpful, of course, for it has perpetuated the illusion that science and ideology (understood as personal preference) are independent of each other, when the real problem lies in the ideology of scientific method itself.

It is easiest to grapple with the connections between method, ideology, and substantive conclusions by examining the works of Malthus, Ricardo, and Marx, for it is relatively easy to grasp the connections in these works and thereby to discern some important and often obscured questions that lie at the heart of any analysis of the population-resources relation.

Malthus

It is sometimes forgotten that Malthus wrote his first *Essay on the Principle of Population* in 1798 as a political tract against the utopian socialist-anarchism of Godwin and Condorcet and as an antidote to the hopes for social progress aroused by the French Revolution. In his introduction, however, Malthus lays down certain principles of method which ought, he argues, to govern discourse concerning such an ambitious subject as the perfectibility of man:

> A writer may tell me that he thinks a man will ultimately become an ostrich. I cannot properly contradict him. But before he can expect to bring any reasonable person over to his opinion, he ought to show that the necks of mankind have been gradually elongating, that the lips have grown harder and more prominent, that the legs and feet are daily altering their shape, and that the hair is beginning to change into stubs of feathers. And till the probability of so wonderful a conversion can be shown, it is surely lost time and lost eloquence to expatiate on the happiness of man in such a state: to describe his powers, both of running and flying, to paint him in a condition where all narrow luxuries would be contemned, where he would be employed only in collecting the necessaries of life, and where, consequently, each man's share of labour would be light, and his portion of leisure ample.
>
> (Malthus 1970: 70)

The method which Malthus advocates is empiricism. It is through the application of this empiricist method that the competing theories of the utopian socialists, the proponents of liberal advancement and the rights of man, and the advocates of 'the existing order of things' can be tested against the realities of the world. Yet, the first edition of the *Essay* is strongly colored by *a priori* deduction as well as by polemics and empiricism. Malthus sets up two postulates: that food is necessary to the existence of man, and that the passion between the sexes is necessary and constant. He places these two postulates in the context of certain conditions; deduces certain consequences (including the famous law through which population inevitably places pressure on the means of subsistence); and then uses the empiricist method to verify his deductions. Thus Malthus arrives at a conception of method which we may call 'logical empiricism'. This method broadly assumes that there are two kinds of truths which we may call 'logical truths' (they are correct deductions from certain initial statements) and 'empirical truths' (they are correct and verifiable factual statements which reflect observation and experiment). Logical truths may be related to empirical truths by uniting the two kinds of statements into a hypothetico-deductive system. If empirical observation indicates that certain of the derived statements are 'factually true', then this is taken to mean that the system of statements as a whole is true, and we then have a 'theory' of, for example, the population–resources relationship. Malthus constructs a crude version of such a theory.

Another feature of empiricism is worthy of note. Empiricism assumes that objects can be understood independently of observing subjects. Truth is therefore assumed to lie in a world external to the observer whose job is to record and faithfully reflect the attributes of objects. This logical empiricism is a pragmatic version of that scientific method which goes under the name of 'logical positivism', and is founded in a particular and very strict view of language and meaning.

By the use of the logical empiricist method, Malthus arrives at certain conclusions supportive of those advanced by the advocates of 'the existing order of things', rejects the utopianism of Godwin and Condorcet, and rebuffs the hopes for political change. The diminution in polemics and the greater reliance on empiricism in the subsequent editions of the *Essay* may in part be regarded as a consequence of Malthus's basic discovery that scientific method of a certain sort could accomplish, with much greater credibility and power than straight polemics, a definite social purpose. The resort to empiricism was facilitated in turn by the growing body of information concerning the growth and condition of the world's population – a prime source, for example, was the work of the geographer Alexander von Humboldt (1811).

Having shown that the 'power of population is indefinitely greater than the power of the earth to produce subsistence', and that it is a 'natural law' that population will inevitably press against the means of subsistence, Malthus then goes on to discuss the positive and preventive checks through which population is kept in balance with the means of subsistence. The subsequent evolution in Malthus's ideas on the subject are too well-known to warrant repetition here. What is often forgotten, however, is the class character with which he invests it. Glacken, for example, who treats Malthus in the penultimate chapter of his monumental study, *Traces on the Rhodian Shore* (1967), ignores this aspect to Malthus entirely.

Malthus recognizes that 'misery has to fall somewhere' and maintains that the positive checks will necessarily be the lot of the lower classes (Malthus 1970: 82). Malthus thereby explains the misery of the lower classes as the result of a natural law which functions 'absolutely independent of all human regulation'. The distress among the lowest classes has, therefore, to be interpreted as 'an evil so deeply seated that no human ingenuity can reach it' (Malthus 1970: 101). On this basis Malthus arrives, 'reluctantly', at a set of policy recommendations with respect to the poor laws. By providing welfare to the lowest classes in society, aggregate human misery is only increased; freeing the lowest classes in society from positive checks only results in an expansion of their numbers, a gradual reduction in the standards of living of all members of society, and a decline in the incentive to work on which the mobilization of labor through the wage system depends. He also argues that increasing subsistence levels to 'a part of society that cannot in general be considered as the most valuable part diminishes the shares that would otherwise belong to more industrious and worthy members, and thus forces more to become dependent' (Malthus 1970: 97).

From this Malthus draws a moral:

> Hard as it may appear in individual instances, dependent poverty ought to be held disgraceful. Such a stimulus seems to be absolutely necessary to promote the happiness of the great mass of mankind, and every general attempt to weaken this stimulus, however benevolent its apparent intention will always defeat its own purpose . . .
>
> I feel no doubt whatever that the parish laws of England have contributed to raise the price of provisions and to lower the real price of labour. They have therefore contributed to impoverish that class of people whose only possession is their labour. It is also difficult to suppose that they have not powerfully contributed to generate that carelessness and want of frugality observable among the poor, so contrary to the disposition to be remarked among petty tradesmen and small farmers. The labouring poor, to use a vulgar expression, seem always to live from hand to mouth. Their present wants employ their whole attention, and they seldom think of the future. Even when they have an opportunity of saving, they seldom exercise it, but all that is beyond their present necessities goes, generally speaking, to the ale-house. The poor laws of England may therefore be said to diminish both the power and the will to save among the common people, and thus to weaken one of the strongest incentives to sobriety and industry, and consequently to happiness.
>
> (Malthus 1970: 98)

Thus, Malthus arrives at what we have now come to know as the 'counter-intuitive solution' – namely, that the best thing to do about misery and poverty is to do nothing, for anything that is done will only exacerbate the problem. The only valid policy with respect to the lowest classes in society is one of 'benign neglect'. This policy is further supported by a certain characterization of 'typical' behaviors exhibited among the lower classes. Arguments such as these are still with us. They appear in the policy statements by Jay Forrester, Edward Banfield, Patrick Moynihan and others. In fact, welfare policy in the United States at the present time is dominated by such thinking.

Malthus's approach to the lower classes has, if it is to be judged correctly, to be set against his view of the roles of the other classes in society, principally those of the industrial and landed interests. These roles are discussed more analytically in *The Principles of Political Economy*. Here he recognizes that there is a problem to be solved in accounting for the accumulation of capital in society. The capitalist saves, invests in productive activity, sells the product at a profit, ploughs the profit back in as new investment, and commences the cycle of accumulation once more. There is a serious dilemma here, for the capitalist has to sell the product to someone if a profit is to be achieved, and the capitalist is saving rather than consuming. If the capitalist saves too much and the rate of capital accumulation increases too rapidly, then long before subsistence problems are encountered, the capitalists will find expansion checked by the lack of effective demand for the increased output. Consequently, 'both capital and population may be at the same time, and for a period of great length, redundant, compared to the effective demand for produce' (Malthus 1968: 402).

Malthus placed great emphasis upon the effective demand problem and sought to convince his contemporary Ricardo that in practice: 'the actual check to production and population arises more from want of stimulant than want of power to produce' (Keynes 1951: 117). Ricardo was not persuaded, and the idea of effective demand in relationship to capital accumulation and wage rates remained dormant until Keynes resurrected it in his *General Theory of Employment, Interest and Money*.

Malthus's solution to the problem of effective demand is to rely upon the proper exercise of the power to consume on the part of those unproductive classes – the landlords, state functionaries, and the like – who were outside of the production process. Malthus took pains to dissociate himself from any direct apologetics for conspicuous consumption on the part of the landed gentry. He was merely saying that if the capitalist, who was not giving in to what Adam Smith calls 'mankind's insatiable appetite for trinkets and baubles', was to succeed in the task of capital accumulation, then someone, somewhere, had to generate an effective demand:

> It is unquestionably true that wealth produces wants; but it is a still more important truth that wants produce wealth. Each cause acts and reacts upon the other, but the order, both of precedence and importance, is with the wants which stimulate industry . . . The greatest of all difficulties in converting uncivilized and thinly peopled countries into civilized and populous ones, is to inspire them with the wants best calculated to excite their exertions in the production of wealth. One of the greatest benefits which foreign commerce confers, and the reason why it has always appeared an almost necessary ingredient in the progress of wealth, is its tendency to inspire new wants, to form new tastes, and to furnish fresh motives for industry. Even civilized and improved countries cannot afford to lose any of these motives.

> (Malthus 1968: 403)

Effective demand, located in the unproductive classes of society, and stimulated by need creation and foreign trade, was an important and vital force in stimulating both the accumulation of capital and the expansion of employment. Labor might be unemployed, consequently, simply because of the failure of the upper classes to consume. This theory of effective demand does not sit easily with the theory of population. For one thing, it appears contradictory to assert via the theory of population that the power to consume be withheld from the lowest classes in society while asserting, through the theory of effective demand, that the upper classes should consume as much as possible. Malthus attempts to resolve this contradiction by arguing that the upper classes do not increase their numbers according to the principle of population – they consume conspicuously and regulate their numbers by prudent habits generated out of a fear of a decline in their station in life. The lowest classes imprudently breed. The law of population is consequently disaggregated into one law for the poor and another law for the rich. But Malthus also has to explain why an effective demand cannot be generated by an increasing power to consume on the part of the laboring classes. Such a possibility Malthus quickly dismisses as illogical for: 'no one will ever employ capital merely for the sake of the demand occasioned by those who work for him' (Malthus 1968: 404).

He adds that the only case in which this could occur would be if the laborers 'produce an excess of value above what they consume.' He dismisses this possibility entirely. But even Ricardo, in annotating this passage, asks quite simply 'why not?' and writes out a simple case to prove his point (Ricardo 1951b: 429). And, of course, it is this idea, which Malthus rejects out of hand, that forms the foundation of Marx's theory of surplus value, out of which the Marxist theory of relative surplus population stems.

Internal to Malthus' own work there is a central contradiction. On the one hand, the 'natural law' of population asserts a doctrine of inevitable misery for the mass of mankind, while the theory of effective demand points to social controls to the employment of both capital and labor. Zinke suggests that Malthus did not need to reconcile these conflicting positions, for the principle of population applies in the long run, while the theory of effective demand is an explanation for short-run cyclical swings [43]. Malthus does not appear to have thought this way about it. In the *Summary View of the Principle of Population*, published in 1830, Malthus attempts to reconcile these divergent views. Here he admits that 'the laws of private property, which are the grand stimulants to production, do themselves so limit it as always to make the actual produce of the earth fall very considerably short of the power of production' (Malthus 1970: 245).

He then goes on to point out that under a system of private property 'the only effectual demand for produce must come from the owners of property', and that the control of effective demand so intervenes with respect to the principle of population that it prevents the visitation of misery on all sectors of mankind and 'secures to a portion of society the leisure necessary for the progress of the arts and sciences', a phenomenon that 'confers on society' a most signal benefit'. Claims for social reform, and particularly any challenges to the principle of private property, are misplaced. To do away with a society based on competitive individualism regulated through the institutions of private property is to permit the principle of population to operate unchecked, an eventuality that will plunge all of mankind into a state of misery. The laws of private property, insofar as they have restricted the opportunities for the laboring classes, have artificially checked the operation of the principle of population and thereby reduced the aggregate misery of mankind. Malthus thus reconciles the principle of population with the theory of effective demand:

> It makes little difference in the actual rate of increase of population, or the necessary existence of checks to it, whether the state of demand and supply which occasions an insufficiency of wages to the whole of the labouring classes be produced prematurely by a bad structure of society, and an unfavourable distribution of wealth, or necessarily by the comparative exhaustion of the soil. The labourer feels the difficulty in the same degree and it must have nearly the same results, from whatever cause it arises.
>
> (Malthus 1970: 247)

Malthus was, in principle, a defender of private property arrangements, and it is this ideology that underlies his formulation of the principle of population as well as the theory of effective demand. Private property arrangements inevitably mean an uneven distribution of income, wealth, and the means of production in society. Malthus accepts some such distributional arrangement and accepts its class character. Specific distributional arrangement may be judged good or bad, but there was no way in which a rational society could be ordered which did not incorporate necessary class distinctions. Malthus bolstered his arguments with analysis and materials blended together, particularly with respect to the theory of population, by appeal to a method of logical empiricism. In his writings on political economy, however, Malthus frequently made use of a method more characteristic of Ricardo. In part the contradictory character of much of Malthus's writings on population and effective demand stems from the disjunction of method used to examine the two phenomena. At this point, therefore, we must turn to that method of investigation most clearly exhibited in the cleanly spelled-out analytics Ricardo.

Ricardo

Ricardo accepted Malthus's principle of population without any reservations and, it must be added, quite uncritically. But the population principle plays a quite different role and is also treated according to a quite different methodology in Ricardo's work. Ricardo's method was to abstract a few basic elements and relationships out of a complex reality and to analyze and manipulate these idealized elements and relationships in order to discern the structure of the system under consideration. In this manner Ricardo built an abstract model of economic allocation through the market mechanism – a working model of capitalist society – that had little need for an empirical base. The function of such a model was to provide a tool for analysis which would both explain and predict change. Ricardo was not an empiricist in the sense that Malthus was in the *Essay on Population*, and he used facts sparingly, largely by way of illustration rather than with the intent to verify theory. The success and legitimacy of such a method depends, of course, entirely upon the reasonableness of the abstractions made. It is important to look, therefore, at the nature of the abstractions and idealizations built into Ricardo's model in order to understand both his substantive conclusions and his treatment of the population-resources problem.

At the heart of Ricardo's system we find a basic assumption concerning the nature of economic rationality: 'economic man' is the model of rationality to which all human beings ought to aspire. Ricardo was, consequently, a normative rather than an empirical (positive) thinker. More deeply buried in Ricardo's work, however, is a doctrine of social harmony achieved through economically rational behavior in the market place. This doctrine of social harmony is frequently found in the political economy of the period, and its appearance in Ricardo's work is not unconnected with the use of an analytic, model-building methodology. A set of elements and relationships linked into a logical structure is bound to be internally consistent and to be internally harmonious. The model also generates equilibrium-type solutions to problems when it is subjected to manipulation and analysis. It is with respect to the social harmony concept that Ricardo's work contrasts most markedly with that of Malthus and Marx. The latter's work is expressive of the theme of class conflict, throughout, whereas in Malthus's work, the sense of class conflict is confused with social harmony (particularly in *The Principles of Political Economy*) as Malthus seeks to combine results arrived at by means of logical empiricism with those arrived at by means of an abstract model of the economy. Class conflict can scarcely be found in the harmonious analytics of Ricardo's market system, although the analytical results are used for class purposes, namely, the defeat of the landed interest and the subservience of wage labor to the interests of the industrial entrepreneur.

Under these conditions it is surprising to find that Ricardo so easily accepted Malthus's principle of population. In part, the simplicity of Malthus's deductive argument must have appealed to him, but there is a much more significant reason for Ricardo's wholehearted endorsement of the principle. Only by means of it could Ricardo keep his system harmonious and in equilibrium. The analytic problem for Ricardo was to explain the equilibrium wage rate. Wages, he argued, were basically determined by two factors: scarcity and the costs of subsistence. In Ricardo's system labor was regarded abstractly as a commodity like any other, and a growing demand for it ought to elicit a supply so that wages would, in the long-run, tend to the level of a 'natural wage' set by the costs of subsistence. The mechanism that Ricardo appropriated from Malthus to achieve the balance between the supply and demand for labor was, of course, the principle of population, through which the laboring population would automatically increase their numbers:

> When, however, by the encouragement which high wages give to the increase of population, the number of labourers is increased, wages again fall to their natural price, and indeed from a re-action sometimes fall below it.
> (Ricardo 1951a: 94)

In the short run and under favorable circumstances, the rate of accumulation of capital could exceed that of the power of population to reproduce, and during such periods wages would be well above their 'natural' price (Ricardo 1951a: 98). But such periods are bound to be short-lived. Also, when a population presses against the means of subsistence, 'the only remedies are either a reduction of people or a more rapid accumulation of capital'. Consequently, the laws determining wages and 'the happiness of far the greatest part of every community' were dependent upon a balanced relationship between the supply of labor, via the principle of population, and the accumulation of capital. Population, Ricardo argued, 'regulates itself by the funds which are to employ it, and therefore always increases or diminishes with the increase or diminution of capital' (Ricardo 1951a: 78). Even Malthus, however, objected to this use of his population principle, observing that it took at least sixteen years to produce a laborer, and that the population principle was far more than just an equilibrating mechanism (Malthus 1968: 319–20).

Ricardo accepted that:

> the pernicious tendency of the poor laws is no longer a mystery since it has

been fully developed by the able hand of Mr. Malthus and every friend of
the poor must adamantly wish for their abolition.

(Ricardo 1951a: 106)

Like Malthus he argues that:

> The principle of gravitation is not more certain than the tendency of such
> laws to change wealth and power into misery and weakness; to call away
> the exertions of labour from every object, except that of providing mere
> subsistence; to confound all intellectual distinction; to busy the mind in
> supplying the body's wants; until at last all classes should be infected with
> the plague of universal poverty.
>
> (Malthus 1951a: 108)

Further, he warns that: 'if we should attain the stationary state, from
which I trust we are yet far distant, then the pernicious nature of these
laws become more manifest and alarming' (Ricardo 1951a: 109).

Ricardo's evocation here of an ultimate stationary state is of interest.
The analytic model-building methodology that he employed naturally
suggests, as we have seen, harmony and equilibrium, and it is understandable that Ricardo should infer from his model that there must
inevitably be some kind of equilibrium or stationary state. (J. S. Mill came
to the same sort of conclusion using a similar methodological framework
(Mill 1965: 752–7).) Ricardo is here arguing also that under such an
equilibrium condition, in which the demand and supply of labor are
equated and the prospects for further capital accumulation eliminated,
there would appear to be a choice between conditions of universal poverty
(everybody receiving a mere subsistence wage) or conditions in which
rational thought and civilization itself could survive, at least among an
elite. Ricardo is also suggesting that social welfare provision will become
particularly pernicious in non-growth situations. Again, this argument is
still with us and we will return to it later.

Ricardo found Malthus's arguments with respect to effective demand
'quite astonishing' however, and commented that:

> A body of unproductive labourers are just as necessary and useful with a
> view to future production as a fire which should consume in the manufac-
> turer's warehouse, the goods which those unproductive labourers would
> otherwise consume.
>
> (Ricardo 1951b: 421)

Ricardo would have no truck with Malthus's defense of the landed
interest and it is clear from his remarks and policies with respect to the

corn laws, rent, and the like, that Ricardo's sympathies lie entirely with the industrial entrepreneur who alone, in Ricardo's system, epitomized economic rationality. Ricardo was in fact offended by the role the landed interest played, and since he discounted the problem of effective demand entirely, Ricardo came to regard the landed interest as a mere barrier to progress and to the achievement of social harmony.

Ricardo's model building analytics permitted him to argue positively for change. He was not deterred by empirical evidence, and he had no sense of debt to history. His normative analytics allowed him to see the possibility for changing and improving reality, rather than just understanding and accepting it. Like August Lösch (another great normative thinker) Ricardo could take the view that 'if my model does not conform to reality then it is reality that is wrong' (Lösch 1954: 363). Ricardo could project upon the world a working model of capitalist society constructed in the image of an idealized social harmony achieved through the benificence of rational economic man. Ricardo sought to change reality to fit this image, and in the process he played an important and vital role in furthering the progress of industrialization in nineteenth-century England.

Marx

Marx argues that both Ricardo and Malthus were projecting ideological assumptions without admitting or even perhaps being aware of them:

> [Malthus's theory] suits his purpose remarkably well – an apologia for the existing state of affairs in England, for landlordism, 'State and Church' ... parsons and menial servants, assailed by the Ricardians as so many useless and superannuated drawbacks of bourgeois production and as nuisances. For all that, Ricardo championed bourgeois production insofar as it signified the most unrestricted development of the social productive forces ... He insisted upon the historical justification and necessity of this stage of development. His very lack of a historical sense of the past meant that he regarded everything from the historical standpoint of his time. Malthus also wanted to see the freest possible development of capitalist production ... but at the same time he wants it to adapt itself to the 'consumption needs' of the aristocracy and its branches in State and Church, to serve as the material basis for the antiquated claims of the representatives of interests inherited from feudalism and the absolute monarchy. Malthus wants bourgeois production as long as it is not revolutionary, constitutes no historical factor of development, but merely creates a broader and more comfortable basis for the 'old' society.
>
> (Marx 1972: 52–3)

The contrasts between Malthus, Ricardo, and Marx are usually portrayed in terms of their substantive views on such issues as the population-resources problem. The more fundamental contrast, however, is surely one of method. Marx's method is usually called 'dialectical materialism', but this phrase conveys little and conceals a lot. Fully to understand it requires some understanding of German critical philosophy and in particular that branch of it which most fully developed a non-Aristotelian view of the world – the most eminent representatives in this tradition being Leibniz, Spinoza, and Hegel. The nature of this non-Aristotelian view requires exposition.

Marx's use of language is, as Ollman has pointed out, relational rather than absolute (Ollman 1971). By this he means that a 'thing' cannot be understood or even talked about independently of the relations it has with other things. For example, 'resources' can be defined only in relationship to the mode of production which seeks to make use of them and which simultaneously 'produces' them through both the physical and mental activity of the users. There is, therefore, no such thing as a resource in abstract or a resource which exists as a 'thing in itself'. This relational view of the world is fundamentally different from the usual and familiar Aristotelian view (characteristic of logical empiricism or Ricardian type model building) in which things are thought to have an essence of some sort and are, therefore, regarded as definable without reference to the relationships they have to other things.

On this basis Marx evolves certain fundamental assumptions regarding the way in which the world is structured and organized. Ollman suggests that: 'The twin pillars of Marx's ontology are his conception of reality as a totality of internally related parts, and his conception of these parts as expandable relations such that each one in its fullness can represent the totality' (Ollman 1973: 495). There are different ways in which we can think of such a totality. We may think of it as an aggregate of elements – a mere sum of parts – which enter into combination without being fashioned by any pre-existing relationships within the totality. The totality can alternatively be viewed as something 'emergent'; it has an existence independent of its parts while it also dominates and fashions the parts contained within it. But Marx's non-Aristotelian and relational view permits him a third view of the totality in which it is neither the parts nor the whole, but the relationships within the totality which are regarded as fundamental. Through these relationships the totality shapes the parts to preserve the whole. Capitalism, for example, shapes activities and elements within itself to preserve itself as an ongoing system. But conversely, the elements are also continually shaping the totality into new configurations as conflicts and contradictions within the system are of necessity resolved.

Marx rarely used the word totality to refer to everything there is. He usually focused on the 'social' totality of human society, and within this totality he distinguished various structures. Structures are not 'things' or 'actions', and we cannot establish their existence through observation. The meaning of an observable act, such as cutting a log, is established by discovering its relation to the wider structure of which it is a part. Its interpretation will depend upon whether we view it in relation to capitalism or socialism, or whether we place it in relation to some quite different structure, such as the ecological system. To define elements relationally means to interpret them in a way external to direct observation; hence the departure from empiricism accomplished by relational modes of thought.

Within the social totality Marx distinguishes various structures (Godelier 1972). The 'economic basis' of society comprises two structures: the forces of production (the actual activities of making and doing), and the social relations of production (the forms of social organization set up to facilitate making and doing). Marx thus distinguished between a technical division of labor and a social division of labor. In addition, there are various superstructural features: the structures of law, of politics, of knowledge and science, of ideology, and the like. Each structure is regarded as a primary element within the social totality and each is capable of a certain degree of autonomous development. But because the structures are all interrelated, a perpetual dynamism is generated out of the conflicts and interactions among them. For example, Marx sees a major contradiction between the increasing socialization of the forces of production (through the intricacies of the division of labor) and the private-property basis of consumption and ownership in capitalist society. Within this system of interacting structures, however, Marx accorded a certain primacy of place to the economic basis. In arguing thus, Marx usually appealed to the fact that man has to eat in order to live and that production – the transformation of nature – therefore has to take precedence over the other structures in a conflict situation. There is a deeper reason for the significance which Marx attached to the economic basis; it is here that the relationship between the natural and social aspects of life become most explicit.

Marx's conception of the man-nature relation is complex (Schmidt 1971). At one level the human being is seen as a part of nature – an ensemble of metabolic relations involving constant sensuous interaction with a physical environment. At another level, human beings are seen as social – each as an ensemble of social relations (Marx 1964) – and capable of creating forms of social organization which can become self-regulating and self-transforming. Society thereby creates its own history by transforming itself, but in the process the relationship with nature is also transformed. Under capitalism, for example:

> Nature becomes for the first time simply an object for mankind, purely a matter of utility; it ceases to be recognized as a power in its own right; and the theoretical knowledge of its independent laws appears only as a stratagem designed to subdue it to human requirements, whether as the object of consumption or as the means of production. Pursuing this tendency, capital has pushed beyond national boundaries and prejudices, beyond the deification of nature and the inherited self-sufficient satisfaction of existing needs confined within well-defined bounds and [beyond] the reproduction of traditional ways of life. Capital is destructive of all this and permanently revolutionary, tearing down the obstacles that impede the development of productive forces, the expansion of need, the diversity of production and the exploitation and exchange of natural and intellectual forces.
>
> (Marx 1973: 410)

Marx saw the capitalist law of accumulation always pushing society to the limits of its potential social relations and to the limits of its natural resource base, continuously destroying the potential for 'the exploitation and exchange of natural and intellectual forces'. Resource limitations could be rolled back by technological change, but the tide of capitalist accumulation quickly spreads up to these new limits.

Marx also argued that capitalism had successfully brought society to the point where mankind could be free of nature in certain important material respects. Human beings are now in a position to *create* nature rather than mindlessly to alter it. Through the creation of nature – a creation that has to proceed through a knowledge and understanding of nature's own laws – human beings could be freed to discover their own essentially human nature within the system of nature. There is, for Marx, an enormous difference between this unalienated creation of nature and the mindless exploitation under capitalism which, in the haste to accumulate, is always concerned, as Engels has it, 'only about the first tangible success; and then surprise is expressed that the more remote effects of actions directed to this end turn out to be of a quite different, mainly of an opposite, character' (Engels 1940: 296).

In the final analysis, the conflict and contradiction between the system of nature and the social system could be resolved only by the creation of an appropriate and entirely new form of human practice. Through such a practice, human beings will 'not only feel but also know their unity with nature' and thereby render obsolete 'the senseless and anti-natural idea of a contradiction between mind and matter, man and nature, soul and body' (Engels 1940: 293).

Marx's methodology allows that knowledge and the processes of gaining understanding are internal to society. Subject and object are not regarded as independent entities but as relationships one to the other. This conception is very different indeed from that of traditional empiricism in which the subject is presumed to be 'instructed by what is outside of him', or from that of a priorism and innatism (clearly implied in Ricardo's method) in which the subject 'possesses from the start endogenous structures which it imposes on objects' (Piaget 1972: 19). Marx in fact fashions a methodology similar to the contructivism advanced by Piaget:

> Whereas other animals cannot alter themselves except by changing their species, man can transform himself by transforming the world and can structure himself by constructing structures; and these structures are his own, for they are not entirely predestined either from within or without.
>
> (Piaget 1970: 118)

The subject is thus seen as both structuring and being structured by the object. As Marx puts it, 'by thus acting on the external world and changing it, [man] at the same time changes his own nature' (Marx 1967: vol. 1, 175).

The thinking subject can create ideas in the imagination. But ideas have at some stage to leave the realms of abstract knowledge and to enter into human practice if they are to be validated. Once incorporated into human practice, concepts and ideas can become (via technology) a material force in production and can alter the social relations of production (through the creation of new modes of social organization). Although many ideas remain barren, some do not: 'at the end of every labour process we get a result that already existed in the imagination of the labourer at its commencement.'

Ideas are therefore regarded as social relations through which society can be structured and reconstructed. But concepts and categories are also produced under specific historical conditions which are in part internal to knowledge (the categories of thought handed down to us) and in part a reflection of the world in which knowledge is produced. The categories of thought available to us are, as it were, our intellectual capital which it is open to us to improve (or destroy). If, however, ideas are social relations, then it follows that we can gain as much insight into society through a critical analysis of the relations ideas express, as we can through a study of society as object. The analysis of ideas in Marx's work is as much directed to understanding the society that produced them as it is to understanding what it is they tell us about the reality they purport to describe. Marx is, thus, adopting a methodological framework that is perpetually revolving around the question: what is it that produces ideas and what is it that these ideas serve to produce?

Marx's substantive conclusions on the 'population problem' are in part generated out of a vigorous criticism of writers such as Malthus and Ricardo. Marx set out to transform the categories handed down to him, for he saw that to do so was necessary if the realities of life were to be transformed. Marx traced the structure of Malthus's and Ricardo's thought back to their respective theories of value. Out of a criticism of these and other theories of value, Marx arrived at the theory of surplus value. Surplus value, he argued, originated out of surplus labor, which is that part of the laborer's working time that is rendered gratis to the capitalist. In order to obtain employment, a laborer may have to work ten hours. The laborer may produce enough to cover his own subsistence needs in six hours. If the capitalist pays a subsistence wage, then the laborer works the equivalent of four hours free for the capitalist. This surplus labor can be converted through market exchange into its money equivalent: surplus value. And surplus value, under capitalism, is the source of rent, interest, and profit. On the basis of this theory of surplus value, Marx produces a distinctive theory of population.

If surplus value is to be ploughed back to produce more surplus value, then more money has to be laid out on wages and the purchase of raw materials and means of production. If the wage rate and productivity remain constant, then accumulation requires a concomitant numerical expansion in the labor force: 'accumulation of capital is, therefore, increase of the proletariat' (Marx 1967: vol. 1, 614). If the labor supply remains constant, then the increasing demand for labor generated by accumulation will bring about a rise in the wage rate. But a rise in the wage rate means a diminution of surplus value, falling profits, and, as a consequence, a slower rate of accumulation. But:

> this diminution can never reach the point at which it would threaten the system itself . . . Either the price of labour keeps on rising, because its rise does not interfere with the progress of accumulation . . . Or accumulation slackens in consequence of the rise in the price of labour, because the stimulus of gain is blunted. The mechanism of the process of capitalist production removes the very obstacles that it temporarily creates.
> (Marx 1967: vol. 1, 619)

Under these conditions, the 'law of capitalist production' that is at the bottom of the 'pretended natural law of population' reduces itself to a relationship between the rate of capitalist accumulation and the rate of expansion in the wage-labor force. This relationship is mediated by technical change, and the increasing social productivity of labor can also be used as 'a powerful lever of accumulation' (Marx 1967: vol. 1, 621). The use of this lever permits an expansion of surplus value through a growing substitution of capital for labor in the production process. Marx then proceeds to show how these processes combine to create a 'law of population peculiar to the capitalist mode of production', adding that 'in fact every special historic mode of production has its own special laws of population, historically valid within its limits alone' (Marx 1967: vol. 1, 632–3). Here we can see a major departure from the thought of both Malthus and Ricardo who attributed to the law of population a 'universal' and 'natural' validity.

Marx largely confines attention to the law of population operative under capitalism. He points out that the laboring population produces both the surplus and the capital equipment, and thereby produces the means 'by which it itself is made relatively superfluous' (Marx 1967: vol. 1, 632). He then goes on to say:

> If a surplus labouring population is a necessary product of accumulation or of the development of wealth on a capitalist basis, this surplus population becomes, conversely, the lever of capitalist accumulation, nay a condition of existence of the capitalist mode of production. It forms a disposable industrial reserve army, that belongs to capital quite as absolutely as if the latter had bred it at its own cost. Independently of the limits of the actual increase of population, it creates for the changing needs of the self-expansion of capital, a mass of human material always ready for exploitation.
> (Marx 1967: vol. 1, 632)

This relative surplus population has however, another vital function: it prevents wages rising and thereby cutting into profits:

> The industrial reserve army, during the periods of stagnation and average prosperity, weighs down the active labour army; during the periods of overproduction and paroxysm, it holds its pretensions in check. Relative surplus population is therefore the pivot around which the law of supply and demand of labour works. It confines the field of action of this law within the limits absolutely convenient to the activity of exploitation and to the domination of capital.
> (Marx 1967: Vol. 1, 632)

The production of a relative surplus population and an industrial reserve army are seen in Marx's work as historically specific, as internal to the capitalist mode of production. On the basis of his analysis we can predict the occurrence of poverty no matter what the rate of population change. Marx explicitly recognizes, however, that a high rate of capital accumulation is likely to act as a general stimulus to population growth; it is likely that laborers will try to accumulate the only marketable commodity they possess, labor power itself (Marx 1967: vol. 3, 218). Marx was not arguing that population growth per se was a mechanical product of the law of capitalist accumulation, nor was he saying that population growth per se did not affect the situation. But he was arguing very specifically, contra the position of both Malthus and Ricardo, that the poverty of the laboring classes was the inevitable product of the capitalist law of accumulation. Poverty was not, therefore, to be explained away by appeal to some natural law. It had to be recognized for what it really was: an endemic condition internal to the capitalist mode of production.

Marx does not talk about a population problem but a poverty and human exploitation problem. He replaces Malthus's concept of overpopulation by the concept of a relative surplus population. He replaces the inevitability of the 'pressure of population on the means of subsistence' (accepted by both Malthus and Ricardo) by an historically specific and necessary pressure of labor supply on the means of employment produced internally within the capitalist mode of production. Marx's distinctive method permitted this reformulation of the population–resources problem, and put him in a position from which he could envisage a transformation of society that would eliminate poverty and misery rather than accept its inevitability.

Methodology and the population–resources relation

The contrasts between Malthus, Ricardo and Marx are instructive for a variety of reasons. Each makes use of a distinctive method to approach the subject material. Marx utilizes a non-Aristotelian (dialectical) framework which sets him apart from Ricardo and Malthus who, in turn, are differentiated from each other by the use of abstract analytics and logical empiricism, respectively. Each method generates a distinctive kind of conclusion. Each author also expresses an ideological position, and, at times, it seems as if each utilizes that method which naturally yields the desired result. The important conclusion, however, is that the method adopted and the nature of the result are integrally related.

It is surprising, therefore, to find so little debate or discussion over the question of method for dealing with such a complex issue as the population–resources relation. Here the ethical neutrality assumption appears to be a major barrier to the advance of scientific enquiry, for if it is supposed

that all scientific methods are ethically neutral, then debates over methodology scarcely matter. The materials on the population-resources relation
published in recent years suggest that the Aristotelian legacy is dominant:
we still usually 'think Aristotle' often without knowing it. Yet the
Aristotelian cast of mind seems ill-suited for dealing with the population-
resources relation, and so there has been a methodological struggle
internal to the Aristotelian tradition to overcome the limitations inherent
in it. There has been, as it were, a convergence toward Marx without overthrowing the Aristotelian trappings. Marx accepts that the appropriate
method to deal with the population-resources relation has to be holistic,
system-wide in its compass, capable of handling dynamics (feedbacks in
particular), and, most important of all, *internally dynamic* in that it has to
be capable of producing new concepts and categories to deal with the
system under investigation and, through the operationalization of these
new concepts and categories, change the system from within. It is this
last feature that gives to Marx's work its dialectical quality. Most contemporary investigations of the population-resources relation recognize
all of Marx's requirements save the last, and rely upon systems theory for
their methodological foundation. Systems-theoretic formulations are
sophisticated enough (in principle) to do everything that Marx sought to
do except to transform concepts and categories dialectically, and thereby
to transform the nature of the system from within. Some examples bear
out this point.

Kneese et al. (1970) adopt what they call a 'materials balance' approach
to the population-resources relation which is, in effect, a two-stage input-
output model. The first stage describes the flows within the economy; the
second stage describes the flows within the ecological system; and the two
systems are linked by the physical principle that matter can neither be
created nor destroyed. The model is descriptive in the sense that the coefficients have to be estimated from empirical data, but experimentation on
the model is possible by examining the sensitivity of results to changes in
the coefficients.

In the study by Meadows et al. (1972), methods derived from systems
dynamics are used; a system of difference equations is simulated to
indicate future outcomes of population growth, industrial expansion,
resource use (both renewable and non-renewable), and environmental
deterioration. The system in this case incorporates feedbacks (both
positive and negative) and is, in contrast to that of Kneese et al., oriented
to development through time. The Meadows model has come in for a great
deal of criticism and a team from the University of Sussex (Cole et al.
1973) has examined the model in detail. They reformulated it in certain
important respects; showed some of the problems inherent in the data

used to estimate the equations; and concluded that some unnecessarily pessimistic assumptions were injected into the Meadows model.

The essential point to note, however, is that *all* of these formulations lead to neo-Malthusian conclusions: strongly voiced in the Meadows model; somewhat muted in the case of Kneese et al. (who speak of the *new* Malthusianism); and long-run in the case of the Sussex team's investigation (rather like Ricardo they seem to suggest that the stationary state is inevitable but a long way off).

The neo-Malthusian results of these studies can be traced back to the Aristotelian form in which the question is posed and the answers constructed. And it is, of course, the ability to depart from the Aristotelian view that gets Marx away from both the short-run and long-run inevitabilities of neo-Malthusian conclusions. Marx envisages the production of new categories and concepts, of new knowledge and understanding, through which the relationships between the natural and social system will be mediated. This relational and dialectical view of things comes closest to impinging upon traditional concerns with respect to the problem of technological change. It has, of course, long been recognized that Malthus was wrong in his specific forecasts because he ignored technological change. Ricardo saw the possibilities of such change, but in the long run he saw society inevitably succumbing to the law of diminishing returns. The difference between the Meadows model and the Sussex team's refashioning of it is largely due to the pessimism of the former and the optimism of the latter. In all of these cases, technological change is seen as something external to society: an unknown that cannot be accounted for. But for Marx technological change was both internal to and inevitable within society; it is the product of human creativity, and stems from the inevitable transformation of the concepts and categories handed down to us. Only if we let ourselves be imprisoned within the system of knowledge handed down to us will we fail to innovate. Further, it is unnecessarily restrictive to think that human inventiveness and creativity apply only in the sphere of technology – human beings can and do create social structures as well as machines. This process Marx regards as essential and inevitable precisely because man could and would respond to the necessities of survival. The only danger lies in the tendency to place restrictions on ourselves and, thereby, to confine our own creativity. In other words, if we become the prisoners of an ideology, prisoners of the concepts and categories handed down to us, we are in danger of making the neo-Malthusian conclusions true, of making environmental determinism a condition of our existence.

It is from this standpoint that Marx's method generates quite different perspectives and conclusions from those generated by simple logical empiricism, Ricardian-type normative analytics, or contemporary systems theory. Let me stress that I am not arguing that the latter methods are illegitimate or erroneous. Each is in fact perfectly appropriate for certain domains of enquiry. Logical empiricism has the capacity to inform us as to what is, given an existing set of categories. Insofar as we make use of this method, we are bound to construct what I have elsewhere called a status quo theory (Harvey 1973). The Aristotelian manner in which normative, analytical model building proceeds yields 'ought-to' prescriptive statements, but the categories and concepts are idealized, abstracted, and *stationary* tools imposed upon a changing world. Systems theory is a more sophisticated form of modelling relying upon various degrees of abstraction and a varying empirical content. Dialectical materialism, in the manner that Marx used it, is 'constructivist' in that it sees change as an internally generated necessity that affects categories of thought and material reality alike. The relationships between these various methods are complex. The methods are not, obviously, mutually exclusive of each other; but different methods appear appropriate for different domains of enquiry. And it is difficult to see how anything other than a relational, constructivist, and internally dynamic method can be appropriate for looking into the future of the population-resources relation, particularly when it is so evident that knowledge and understanding are such important mediating forces in the construction of that future. Results arrived at by other means may be of interest, only if they are set within the broader interpretive power provided by Marx's method. All of this would be a mere academic problem (although one of crucial significance) were it not for the fact that ideas are social relations, and the Malthusian and neo-Malthusian results arrived at (inevitably) by means of other methods are projected into the world where they are likely to generate immediate political consequences. And it is to these consequences that we now turn.

The political implications of population-resources theory

At the Stockholm Conference on the Environment in 1972, the Chinese delegation asserted that there was no such thing as a scarcity of resources and that it was meaningless to discuss environmental problems in such terms. Western commentators were mystified and some concluded that the Chinese must possess vast reserves of minerals and fossil fuels, the discovery of which they had not yet communicated to the world. The Chinese view is, however, quite consistent with Marx's method and should be considered from such a perspective. To elucidate it we need to bring into our vocabulary three categories of thought: subsistence, resources, and scarcity.

Subsistence

Malthus appears to regard subsistence as something absolute, whereas Marx regards it as relative. For Marx, needs are not purely biological; they are also socially and culturally determined (Orans 1966). Also, as both Malthus and Marx agree, needs can be created, which implies that the meaning of subsistence cannot be established independent of particular historical and cultural circumstances if, as Marx insisted, definitions of social wants and needs were produced under a given mode of production rather than immutably held down by the Malthusian laws of population. Subsistence is, then, defined internally to a mode of production and changes over time.

Resources

Resources are materials available 'in nature' that are capable of being transformed into things of utility to man. It has long been recognized that resources can be defined only with respect to a particular technical, cultural, and historical stage of development, and that they are, in effect, technical and cultural appraisals of nature (Firey 1960; Spoehr 1956).

Scarcity

It is often erroneously accepted that scarcity is something inherent in nature, when its definition is inextricably social and cultural in origin. Scarcity presupposes certain social ends, and it is these that define scarcity just as much as the lack of natural means to accomplish these ends (Pearson 1957). Furthermore, many of the scarcities we experience do not arise out of nature but are created by human activity and managed by social organization (the scarcity of building plots in central London is an example of the former; the scarcity of places at university is an example of the latter). Scarcity is in fact necessary to the survival of the capitalist mode of production, and it has to be carefully managed, otherwise the self-regulating aspect to the price mechanism will break down (Harvey 1973).

Armed with these definitions, let us consider a simple sentence: 'Overpopulation arises because of the scarcity of resources available for meeting the subsistence needs of the mass of the population.' If we substitute our definitions into this sentence we get: 'There are too many people in the world because the particular ends we have in view (together with the form of social organization we have) and the materials available in nature, that we have the will and the way to use, are not sufficient to provide us with those things to which we are accustomed.' Out of such a sentence all kinds of possibilities can be extracted:

1. we can change the ends we have in mind and alter the social organization of scarcity;
2. we can change our technical and cultural appraisals of nature;
3. we can change our views concerning the things to which we are accustomed;
4. we can seek to alter our numbers.

A real concern with environmental issues demands that all of these options be examined in relation to each other. To say that there are too many people in the world amounts to saying that we have not the imagination, will, or ability to do anything about propositions (1), (2), and (3) . In fact (1) is very difficult to do anything about because it involves the replacement of the market exchange system as a working mode of economic integration; proposition (2) has always been the great hope for resolving our difficulties; and we have never thought too coherently about (3) particularly as it relates to the maintenance of an effective demand in capitalist economies (nobody appears to have calculated what the effects of much reduced personal consumption will have on capital accumulation and employment).

I will risk the generalization that nothing of consequence can be done about (1) and (3) without dismantling and replacing the capitalist market exchange economy. If we are reluctant to contemplate such an alternative and if (2) is not performing its function too well, then we have to go to (4). Much of the debate in the western world focuses on (4), but in a society in which all four options can be integrated with each other, it must appear facile to discuss environmental problems in terms of naturally arising scarcities or overpopulation – this, presumably, is the point that the Chinese delegation to the Stockholm Conference was making.

The trouble with focusing exclusively on the control of population numbers is that it has certain political implications. Ideas about environment, population, and resources are not neutral. They are political in origin and have political effects. Historically it is depressing to look at the use made of the kind of sentence we have just analyzed. Once connotations of absolute limits come to surround the concepts of resource, scarcity, and subsistence, then an absolute limit is set for population. And what are the political implications (given these connotations) of saying there is 'overpopulation' or a 'scarcity of resources'? The meaning can all too quickly be established. Somebody, somewhere, is redundant, and there is not enough to go round. Am *I* redundant? Of course not. Are *you* redundant? Of course not. So who is redundant? Of course, it must be *them*. And if there is not enough to go round, then it is only right and proper that *they*, who contribute so little to society, ought to bear the brunt of the burden. And if we hold that there are certain of *us* who, by virtue of our skills, abilities, and attainments, are capable of 'conferring a signal benefit upon mankind' through our contributions to the common good and who, besides, are the purveyors of peace, freedom, culture, and civilization, then it would appear to be our bound duty to protect and preserve ourselves for the sake of all mankind.

Let me make an assertion. Whenever a theory of overpopulation seizes hold in a society dominated by an elite, then the non-elite invariably experience some form of political, economic, and social repression. Such an assertion can be justified by an appeal to the historical evidence. Britain shortly after the Napoleonic Wars, when Malthus was so influential, provides one example. The conservation movement in the US at the turn of this century was based on a gospel of efficiency that embraced natural resource management and labor relations alike. The combination of the Aryan ethic and the need for increased *lebensraum* produced particularly evil results in Hitler's Germany. The policy prescriptions that frequently attach to essays on the problems of population and environment convey a similar warning. Jacks and Whyte (1939), writing in the twilight years of the British Empire, could see only one way out of the scarcity of land resources in Africa:

> A feudal type of society in which the native cultivators would to some extent be tied to the lands of their European overlords seems most generally suited to meet the needs of the soil in the present state of African development … It would enable the people who have been the prime cause of erosion [the Europeans] and who have the means and ability to control it to assume responsibility for the soil. At present humanitarian considerations for the natives prevent Europeans from winning the attainable position of dominance over the soil.
>
> (Jacks and Whyte 1939: 276)

Such direct apologetics for colonialism sound somewhat odd today.

Vogt, whose book *The Road to Survival* appeared in 1948, saw in Russian overpopulation a serious military and political threat. He argued that the Marshall Plan of aid to Europe was the result of an unenviable choice between allowing the spread of communism and providing international welfare, which would merely encourage population increase. He also points to the expendability of much of the world's population:

There is little hope that the world will escape the horror of extensive famines in China within the next few years. But from the world point of view, these may be not only desirable but indispensable. A Chinese population that continued to increase at a geometric rate could only be a global calamity. The mission of General Marshall to this unhappy land was called a failure. Had it succeeded, it might well have been a disaster.

(Vogt 1948: 238)

It is ironic indeed that this prediction was published in the very year that Mao Tse-tung came to power and sought, in true dialectical fashion, to transform China's problem into a solution through the mobilization of labor power to create resources where there had been none before. The resultant transformation of the Chinese earth (as Buchanan (1970) calls it) has eliminated famine, raised living standards, and effectively eliminated hunger and material misery.

It is easier to catch the political implications of overpopulation arguments in past eras than it is in our own. The lesson which these examples suggest is simply this: if we accept a theory of overpopulation and resource-scarcity but insist upon keeping the capitalist mode of production intact, then the inevitable results are policies directed toward class or ethnic repression at home and policies of imperialism and neo-imperialism abroad. Unfortunately this relation can be structured in the other direction. If, for whatever reason, an elite group requires an argument to support policies of repression, then the overpopulation argument is most beautifully tailored to fit this purpose. Malthus and Ricardo provide us with one example of such apologetics. If a poverty class is necessary to the processes of capitalist accumulation or a subsistence wage essential to economic equilibrium, then what better way to explain it away than to appeal to a universal and supposedly 'natural' law of population?

Malthus indicates another kind of apologetic use for the population principle. If an existing social order, an elite group of some sort, is under threat and is fighting to preserve its dominant position in society, then the overpopulation and shortage of resources arguments can be used as powerful ideological levers to persuade people into acceptance of the status quo and of authoritarian measures to maintain it. The English landed interest used Malthus's arguments thus in the early nineteenth century. And this kind of argument is, of course, even more effective if the elite group is in a position to create a scarcity to demonstrate the point.

The overpopulation argument is easily used as part of an elaborate apologetic through which class, ethnic, or (neo-)colonial repression may be justified. It is difficult to distinguish between arguments that have some real foundation and arguments fashioned for apologetic reasons. In general the two kinds of arguments get inextricably mixed up. Consequently, those who think there is a real problem of some sort may, unwittingly, contribute strength to the apologists, and individuals may contribute in good faith to a result which, as individuals, they might find abhorrent.

And what of the contemporary ecology and environmental movement? I believe it reflects all of the currents I have identified, but under the stress of contemporary events it is difficult to sort the arguments out clearly. There are deep structural problems to the capitalist growth process (epitomized by persistent 'stagflation' and international monetary uncertainties). Adjustments seem necessary. The welfare population in the US is being transformed from a tool for the manipulation of effective demand (which was its economic role in the 1960s) into a tool for attacking wage rates (through the work-fare provision), and Malthus's arguments are all being used to do it. Wage rates have been under attack, and policies for depressing real earnings are emerging in both America and in Europe to compensate for falling rates of profit and a slowdown in the rate of capital accumulation. There can be no question that the existing social order perceived itself to be under some kind of threat in the late 1960s (particularly in France and the US, and now in Britain). Was it accidental that the environmentalist argument emerged so strongly in 1968 at the crest of campus disturbances? And what was the effect of replacing Marcuse by Ehrlich as campus hero? Conditions appear to be exactly right for the emergence of overpopulation arguments as part of a popular ideology to justify what had, and what has, to be done to stabilize a capitalist economic system that is under severe stress.

But at the same time there is mounting evidence (which has in fact been building up since the early 1950s) of certain ecological problems that now exist on a worldwide as opposed to on a purely local scale (the DDT example being the most spectacular). Such problems are real enough. The difficulty, of course, is to identify the underlying reason for the emergence of these difficulties. There has been some recognition that consumption patterns induced under capitalism may have something to do with it, and that the nature of private enterprise, with its predilection for shifting costs onto society in order to improve the competitive position of the firm, also plays a role (Kapp 1950). And there is no question that runaway rates of population growth (brought about to a large degree by the penetration of market and wage-labor relationships into traditional rural societies) have also played a role. But in their haste to lay the origin of these problems at the door of 'overpopulation' (with all of its Malthusian connotations), many analysts have unwittingly invited the politics of repression that invariably seem to be attached to the

Malthusian argument at a time when economic conditions are such as to make that argument extremely attractive to a ruling elite.

Ideas are social relations; they have their ultimate origin in the social concerns of mankind and have their ultimate impact upon the social life of mankind. Arguments concerning environmental degradation, population growth, resource scarcities, and the like can arise for quite disparate reasons and have quite diverse impacts. It is therefore crucial to establish the political and social origins and impacts of such arguments. The political consequences of injecting a strongly pessimistic view into a world structured hierarchically along class and ethnic lines and in which there is an ideological commitment to the preservation of the capitalist order are quite terrifying to contemplate. As Levi-Strauss warns in *Tristes Tropiques*:

> Once men begin to feel cramped in their geographical, social and mental habitat, they are in danger of being tempted by the simple solution of denying one section of the species the right to be considered human.
>
> (Levi-Strauss 1973: 401)

Conclusions

Twentieth-century science in the western world is dominated by the tradition of Aristotelian materialism. Within that tradition, logical empiricism, backed by the philosophical strength of logical positivism, has provided a general paradigmatic basis for scientific enquiry. More recently the 'model builders' and the 'systems theorists' have come to play a larger role. All of these methods are destined to generate Malthusian or neo-Malthusian results when applied to the analysis of global problems in the population-resources relation. Individual scientists may express optimism or pessimism about the future, while the results of scientific investigation may indicate the inevitable stationary state to be far away or close at hand. But, given the nature of the methodology, all the indicators point in the same direction.

The political consequences that flow from these results can be serious. The projection of a neo-Malthusian view into the politics of the time appears to invite repression at home and neo-colonial policies abroad. The neo-Malthusian view often functions to legitimate such policies and, thereby, to preserve the position of a ruling elite. Given the ethical neutrality assumption and the dominant conception of scientific method, all a ruling elite has to do to generate neo-Malthusian viewpoints is to ask the scientific community to consider the problems inherent in the population-resources relation. The scientific results are basically predetermined, although individual scientists may demur for personal 'subjective' reasons.

It is, of course, the central argument of this paper that the only kind of method capable of dealing with the complexities of the population-resources relation in an integrated and truly dynamic way is that founded in a properly constituted version of dialectical materialism.

This conclusion will doubtless be unpalatable to many because it *sounds* ideological to a society of scholars nurtured in the belief that ideology is a dirty word. Such a belief is, as I have pointed out, ideological. Further, failure to make use of such a method in the face of a situation that all regard as problematic, and some regard as bordering on the catastrophic, is to court ignorance on a matter as serious as the survival of the human species. And if ignorance is the result of the ideological belief that science is, and ought to be, ideology-free, then it is a hidden ideology that is the most serious barrier to enquiry. And if, out of ignorance, we participate in the politics of repression and the politics of fear, then we are doing so largely as a consequence of the ideological claim to be ideology-free. But then, perhaps, it was precisely that participation that the claim to be ideology-free was designed to elicit all along.

CHAPTER 4

On countering the Marxian myth – Chicago-style

First published in Comparative Urban Research, *1978.*

The invitation to engage in a *dialogue* between 'urban sociologists in the Chicago School tradition and Marxian challengers to that tradition' sounds, at first blush, like an invitation to the sheep to come sit down and parley with the wolves. It certainly suggests a reduction of the level of combat from a grand gladiatorial contest of words and ideologies to a much quieter, and perhaps more subtle, level of polemical jousting. What is interesting about the idea is that an analysis of the pitfalls which pockmark the approach to the conference room can tell us a great deal about the real differences which arise as 'scientists' from radically different traditions seek to understand the same material phenomena.

Consider, first, the difficulty of establishing the rules for debate. We might agree at the outset, for example, that polemical namecalling is not very useful. The problem with such a rule is that we have to know when a category is a perjorative name. I am called a 'Marxian challenger'. I object to that for two reasons. First, I regard myself primarily as a *scientist* seeking a comprehensive understanding of the world in which we live. I have turned to the Marxian categories because they are the only ones I have so far come across which allow me to make sense of events. Secondly, I know only too well that the name 'Marx' and the epithets 'Marxist' and 'Marxian' are so colored with a history of cold-war invective and McCarthyite witch-hunting that to be so called is to have the majority of any professional audience in the United States turn away from me before I even begin. What chance do I have, then, pitted against the respectable-sounding 'urban sociologists in the Chicago School tradition'? I could try to even things up a bit by seeking somehow to conjure up associations between Chicago urban sociology and, say, Al Capone. But even were I to succeed in such an odd enterprise, it would scarcely redress the balance because Al Capone is, in some respects, an American folk hero in ways that Karl Marx most definitely is not. Or I could take revenge for being dumped into the 'Marxian' category by referring to the Chicago School

68

urban sociologists as 'bourgeois social scientists'. Such a characterization is perfectly valid from the standpoint of the 'Marxian tradition', but it is surprising how upset our opponents get when so described. My revenge would, of course, be shortlived because, next to the dreaded name of Marx himself, the word 'bourgeois', in American thought at least, is a big giveaway which amounts almost to 'red-baiting' (or at least 'red-tainting') anyone who uses it except for fun in quotation marks.

I draw attention to these problems because a necessary precondition for dialogue between 'Marxian challengers' and 'bourgeois social scientists' is the creation of a basic vocabulary with commonly-agreed meanings. In its most elemental form, the struggle between the two traditions is a struggle to establish a hegemonic system of concepts, categories and relationships for understanding the world. It is a struggle for language and meaning itself. And it is not simply the literal meanings which attach to jargon words that we are concerned with, for we need to take cognizance, too, of the multiple connotations and the weight of historical content which particular words contain. Again, the 'Marxian challengers' seem to start with a handicap since they are expected to acquire a thorough understanding of 'bourgeois' terms while most 'bourgeois social scientists' engage even in polemics about Marx without, it seems, paying the least attention to the integrity of the Marxian meanings.

Since I have sat on both sides of the fence (acquiring my tenure, conveniently enough, on the strength of thoroughly 'bourgeois' work), I can testify to some of the extraordinarily complex problems which arise when confronting the Marxian meanings from out of a 'bourgeois' positivist and analytical tradition. It has taken me almost seven years to acquire even a limited fluency in the use of the Marxian concepts, and I am always being surprised by the continued unfolding of new patterns of meanings and relationships. I point this out not to reject any commentary on Marx except by those who have gone through certain 'rites of passage' to the received wisdom, but to emphasize the sheer difficulty of dialogue under such conditions. I would add that this difficulty is not something inherent in the Marxian representation of affairs – indeed there is something very natural and common-sensical about the Marxian representation. It arises because most of us are 'trained' for most of our lives in an empiricist, analytical tradition which brooks no trace of dialectical thought or relational meanings. My own experience is, quite simply, that the more highly trained scholars are in 'bourgeois' social science, the more they have difficulty in acquiring even a sense of what the Marxian representation is all about.

This difficulty does not arise when looked at from the other side of the fence. For those working in the Marxian tradition, an understanding of bourgeois categories is a necessity. The matrix of bourgeois thought, understood in its own terms, provides a vital and malleable raw material which, when fashioned with the aid of the awesome critical tools which Marx provides, can reveal a transformed and *more valid* understanding of the world. At this juncture I do not propose to examine this process in detail, because I only wish to establish one crucial point. The Marxian tradition already contains a 'dialogue' of sorts between bourgeois and Marxian representations simply because the former provides much of the raw material which, when transformed, provides the latter with its own understanding. Put another way, a dialectical relation must, in the Marxian view, *necessarily* exist between bourgeois and Marxist thought; the former is a representation of the world drawn up from the standpoint of *capital* while the latter is a representation of the world drawn up in terms of the opposition of *labor*.

Bourgeois social theory typically denies such a class basis for knowledge and purports to define some 'objective', 'neutral' understanding of the world, free from class bias. To this end it has fashioned a certain kitbag of tools and methods and theoretical frameworks, as well as a whole corpus of categories, concepts, and relationships. In its own terms, bourgeois social science appears to be a reasonably successful endeavor in the sense that a lot of people make a living at it while bourgeois society seems to be reasonably well-persuaded of its utility. Why on earth, then, would bourgeois social scientists wish to engage in a dialogue with their 'Marxian challengers'? The idea of a dialectical relation between the two traditions is totally foreign to the bourgeois viewpoint, in part because the practice of the dialectic is lacking and in part because to concede the notion of a fundamental opposition between the two comes dangerously close to conceding what the bourgeois social scientist is least willing to concede, namely, the class basis of knowledge. The relationship between the two traditions is, therefore, very different depending upon which side one is on.

Having sat on both sides of this fence, I do have some feeling for why the bourgeois social scientist can and does flirt with the Marxian challenge. Inherent in the bourgeois tradition is a notion of 'free intellectual enquiry' which represents, at its best, certain conditions of work (freedom from external dominance) and an internalization of the idea of 'the pursuit of truth' as a matter of personal conscience. There has always been a certain tension, therefore, between bourgeois thought and 'the powers that be' in society. From the latter's standpoint, the dynamics of capitalism require a good deal of scientific, technical, organizational and social inventiveness, and there is, therefore, a certain trade-off between permitting the conditions of free intellectual enquiry to flourish and suppressing those strains of thought which are anathema to the preservation of the existing social order. From time to time, the 'powers that be' organize a massive house-cleaning of sorts and 'clean out the reds from under the bed', as they did in the McCarthy era in the United States and as they are currently doing most systematically in West Germany. But the notion of free intellectual enquiry persists; and, in the course of events, individual bourgeois scholars may pick up Marx, read it, understand it, and even become convinced by it. However, individual actions of this sort do not a social movement make. To explain phases of wholesale flirtation with Marx we must appeal to a different kind of explanation.

The analytic and empirical stance which dominates in bourgeois social science inevitably leads to an excessive fragmentation of knowledge. The fragmentations begin with the formation of disciplines and sub-disciplines and proceed to 'areas of specialization' right the way on down to minutiae. This technical division of labor has its social parallel in the formation of professional associations, interest groups and specialized committees, which frequently treat a particular subject matter as a special 'turf', as their own preserve, hanging out all manner of 'keep off' and 'no trespassing' signs. Mutual respect for 'property rights' to areas of enquiry leads also to a certain self-censorship as individual researchers come to accept the view that there are vast areas of enquiry that they are not competent to traverse. Economists say that political and social questions lie outside of their jurisdiction, sociologists say economic and political questions are not within their proper domain, and the like. The technical division of labor is not unproductive of new understandings, and provided the social divisions are appropriate, they may also be useful. But when the technical and social divisions become ossified, the problems begin, particularly in a society which demands dynamism in thought as well as in production. A disciplinary framework which broadly emerged at the beginning of this century and which has set itself in concrete ever since is hardly likely to remain consistent for long in relation to a society undergoing fundamental changes.

The fragmentations and specializations also produce a sense of diminishing returns after a while. If we search the literature, there must, by now, be literally thousands upon thousands of hypotheses which have been proven 'true' at the 0.05 level of significance. How on earth are we to make sense of all of these true statements? Excessive analytic fragmentation and empirical specialization inevitably spawns a desire for synthesis, and the 'grand synthesisers' and 'systematisers' must have their day in the sun.

Consider, from this standpoint, the disciplinary specialty we call 'urban sociology'. Such a title indicates two immediate limitations to enquiry: the 'urban' and the 'sociological'. The isolation of the 'urban' as a distinctive epistemological object of enquiry is problematic even within the framework of bourgeois social science. It is difficult to contain the 'urban' within strict bounds. There is a certain ambiguity at the edges (where does the urban begin and end, both physically and conceptually?) while relationships to other epistemological objects, such as the 'rural', cannot be avoided in any deep study of phenomena such as migration, poverty, and the like. Yet in spite of these difficulties, the urban is constituted as a distinctive epistemological object of enquiry in bourgeois social science, and researchers happily delve away, sometimes seemingly oblivious of all else, within this restricted domain.

The second limitation is the 'sociological'. This, too, is problematic even within the bourgeois framework of study, because the sociological, economic, psychological 'factors' in material life have an unhappy habit of running into each other. But again, when times are quiescent, the limitation of the sociological comes broadly to be accepted, and specialized workers beaver away to their hearts' content, not only within the urban but within certain bounds as to what kinds of relationships are to be examined amongst what kinds of social groupings.

There is a third limitation which arises out of the empirical stance of bourgeois social science. This is the limitation within space and time. The uniqueness of position in space and time, considered as absolute frameworks for locating objects and events, has the habit of asserting itself in spite of the supposed generic and low-seeking posture of positivist social science. As a consequence, articles on 'ethnicity in Sydney in the 1960s', 'kinship in Dar-es-Salaam in the postcolonial era', or 'family relations in Flint, Michigan, in the depression of the 1930s' become the standard bread-and-butter product and diet of the urban sociologist.

Such standard fare rarely satisfies the gourmet and can become quite indigestible after a while to even the most unimaginative and insatiable urban sociologist. The fragmentations and limitations, the product of a well-meaning division of labor, can, under such circumstances, come to be construed as barriers, rather than as aids, to understanding. The urge to see the urban in some broader framework, to overcome the limitations of the 'purely sociological' perspective, and to reach out to comparative studies across space and time, is never entirely absent. The deeper the fragmentations, the more the moves towards synthesis and comparison become imperative. Within the autonomous development of urban sociology, there is a fundamental opposition – dare I say a 'dialectical movement' – between the quest for specificity within limited horizons and the quest for generality within a broader universe of discourse.

But the autonomy which exists within urban sociology is only a *relative* autonomy. Somewhere along the line the urban sociologist has to prove his or her utility by coming up with some useful ideas relevant to the problems of the time. And no matter what the personal beliefs or political affiliation of individual sociologists, the profession of 'urban sociology' as a whole is accountable to the 'powers that be' which allocate resources to it. It would be idle to pretend in North America that these powers that be represent all elements in society equally and that in this the 'working class' gets as fair a shake as 'big corporate capital'. The 'social relevance' of 'urban sociology' is, of course, a very hazy notion, but at the bottom of it lies (1) a set of problems which are real enough to cause serious concern to (2) 'the powers that be' in society.

At this point, we have to consider the relation between the structure of the problems and the structure of the science seeking answers to these problems. If the problems are fragmented, then a fragmented approach based on a compartmentalized division of labor may prove adequate and successful to providing the solutions. The social science of the 1950s had much of this quality to it. But it is entirely another matter when the problems build into more complex configurations as they did in the 1960s. Within the US, problems of social inequality, minority group oppression, and social unrest had a strongly urban expression so that the social explosions could be characterized as a distinctively urban crisis. The crisis appeared localized within the urban, even if rather complex. Limited 'urban solutions' could be pursued. This search increasingly led to the conclusion that this limited view was false – everything appeared to be relating to everything else. The structure of the problem led people to reach out for broader theoretical frameworks. Yet the institutionalization of the division of labor was such that urban economists could still write learned tracts on municipal finance as if social inequality or racism were non-existent, while sociologists could write on the latter topics as if the imperative economic logic of municipal finance had no meaning. The response of the 'powers that be' was a typical combination of stick and carrot. Money flowed to a variety of inter-, multi-, cross- and even metadisciplinary studies of the urban problem as well as to those who, armed with a variety of new techniques such as systems analysis, could promise the grand synthetic solution. The results were scarcely gratifying and the solutions at best transitory.

The gathering crisis of western capitalism, dictated by its own essential logic, was propelling events in ways which bourgeois social science could not grasp. Economists were bewildered by stagflation, and once the empirical generality of the celebrated Phillips curve was shown no longer to be operative in the conditions of the 1970s, they lost any sense of coherent policy solutions. Events such as those which overtook New York City in 1975 clearly indicated to the 'urban sociologist' that solutions which did not incorporate the facts of municipal finance in the context of a financial system which was global in scope were scarcely appropriate. There has never been a major crisis in capitalist society which has not simultaneously meant a crisis for its distinctive form of social science. And so it is in the troubled and crisis-ridden 1970s. Understandably, the 'powers that be' at such conjunctures grow impatient with economists who pass the buck to sociologists who promptly pass it on to the political scientist, all of whom claim that they have no competence to understand the workings of the system as a whole.

Capitalist crises translate into crises in bourgeois social science not simply because the latter fragments in ways inappropriate to understanding the former. Bourgeois social science inclines, by virtue of being bourgeois, to interpret social affairs in terms of competing interests and functions within a social totality which is perceived to be either actually or potentially harmonious in its workings. Pluralist political theories, neoclassical economics, and functionalist sociology all have that in common. Understood from the Marxian perspective, bourgeois science must do that because it must assert the potentiality for harmony between capital and labor. Such a social science has no profound theory of crises and does not even perform well when it comes to questions of social change. When crises erupt, they are either greeted with stunned silence or attributed to some malignant external force (accidents, wars, famines, pestilence or seemingly arbitrary acts like Arab intransigence on oil prices). The political economists, Marx once observed, are reduced at times of crisis merely to saying that all would be well if the economy would only perform according to their textbooks.

At such conjunctures, the Marxian theory takes on a peculiar fascination for a number of reasons. First of all, the Marxian theory is pre-eminently a theory of crisis. Marxian theory sees historical movement as founded in a deep and pervasive struggle between competing and opposing forces which are anything but harmonious with each other (except by accident!). The polarizations which occur in the course of a crisis are the crucible out of which new social configurations emerge, with one power asserting its power of domination over another. When crises erupt, a theory which deals directly with crisis formation begins to be taken seriously both by the populace at large and by those working in the sheltered groves of academia. Capitalist crisis generates a resurgence of interest in the Marxian theory to the same degree that it provokes a crisis within bourgeois social science.

But there are other attractions for the bourgeois social scientist provided that the fog of misconceptions and prejudices about Marx can be cleared away. To illustrate the argument I will dive to the philosophical and epistemological level to try and show something of what the Marxian theory is and what it is not.

The first point to make is that the Marxian theory is holistic and works with a particular sense of how the parts relate to the totality. The totality is regarded neither as an aggregate of elements nor as something that has a meaning independent of its parts, but as a 'totality of internally related parts' each of which can be conceived of as 'an expandable relation such that each one in its fullness can represent the totality'. Having sought to elaborate on the meaning of this in the final chapter of *Social Justice and the City*, I will not repeat myself here. Put simply, the Marxian method accepts fragmentation and separation for purposes of analysis only on the condition that the integrity of the relation between the whole and the part is maintained intact. The Marxian theory thus starts with the proposition that everything relates to everything else in society and that a particular object of enquiry must necessarily internalize a relation to the totality of which it is a part. The focus of the enquiry is, then, on the *relations* of the epistemological object to the totality. The purpose of isolating a particular epistemological object for enquiry – as Marx does when he commences his analysis of capitalism by an examination of the commodity – is to discover the relations within it that reveal the real nature of the capitalist mode of production. Marx starts out to do directly what the urban researchers of the 1960s found themselves compelled to do: understand the manner in which everything relates to everything else.

Yet the open and direct relational and dialectical style of investigation which Marx practices is extraordinarily difficult for the bourgeois social scientist to grasp. The relational aspect simply acknowledges that the meaning of an object or action (digging a ditch, say) cannot be understood without an understanding of the whole social framework of which it is a part (the action of digging changes its meaning according to whether the digging is done under the social relations of slavery, wage labor, collective production, and so on). The meaning is internalized *within* the action, but we can discover what the action internalizes only by a careful study and reconstruction of the *relations* it expresses to surrounding events and actions.

The same argument applies to the analysis of the 'urban'. We can find 'cities' at various times and places, but the category 'city' or 'urban' changes its meaning according to the context in which we find it. The self-same set of buildings will assume a different meaning under capitalism compared to feudalism or socialism. The surface appearance may be the same but the underlying essential meaning changes. To speak of the 'urban' as if it had a universal meaning is, from the Marxian perspective, to engage in a reification which can only 'mystify' (if I may use another word which has a technical meaning in the Marxian critique but which usually provokes either anger or mirth amongst bourgeois social scientists). The 'urban problematic' from the Marxian standpoint is to uncover the changing meaning of both the word and what it represents in the context of the social dynamic of history.

The dialectical aspect of Marxist thought focuses upon contradiction. Given the relational meanings, we must expect the contradictions to become internalized within particular objects or events and therefore fundamental to understanding meaning. The action of the laborer digging the ditch may express a certain energy which springs from pride in work, from a sense of shaping nature to human need, at the same time as it expresses a certain listlessness which derives from the alienation inherent in the conditions of wage labor under the social relations of capitalism. The 'urban' likewise expresses numerous contradictions. Consider, for example, the 'urban' in its purely physical aspect. On the one hand, the built environment of the capitalist city can be viewed as the crowning glory of human achievement, as a testimony to the power of human labor over nature, as a manifestation of the ability to *create* whole landscapes, albeit in the image of capitalist social relations. On the other hand, the weight of the past which such a physical landscape expresses can act as a prison for daily life and as a barrier to future development. The dominance of 'dead' past labor is asserted over the creative powers of living labor. Once put in motion, this idea can help us identify a whole set of contradictions within the 'urban'. Capitalism struggles to create a physical landscape appropriate to its needs and purposes (both in production and consumption) at one point in time, only to find that what it has created becomes antagonistic to its needs at a future point in time. Part of the dynamic of capitalist accumulation is the necessity to build whole landscapes only to tear them down and build anew in the future.

This notion of contradiction in the Marxian view of the urban problematic is not at all abstract. On the contrary, it is very real. Consider what is currently happening to the built environment of New York City compared to the rate of investment in physical infrastructure in, say, Houston. And how else can we explain all of those mechanisms which have been created in the US for organizing on a systematic basis the restructuring of built environments to new needs: urban renewal programs, 'red-lining practices' by the financial institutions, and the like? From the Marxian standpoint, contradictions within society inevitably spawn a variety of attempts to deal with them. And the 'solutions' inevitably become a part of 'the problem' (consider the history of housing policy through the 1968 act to the Nixon moratorium, for example).

The Marxian emphasis upon relations and contradictions within a totality yields, when properly executed, a unity of analysis and synthesis. The initial response of the social scientist trained in the bourgeois framework of thought is to misinterpret this simply because of the difficulty posed by the relational and dialectical way of proceeding. The misinterpretation may be quite fruitful for bourgeois social science. Sociologists often interpret Marx as trying to import an 'economic' viewpoint into sociology, and they may indeed broaden their perspective as a response. Along with this goes a tendency to see the Marxian theory as just another theory amongst a whole bunch of possible theories which might be used to explain some phenomena but not others. The Marxian theory can be dusted off for use when the problem is right for it, and typically the sociologist does so when the 'economic factor' is important. The sociologist usually adds, quite emphatically, that 'Of course, it is inappropriate to try and reduce all issues to questions of political economy as Marxists are wont to do.' The odd thing is that my contact with economists yields a different version of the same story. They see the Marxists as seeking to introduce a sociological component into economic analysis – a component which some may concede is very helpful under certain conditions – but they are also usually quick to point out that 'All problems cannot be reduced to questions of class relations as the Marxists invariably imply.' The Marxists are either being just plain contrary in trying to turn economists into sociologists and sociologists into economists, or else they are seeking to do something quite different. So how can we explain what the Marxists are up to?

In the first place, disciplinary boundaries make no sense whatsoever from the Marxian standpoint. The technical division of labor is obviously necessary but its social representation is to be rejected. But at this point the Marxian challengers encounter a peculiar difficulty. We live in a world in which the bourgeois framework for organizing knowledge is hegemonic. Individually, we must appear expert in some discipline and to some degree conform to its rules if we are to be listened to or even to gain employment.[1] The Marxian challenge has therefore to be mounted within

---

[1] I am particularly sensitive to this problem on a number of counts. My employment prospects are almost entirely enclosed within the professional framework of geography; yet my colleagues in this field typically dismiss my work as 'not geography', but political economy, sociology, and the like. But I do not possess the professional credentials to be considered a bona fide critic in these fields. I notice, for example, that *Social Justice and the City* was not thought worthy of review in that prestigious Chicago-based journal of

the existing framework for knowledge which makes it appear that there is a 'Marxist sociology' or a 'Marxist economics' when in truth there is just Marxian analysis and that is that. The Marxian challenge thus attempts to be subversive of all disciplinary boundaries.

In the second place, Marx did not disaggregate the world into 'economic', 'sociological', 'political', 'psychological', and other factors. He sought to construct an approach to the totality of relations within capitalist society. There are many aspects of this approach which are problematic, however, and there are plenty of controversies within the Marxian tradition as a consequence. There are various schools of thought (including one which is very 'economistic' and 'reductionist') as well as a variety of good and bad works in the Marxian tradition. Obviously, it is difficult for the bourgeois social scientist to discriminate between the various schools of thought and the good and bad work; from outside they tend to all look alike. It takes great patience and a fair amount of sophistication to wade through the mass of Marxian argument, and on the first run-through the bourgeois social scientist is quite naturally going to hang onto the bourgeois categories to try and make sense of things. The inevitable result is misinterpretation. Consider, for example, the controversy within the Marxian tradition over the relations between 'the economic base' (comprising the 'productive forces' and the 'social relations' of production) and the 'super-structural' forms of politics, ideology, consciousness, law, institutions, and the like. The arguments here are myriad and complex, but the one reduction which is disastrous for preserving the integrity of the Marxian meaning, is to equate the 'economic base' in the Marxian theory with the 'economic factor' as bourgeois social science typically treats it. The similar-sounding phrases express quite different meanings.[2]

All of this may help to establish what the Marxian theory is not. It is more difficult to state clearly what the Marxian theory *is* because, in the final analysis, the Marxian theory boils down to concrete practices, some of which are carried on in the academic arena. So the best way for me to proceed at this point is to illustrate, very briefly, the thinking which underlies my current academic practice as I struggle to grasp the real meaning of the 'urban question'. The presentation must perforce be brief and unduly schematic, but I am prepared to take the risk of misinterpretation in the interests of dialogue.

---

sociology. And I have also come to recognize that no matter what I do, the work of Castells poses a far more serious challenge to sociologists simply because it is a challenge mounted from within the field and a challenge which seems to demand a reconstitution of that field of study.

[2] Some of the best insights on this can be gained from a careful reading of Ollman (1971).

To begin with, I confine myself to an attempt to understand the *capitalist* forms of urbanization. This emphasis is a necessary reflection of the epistemological position that the 'urban' has a specific meaning under capitalism which cannot be carried over without a radical transformation of meaning into other social contexts.[3]

Within the framework of capitalism, I hang my interpretation of the 'urban' on the twin themes of *accumulation* and *class struggle*. The two themes are integral to each other and have to be regarded as different sides of the same coin – different windows from which to view the totality of capitalist activity. The class character of capitalist society means the domination of labor by capital. Put more concretely, a class of capitalists is in command of the work process and organizes that process for the purposes of producing profit. The laborer, on the other hand, has command only over his or her labor power which must be sold as a commodity on the market. The domination arises because the laborer must yield the capitalist a profit in return for a living wage. All of this is terribly simplistic, of course, and actual class relations and an actual system of production (comprising production, services, necessary costs of circulation, distribution and exchange, and so on) are much more complicated. The essential Marxian insight, however, is that profit arises out of the domination of labor by capital but that the capitalists as a class must, if they are to reproduce themselves, expand the basis for profit. We thus arrive at a conception of a society founded on the principle of 'accumulation for accumulation's sake, production for production's sake'. The theory of accumulation which Marx constructs in *Capital* amounts to a careful enquiry into the dynamics of accumulation and an exploration of its contradictory character. It sounds (and reads) rather 'economistic', but the other side of the coin of accumulation is that this is simply the means by which the capitalist class reproduces itself at the same time as it reproduces its domination over labor.

We can spin a whole web of argument concerning the 'urban' from an analysis of the contradictory character of capitalist accumulation. Let me first put the contradictions firmly in place. A contradiction arises *within* the capitalist class because individual capitalists, each acting purely in his or her own self-interest in a context of competitive profit-seeking, produce a result which is antagonistic to their own class interest. Marx's analyses suggest that this contradiction creates a persistent tendency towards 'overaccumulation', which is defined as a condition in which too much capital is produced relative to the opportunities to find profitable

---

[3] I attempted to say something about this in Chapter 6 of *Social Justice and the City*, but a much better account has since appeared, Merrington (1975).

employment for that capital. The tendency towards overaccumulation is manifest in periodic crises marked by falling profits, idle productive capacity, over-production of commodities, unemployment, idle money capital, and the like. The second major source of contradiction arises out of the antagonism *between* capital and labor. The relative shares of profits and wages are defined through class struggle. When capital is omnipotent, competition between capitalists tends to drive the wage rate down to the point where capitalists destroy the capacity to realize the values they produce in the market by an excessive reduction in the purchasing power of labor. When labor is very strong, it can hold down profits and check the rate of accumulation, which means a reduction in the rate of expansion of job opportunities, and with technological change the employment opportunities may even diminish. One-sidedness in the class struggle, therefore, produces 'crises of disproportionality' either for capital or for labor. The third set of contradictions arises out of the often antagonistic relation between the capitalist production system and non- or pre-capitalist sectors which may exist within capitalist economies (domestic sectors, peasant sectors, and so on) or be largely external to them (as in some Third-World countries or socialist countries). And finally, we should add the contradiction which inevitably arises between the dynamics of capital and the natural resource base as it is defined in capitalistic terms.

These various 'contradictions' give rise to periodic crises within the capitalist production system. These crises serve to 'rationalize' the system, driving out inefficient enterprises, reducing the power of labor to resist technological change or to command a high wage, bringing non- or pre-capitalistic sectors to heel (often through political force), and so on. Yet to resolve these difficulties in capitalistic terms means to create, in the course of the crisis, conditions for a heightening of class struggle, for a heightening of political awareness and, as a consequence, a growing awareness of the need to explore the socialist alternative.

Let us now turn to consider some of the basic attributes of the capitalist production system because a part of the understanding of the 'urban' that I have to offer comes out of a bringing together of the notion of 'crisis formation and resolution' within the context of the accumulation process. To begin with, the system of production which capital established was founded on a physical separation between a place of work and a place of living. The growth of the factory system, which created this separation, rested on the organization of cooperation and economies of scale in the work process. But it also meant an increasing fragmentation in the division of labor and the seeking of collective economies of scale through agglomeration. All of this implied the creation of a built environment to function as the collective means of production for capital. A part of that built environment must be allocated to the transport of commodities in space, the speed and efficiency of which has a direct impact on the rate of accumulation. As Marx observed, the 'annihilation of space by time' becomes a historic necessity for capital, and with this comes the drive to create configurations of space that are 'efficient' (for capital) with respect to circulation, production, exchange and consumption.[4] Accumulation requires, then, the creation of a physical landscape conducive to the organization of production in all of its aspects (including the specialized functions of exchange, banking, administration, planning and coordination, and the like, which typically possess a hierarchical structure and a particular form of spatial rationality).

But there is also a landscape for consumption, a landscape for living as opposed to working. This is in part created out of the manner in which the bourgeoisie consume their revenues, and it is therefore a particular expression of bourgeois culture with all that this implies. There must also be a landscape for the reproduction of labor power – not only quantitatively, physically and in locations proximate to production activities, but also in terms of those skills, attributes and values which must to some degree be consistent with the capitalist work process. Furthermore, as the purchasing power of labor grows – as it must with accumulation – so the manner in which that purchasing power is expressed in the marketplace mediates the circulation of capital. To an increasing degree, then, the realization of the values produced in the workplace depends on the consumption habits of both the bourgeoisie and labor in the living place. Capital therefore reaches out to dominate living as well as working, and it does so because it must (Harvey 1977). The socialization of labor which takes place in the living place – with all that this implies for attitudes to work, consumption, leisure, and the like – cannot be left to chance. Capital may seek direct forms of domination, as it did in the early model communities and in later experiments such as Pullman, but finds it much more appropriate to seek indirect controls through the mediating power of the state and its associated institutions (educational, philanthropic, religious, and the like). The collectivization of consumption through the state apparatus becomes a necessity for capital with the inevitable consequence that class struggle becomes internalized within the state and its associated institutions. We will see these features again in the context of class struggle in general.

One of the threads which we can prize out from the preceeding argument is the necessity for a specific kind of relation between the processes of accumulation – the means by which the capitalist class reproduces

---

[4] Marx's treatment of these issues has been summarized in Harvey (1975a).

itself – and the creation of a built environment. Whatever else it may entail, the 'urban' implies the creation of such a built environment as a resource system to facilitate capitalist production, exchange and consumption. This provides us with one clear point of contact between the study of the 'urban' and the study of capitalist society. And if our epistemological stance is correct, then the study of the ways in which the built environment is created and of the physical form it assumes will reveal a great deal about the nature of capitalism viewed as a totality. The built environment internalizes within it the contradictory relations inherent in the accumulation of capital. Let us examine this theme for a moment.

The creation of a built environment absorbs a vast amount of productive energy at one point in time to provide a continuous stream of use–value benefits over an extended period of time. The creation of a built environment under capitalism depends upon the existence of a surplus both of capital and labor – surplus here defined in terms of unemployed resources relative to immediate needs. One of the inherent tendencies in the capitalist accumulation process is towards overaccumulation, the production of such surpluses on a periodic basis. This means that the internal dynamic of accumulation periodically creates conditions which are markedly favorable to investment in the built environment. Such periodicity is represented historically in the 'long-swings' in construction activity, in urban building, in investment in transportation, in real estate development, in land speculation, and the like. There is no lack of historical evidence for these long-swings. They have been studied in detail at the international, national and even local levels. (Homer Hoyt's studies of the real-estate cycle in Chicago are surely classic.) The problem with all of these studies is that they lack an adequate theoretical framework to tie the long-swing movements into the dynamics of capitalist accumulations (Thomas 1973; Gottlieb 1976). This link, the Marxian theory can provide.

But this integration also yields certain other insights. The tendency towards overaccumulation is temporarily relieved by the syphoning-off of surplus capital and labor to produce the built environment. But if 'productive' uses cannot be found (directly or indirectly), then the exchange value taken up in the production of the built environment is lost and we have a 'devaluation' of the capital incorporated in it. Such devaluations have often been the precursors to general crises in the accumulation process (consider the role of the collapse of the property sector in late 1973 in triggering the subsequent recession). The periodic devaluation of assets in the built environment is quite a common phenomenon – be it in the form of over-extensions in railroad investments in the nineteenth century, overcapitalization in the urban mass-transit systems at the turn of this century, the real-estate development boom of the 1920s in the US, or the office and development boom of 1969–73 in Britain and the US.

We can push this analysis deeper. Fixed capital in the built environment is both immobile and long-lived. It expresses the power of dead labor over living by committing the latter to certain patterns of use for an extended time within the particularity of spatial location. From this derives the central tension which we have already noted which forces capital to create a landscape, only to have to overcome the barriers which this landscape contains at a later point in time. The need to accelerate the turnover of capital, to overcome space with time, for example, leads capital to create a transport network which is, however, fixed in space. Hence arises the paradox that capital creates fixed spatial systems in order to overcome spatial barriers. With the progress of accumulation, the capital tied up in the transport system itself becomes the barrier to be overcome. Capital, as Marx observed, perpetually creates barriers to its own further development.

Such contradictions must be overcome if capitalism is to survive. Capital is therefore always promoting 'internal revolutions' within the accumulation process – revolutions which are forced through by crises which affect the production and use of the built environment. The ebb and flow of urban investment in both space and time is a product of this permanently revolutionary force which capital itself expresses.

To bourgeois ears, all of this probably sounds very economistic and reductionist. And in any case the 'urban', my opponents will correctly contend, is much more than the 'mere' study of the physical artifact that is the city. My riposte is simple. Bourgeois social science typically engages in reifications, invariably representing social relations as things; consider, for example, how the social relations between capitalists, laborers, and landlords are reduced in bourgeois economic thought to relations between the 'factors of production', land, labor and capital. As a consequence, bourgeois social science often avoids the study of material things (except when it has one of its not infrequent bouts of environmental determinism) because there is not much of interest to be discovered about things studied as things. The whole thrust of the Marxian argument is, of course, to concentrate on the social meaning of things. Starting with the physical artifact that is the city, we can reach out, step by step, into the myriad social relations (between landlords and financiers, building laborers, artisans and capitalist builders, between users and producers, between the state and individuals, between communities and speculators, and so on) and into the extraordinary complexity of interactions, conflicts, coalitions within a framework of institutional arrangements, all of which lead to the creation of this physical landscape. If the aim of the project is to uncover the social meaning of this physical landscape, then the horizons for research are bounded only by the limits of the totality of capitalist society. But just in case my bourgeois opponents are unconvinced by such an assertion, I will switch my 'window' on the world and look at the 'urban' from the standpoint of class struggle.

The central point of tension between capital and labor lies in the workplace and is expressed in struggles over the conditions of labor and the wage rate. These struggles. take place in a context. The law (property rights, contract, combination, and the like), together with the power of the capitalist class to enforce its will via the power of the state, are obviously fundamental, as any casual reading of labor history will abundantly illustrate. But the nature of the demands, the capacity of labor to organize, and the resolution with which these struggles are waged, depend upon a whole set of contextual relations, some of which bring us close to the terrain of traditional bourgeois urban sociology.

The quantity of labor in relation to the needs of capital is crucial, for example. The significance to the supply of labor power lies simply in this: that the greater the labor surplus and the more rapid its rate of expansion, the easier it is to control the struggle in the workplace to the benefit of capital and, therefore, the higher the rate of accumulation can be (other relations remaining constant, of course). Capital therefore has a direct interest in expanding its capacity to mobilize an 'industrial reserve army', either by stimulating the migration of capital or labor (including temporary migrations) or by drawing hitherto 'unused' elements in the population (women and children, labor employed in non-capitalist sectors, and so on) into the labor force. The study of the migration and 'mobilization' processes in capitalist societies can thus be related to accumulation via the whole history of class struggle in the workplace. It was no accident that Ford used newly-landed immigrant labor almost exclusively to man his assembly lines and that United Steel, when faced with labor troubles of its own, turned to black workers from the south to break the strikes.

We also have to consider the costs of reproduction of labor power at a standard of living which is itself set by a whole host of cultural, historical, moral and environmental considerations. A change in these costs or in the definition of the standard of living itself has obvious implications in relation to wage demands. Conversely, capitalists have increasingly looked to inflation as a tool for managing the real wage rate and have thus avoided the violent opposition which direct wage-cutting has often provoked. Furthermore, as capitalism has advanced, so the internal market formed by the purchasing power of labor has become significant to accumulation.

Consequently, the consumption habits of labor – the standard of living once more – become a part of the focus of struggle.

Finally, we have to consider a whole host of qualitative aspects of labor power encompassing not only skills and training, but also attitudes of mind, levels of compliance, the pervasiveness of the 'work ethic' and 'possessive individualism', and the variety of fragmentations within the labor force which derive from the division of labor, occupational status, religious, ethnic, racial features, and the like. The ability and the urge for labor to organize along class lines depends upon the creation and mainte-nance of a sense of class consciousness and class solidarity in spite of these fragmentations. The ideological struggle for the 'hearts and minds' of individuals in the labor force is fundamental to understanding the dynamics of class struggle.

This leads us to the notion of *displaced* class struggle, by which I mean class struggle within and around the contextual relations of class struggle in the workplace. We can trace these displacements to almost every corner of the social totality and extend the idea to deal with class struggle within the institutions of community, within the institutions of the state, and the like. Consider, as an example, the struggles around public education. Charles Dickens had some pertinent things to say about this. That para-digmatic bourgeois, Mr Dombey, regarded public education as a most excellent thing provided it taught the common people their proper place in the world, and in *Hard Times* Dickens constructed a brilliant satirical counterpoint between the factory system and the educational, philan-thropic, and religious institutions designed to cultivate habits of mind amongst the working class that were conducive to the workings of the factory system. Public education as a right was a basic working-class demand, but the bourgeoisie quickly grasped that public education could be mobilized against the interests of the working class. The struggle over social services in general is not only over their provision but over the very nature of what is provided. A national health care system fashioned as collective consumption in the interest of capital accumulation In the 'medical-care industry' or which defines ill-health as inability to go to work (inability to produce surplus value for capital) is very different indeed from a health care system dedicated to the total mental and physical well-being of the individual in a given physical and social context.

The socialization and training of labor and the 'management of human resources' (with all that this phrase implies about the management of labor power as fodder for the production process) are far too important to be left to chance because it is out of the crucible of these relations that the real stuff of class-consciousness is forged. The links and relations are intricate and difficult to unravel. I will briefly consider two facets, both located within the living space rather than within the workplace.

The demand for adequate shelter is clearly high on the list of priorities from the standpoint of labor. The broad lines of class struggle around the 'housing question' have had a major impact upon the shaping of the 'urban' as we now know it. We can trace some of the links to class struggle in the workplace. To begin with, the cost of shelter is an important determinant of the cost of labor power. The more labor has the capacity to press home its wage demands, the more capital will become concerned about the costs of housing and the more it will tend to support programs for cheap subsidized housing construction and the like. Also, since housing is such an important item in the laborer's budget, it becomes a prime target for commodity production and accumulation. In the US, the integration of housing production and accumulation became so complete after 1945 that the former came to function as a Keynesian 'contra-cyclical' regulator for the accumulation process as a whole, at least until the debacle in the construction industry in 1973–6. But housing implies more than mere direct commodity consumption. To begin with, the whole structure of consumption is related to the form which housing provision takes. The dilemmas of potential overaccumulation which faced the US in 1945 were in part resolved through a suburbanization process which created new modes of living in accordance with the needs of capital accumulation. Furthermore, individualized homeownership promoted via a credit system which encourages debt encumbrance, is hardly neutral in relation to class struggle. In the 1930s in the US, individual homeownership was seen as a vital tool for achieving social stability at a time of widespread social unrest. Much of what has happened in the housing field, and the shape of the 'urban' that has resulted, can be explained only in terms of these various displacements of class struggle from the point of production to the place of reproduction and socialization (Harvey 1974, 1975b).

The second example I shall take is even more complex. Consider, in its broad outlines, the history of the bourgeois response to acute threats of civil strife associated with marked spatial concentrations of the working class and the unemployed. The revolutions of 1848, the Paris Commune of 1871, the urban violence which accompanied the great railroad strikes of 1877, and the Haymarket incident in Chicago in the US, clearly demonstrated the revolutionary dangers associated with high concentrations of what Charles Loring Brace called 'the dangerous classes' of society. The danger could, to some degree, be alleviated by following a policy of dispersal so that the poor and the working class could be subjected to what the nineteenth-century urban reformers called 'the moral influence of the suburbs'. Cheap suburban land, cheap housing, and cheap transportation were a part of this solution; the shape of London, for example, was greatly altered by the Cheap Trains Act of 1882, which was a part of the package of bourgeois responses to the panic induced all across Europe by the events of 1871 in Paris. The events in the latter city belied Napoleon III's hopes and confirmed his fears as he sought explicitly by 'urban renewal' to prevent that concentration of workers which could all too easily form the basis for revolutionary action. And what was the bourgeois response to the urban riots of the 1960s in the ghettos of the US? Open up the suburbs, promote low-income and black homeownership, improve access via the transport system . . . the parallels are remarkable (Walker 1976).

The alternative to dispersal is what we now call 'gilding the ghetto', but this, too, is a well-tried and persistent bourgeois response to a structural problem which just will not disappear. As early as 1812, the Reverend Thomas Chalmers wrote with horror of the specter of a tide of revolutionary violence sweeping Britain as working-class populations steadily concentrated in large urban areas (Chalmers 1900). Chalmers saw 'the principle of community' as the main bulwark of defense against this revolutionary tide, a principle which he argued should be deliberately cultivated to persuade all that harmony could be established around the basic institutions of community – a harmony which could function as an antidote to class war. The principle entailed a commitment to community improvement and a commitment to support those institutions (such as the church, civil government, and the like) capable of forging community spirit. From Chalmers through Octavia Hill and Jane Addams, through the 'urban reformers' such as Joseph Chamberlin in England and the 'progressives' in the US, through to model cities and citizen participation, we have a continuous thread of a response to problems of civil strife and social unrest via the principle of community mobilization and improvement (Walker 1976; Harvey 1985a).

But the 'principle of community' which the bourgeoisie has done so much to instill in the course of 'urban' history can be used by the working class in their own self-defense and even as an offensive weapon in the course of class struggle. The institutions of the church in the early years of the industrial revolution were sometimes mobilized to working-class ends, much as they became an instrument for black mobilization in the civil rights movement of the 1960s in the US and currently are used in the Basque country of Spain. The principle of community can therefore be the springboard for class action as well as an antidote to class consciousness. The definition of community as well as the command of its institutions – particularly those of the state – become the focus for class struggle. Yet this struggle can break open into innumerable dimensions of conflict, pitting one element within the bourgeoisie against another and various fragments of the working class against others as the principles of 'turf' and of 'community autonomy' become an essential part of life in capitalist society. The bourgeoisie has frequently sought to divide and rule but has just as frequently found itself caught in its own contradictions as well as in the harvest of conflict which it has helped to sow. The attempt to displace class conflict by the principle of community is replete with its own contradictions as we find 'bourgeois' suburbanites resisting the further accumulation of capital via the extension of the built environment and as we find civil disorder within the 'urban' fabric escalating out of control as ethnic, religious and racial tensions take on their own dynamic. The relations here are intricate and complex – and bourgeois urban sociology frequently can tell us something about the details – but the problem is to find the interpretive framework to make sense of the myriad cross-currents which make up the fabric of contemporary social and political life under the capitalist form of urbanization.

I am not seeking here to set out any rigorous or perfected framework for thought about the 'urban' under capitalism. I am, rather, seeking to illustrate how the twin themes of accumulation and class struggle, when understood from the Marxian perspective, can serve as reference points for understanding almost anything that the urban sociologist would want to understand and a good deal more besides (Harvey 1985a). Since this is the kind of statement that my 'bourgeois' opponents will almost certainly misinterpret, I will conclude by explicating its meaning. I am not claiming some vast systemic overview of society in which to interpret the 'urban'. The essence of the relational and dialectical style of investigation is, as I have already suggested, to explore the totality by way of the internal relations which compose it. Since everything relates to everything else in society, we could, in principle, start our investigations elsewhere, and there is often some virtue in so doing. But some points of departure are more fruitful than others, and those of accumulation and class struggle are the best I know. Yet they are not fixed concepts in the matrix of thought I use but concepts which change their meaning with use. As I reach out from an analysis of accumulation and class struggle to embrace the 'urban question' in all its aspects, so I come to understand more clearly and more fully what the concepts of accumulation and class struggle mean. This I take to be the real dialectic involved in the Marxian way of discovering meaning.

Aristotle once remarked that if only there were a fixed point in outer space, we could construct a lever to move the world. The remark tells us a great deal about the shortcomings of Aristotelian thought. Bourgeois social science is heir to the same shortcomings. It attempts to construct a view of the world from outside, to discover some fixed points (categories or concepts) on the basis of which an 'objective' understanding of the world may be fashioned. The bourgeois social scientist typically seeks to leave the world by way of an act of abstraction in order to understand it. The Marxist, by way of contrast, always seeks to construct an understanding of society from within rather than imagining some point without. The Marxist finds a whole bundle of levers for social change within the contradictory processes of social life and seeks to construct an understanding of the world by pushing hard upon the levers.

To construct knowledge requires an active involvement in the processes of social change. We discover ourselves by striving to change the world, and in the course of the struggle we change both the world and ourselves. Nothing is certain in this struggle, and we may as individuals fail in our practice, lose our way, be repressed or destroyed. But the only sure path to that knowledge which has the capacity to change the world is to engage in the struggle. Herein lies the most serious barrier to the bourgeois understanding of Marxian thought, for to understand that thought means ultimately to practice it, which means, quite simply, that the bourgeois academic will have to cease to be bourgeois and come to the other side of the barricades if he is really to understand what the view from inside, the view from the standpoint of labor, is really all about.

CHAPTER 5

Owen Lattimore: a memoire

First published in Antipode, *1983.*

Owen Lattimore's life spans the twentieth century and his career inter-weaves with two of its more significant events: the revolutionary upheaval that transformed China from 1911 to 1949 and the savage aftershock known as McCarthyism in the United States. An assessment of that biography, long overdue, has much to tell us about the politics of doing geography in a world in political turmoil.

Lattimore, once of The Johns Hopkins University and now an 83-year-old emeritus professor of Leeds University, lives in Cambridge, England. He continues to write reviews and books, gives lectures and travels. He describes himself as a 'radical conservative' and from that standpoint is still more than willing to take issue with American foreign policy. As recently as 1981, a reporter from the *Baltimore Sun* caught up with him in China en route to Ulan Bator for the sixtieth anniversary of Mongolia's revolution (Lattimore is the only westerner to be a member of the Mongolian Academy of Sciences). Lattimore immediately took issue with the US conception of 'the China card' in international affairs and worried that we were becoming 'the best teacher of communism' in El Salvador. He here echoed a lesson he once learned from Mao. 'I once asked Mao,' he told me on the eve of his China trip in 1981, 'how communist conceptions spread so quickly amongst an illiterate peasantry. Mao responded that they had had the two best instructors of communism in the world. I naturally thought he meant Marx and Lenin but Mao just chuckled and said, "No, Chiang Kai-shek and the United States."'[1] The subsequent *Sun* article (8 July 1981) sparked the kind of response that has dogged Lattimore for many years now. 'The *Sun*,' wrote an irate correspondent, 'would have been well advised' to leave Lattimore 'to the obscurity he richly deserves.' Had not the Senate Judiciary Committee of 1952 concluded 'after one of the most exhaustive probes in Senate history,' that 'Owen Lattimore had been since the 1930s a conscious, articulate instrument of the Soviet conspiracy?'

[1] Personal interview, 19 June 1981.

COMMUNIST LABEL PUT ON LATTIMORE

Budenz Tells Senate Group Of I.P.R. Hookup

By WILLIAM KNIGHTON, JR.
[Washington Bureau of The Sun]
Washington, Aug. 22—Louis F. Budenz today testified under oath that Communist party officials "specifically mentioned" Owen Lattimore "as a member of the Communist cell under instructions."

To penetrate the mysteries of that life requires far more sophistication than I am ever likely to possess. For what Lattimore did and said, and what others did to and said of him, cannot be understood independently of the social forces at work on, and the intersecting histories of, two vast and complex continents. It would take an understanding of China and inner Asia at least equivalent to that which Lattimore acquired from a life-time of experience and study together with an understanding of America – particularly of what Hofstadter calls 'the paranoid style in American politics'[2] – that seems to have eluded Lattimore himself. I put the matter that way because if Lattimore is guilty of anything, I suspect it was simply that he knew the realities of China too well and that he sought to convey them to an America which he understood too little, largely in terms of its own surface idealism. His whole deportment during the McCarthy period testifies to his shocked disbelief that such an outrageous violation of the spirit of American constitutionality could so easily be inflicted upon him. In this he was not alone. McCarthyism hit working-class organizations – trade unions, immigrant organizations, neighborhood groups – professionals of all sorts (teachers and lawyers) and savaged the political left (of whatever stripe) for a whole generation. Of that whole social process, Lattimore is but a symbol and cypher (as, indeed, was Senator McCarthy himself). A study of his career is, therefore, a means to an end for, as Lattimore himself insists, history is not the study of great men but the dissection of social movements in all their fullness.[3]

Lattimore was born in Washington, DC on 29 July 1900.[4] Within a year his father, a teacher, had accepted a position in China and it was twenty-eight years before Lattimore was to set foot in the US again. His whole

---

[2] Hofstadter, 1967.
[3] Personal interview.
[4] Lattimore gives his own account of his life and career in the 'preface' to *Studies in Frontier History: Collected Essays 1928–58*. I have drawn heavily on this account supplementing it with some information from *Who's Who*, my own personal interview, and various snippets from press accounts in the Lattimore file in Baltimore's Enoch Pratt Library. A complete bibliography of Lattimore's writings up until 1962 is appended to his *Studies in Frontier History*.

childhood was spent in China. At age twelve he went to Switzerland (two years) and England (five years) to complete his only formal education. At age nineteen he returned to China ('for financial reasons') where he was variously employed as a journalist for an English newspaper in Tientsin and as a buyer and commercial agent for a British firm 'which imported into China everything that the West had to sell, and exported everything that the West would buy.' These were chaotic years in China. Internal disorders, conflict between feuding warlords and the breakdown of Western domination were preparing the way for the Second Chinese Revolution of 1925–7.

As a commercial traveller and buyer working out of Shanghai, Tientsin and Peking, Lattimore was in an unparalleled position, as he later realized, to learn about 'the realities of an historical period' and to evolve an understanding of commodities 'based on experience rather than upon books'. He was, by his own account, bored with his job and intellectually frustrated. He sought escape by building a knowledge of China and the Chinese sufficient 'to be considered a little queer by my fellow Philistines' in the commercial world and by abandoning the safety of the treaty ports to travel without an interpreter, servant, or supplies of western food (then considered a most daring venture) into the interior regions of China. He there began to unravel some of the secrets of how commodities are produced and traded under social conditions that westerners had little or no comprehension of. A difficult trip to Inner Mongolia was a turning-point:

> The bits and scraps that I learned from merchants and caravan men made up my mind for me. I would follow the caravans to the end of the line, and see what there was to be seen. When I got back to Tientsin I resigned from my firm . . .

And so began one of the most extraordinary journeys by any westerner into the heart of inner Asia. After a year of picking up diplomatic niceties and information in Peking, where he also met and married Eleanor Holgate (then Secretary of the Peking Institute of Fine Arts, a writer of proven ability and Lattimore's close companion until her death in 1970), they set off to explore the route through Mongolia into Sinkiang and from thence right through to Kashmir. Lattimore set off in March 1926, but the outbreak of civil war, diplomatic problems, and separation from his wife (who caught up with him by travelling 400 miles by sled)[5] delayed completion of the journey. They arrived in Kashmir in October 1928.

---

[5] Eleanor Lattimore's account can be found in *Turkestan Review* (1934).

The experience was recorded as a book – *The Desert Road to Turkestan* (1928) – and Lattimore thereby claimed the attention of Isaiah Bowman, then President of the American Geographical Society and himself deeply fascinated by the idea of comparative study of frontier regions. And so began an entirely new phase of Lattimore's life. With Bowman's help, he procured a fellowship from the Social Science Research Council to travel in Manchuria. Since he did not have a PhD, or even formal university training, he was supported for a year of preparatory research in the Division of Anthropology at Harvard University. The year in Manchuria was followed by three years of fellowships (from the Harvard Yenching Institute and the Guggenheim Foundation) that allowed him to live in Peking. He published numerous articles and books – *High Tartary* (1930), *Manchuria, Cradle of Conflict* (1932) – and gathered many of the materials and experiences to be embodied in later studies of *The Mongols of Manchuria* (1934), *Inner Asian Frontiers of China* (1940) and *Mongol Journeys* (1941). By the mid-1930s he had proved himself a prolific writer, lacking, perhaps, in academic niceties, but with an unparalleled command over his subject matter built upon a quite unique experiential base.

Lacking other employment prospects when he returned to America in 1933, Lattimore jumped at the chance to be the paid editor of *Pacific Affairs*, a position held from 1934 to 1941. This journal was published by the Institute of Pacific Relations (IPR), an international organization set up in 1932 by religious leaders, scholars and businessmen from various countries in the Pacific region and largely funded by grants from the Rockefeller and Carnegie Foundations. The IPR was organized into autonomous national councils including the US, China, Japan, Britain, France, and the Soviet Union. The task of *Pacific Affairs* was to provide an open forum for debate and discussion on matters of controversy in the Far East. There was, of course, no end of controversy and while independent scholarship was overtly stressed, there was no way to avoid the collision of national interests. To keep scholars from the Soviet Union and the US participating in dialogue was just as difficult as keeping the Japanese and Chinese together during a period of intense geopolitical conflict and open warfare, while the British and French councils also tended to press distinct national interests. Lattimore was faced with a delicate and difficult task during a period of extraordinary

> **I.P.R. 'CONDUIT' TO REDS DENIED BY LATTIMORE**
>
> 'Farfetched,' Professor Calls Suggestion Soviet Got Secrets

political convulsion and geopolitical tension. But he evidently relished it and sought to keep *Pacific Affairs* an open forum which meant, quite simply, making complex compromises with the different national councils and groups. The decisions to publish articles, though based on scholarship, had in some cases to be political decisions as well. The price of continued Soviet participation in IPR, for example, was acceptance of articles that expressed the Soviet line. Lattimore later justified their publication on the grounds that the 'line' deserved expression as part and parcel of the controversy.

The position of editor of *Pacific Affairs* seems to have suited Lattimore's temperament and career objectives admirably. He could continue to live in Peking and devote half his time to research. He was at the center of a network of scholars and political interests that allowed him an unparalleled opportunity to follow the pulse of change in Asia and the cross-currents of conflicting interest and opinions. He was able to solicit articles on topics of particular interest to him (such as climatic change and the social origin of deserts). And he could continue to make frequent visits to Inner Mongolia and other areas of China. In 1937 he managed to get into Yenan Province, then under communist control. He there participated in interviews with Mao Tse-Tung, Chou En-lai and other leaders of the communist movement and observed first hand many of the communists' organization practices.

When, after the Japanese occupation, it became too difficult to live in Peking, he returned to the US where Isaiah Bowman, now President of The Johns Hopkins University, appointed him director of a revived Walter Hines Page School of International Relations, a position Lattimore held from 1938 to 1950. Bowman probably viewed him as the cutting-edge to fashion a geography presence in a university that lacked one, while simultaneously bringing the prestige of a foremost authority on Asiatic affairs to the university as a whole.

Undoubtedly, Lattimore was a controversial figure. As editor of *Pacific Affairs*, 'I was continuously in hot water, especially with the Japanese Council, which thought I was too anti-imperialist, and the Soviet Council, which thought that its own anti-imperialist line was the only permissible one.' He was also working in a field dominated by prima donna academic scholars who sometimes seemed to view him as an untrained upstart and for whose peccadillos Lattimore had little patience. And he was by no means loath to engage directly in controversy, freely speaking his mind in highly politicized environments.

The American Council of the IPR in the 1930s for example, comprised all shades of political opinion from communists to rabid ultra-conservatives. Many of those who subsequently testified against him (Wittfogel,

Colegrove, Utley, for example) did so on the basis of opinions that Latti-more freely expressed to them during the 1930s and 1940s. And some of these opinions can, with the benefit of hindsight, be judged questionable: his characterization in *Pacific Affairs*, for example, of Stalin's show trials as a harbinger of democratization. Others, also with the benefit of hindsight, appear as perceptive judgements on dangerously murky situations: his analysis of communist strength in China, or how and why the Mongolian People's Republic clung to the Soviet orbit, for example. Under such conditions, it was easy for Lattimore to acquire those enemies which Professor Newman (1983) so diligently documents.

The outbreak of war brought new opportunities and challenges. In 1941, six months before Pearl Harbor, Chiang Kai-Shek, at Roosevelt's request, appointed Lattimore as a personal adviser (Bowman probably influenced the appointment). Lattimore remained in that position until the end of 1942 and apparently developed a consider-able admiration for Chiang as an intelligent individual in the midst of conspiring and often corrupt forces.[6] The outbreak of war in the Pacific brought the US into a formal alliance with China, thus eliminating the need for the kind of close informal liaison that Lattimore provided. He returned to America, where in 1943 he became Director of Pacific Operations, Office

CHIANG PICKS LATTIMORE AS POLITICAL AIDE

Appointment Is Recom-mended To Chinese Gen-eral By Roosevelt

of War Information (a US intelligence unit). He returned to his full university position in 1944 but accompanied Vice-President Wallace on a trip to Siberia and China in 1944, acted as an economic consultant on the American Reparations Mission in Japan (1945), and went to Afghanistan as a UN technical assistant in 1950 – it was while he was in Afghanistan that McCarthy first made charges against him and so, guilty or not, ended this kind of governmental activity.

But he appears to have viewed such activity largely as means to an end. He evidently relished the public recognition that came with the Chiang Kai-Shek appointment while the nature of the assignments gave him abundant opportunities to learn and to keep tabs on the situation in Asia. 'I was more fortunate than many,' he writes, 'in that the war years did not completely interrupt my academic interests.' There is little evidence that

---

[6] He still feels that way about Chiang and is currently working on recollections of the period he spend as Chiang's adviser.

he sought to insinuate himself into the councils of government though he was more than willing to participate when asked. His preferred mode was to influence and educate public opinion through a massive outpouring of articles, books, lectures, (the *Baltimore Sun* accounts indicate that Lattimore frequently lectured to packed houses in the city after 1944). *Solution in Asia* (1945) and *The Situation in Asia* (1949) were two well-written and popular statements of his views that most certainly did have an impact on American public opinion. In both, he sought to analyze the internal

CHINA STIRRED BY ASSIGNMENT OF LATTIMORE

Speculation High As To Prospects Of Success For His Mission

Hopkins Man Is Silent On Plans As Adviser To Chiang Kai-Shek

dynamics of Asiatic societies and the geopolitical consequences of past, present, and future great-power policies.

The ideas he presented were coherent but controversial. In *Solution in Asia*, for example, he depicted a world divided into capitalism and collectivism with a third division of mixed and contested influence. He refused to see the Soviet Union as a 'red menace' and argued that, for many of the peoples of Asia, the Soviet Union had a great power of attraction because it stood for 'strategic security, economic prosperity, technological progress, miraculous medicine, free education, equality of opportunity' and, by virtue of all of the above, and in comparison to the social systems then prevailing (in which western notions of freedom of speech and parliamentary democracy had no meaning), it also stood for democracy. It also had the geopolitical advantage of a seemingly indestructible land base (a Mackinder-type proposition) from which to project these attractions onto the rest of Asia The problem for western diplomacy was to come to terms with this influence in such a way as not to divide the unaligned countries into warring camps. Lattimore therefore counseled a politics of negotiation and cooperation with the Soviet Union. For this reason he welcomed the setting-up of the UN, the Bretton Woods Conference and all the other attempts to create supranational institutions.

He made similar points about the internal politics of China. The communists, he argued, were much more profoundly entrenched and much more attractive to the peasantry than most Americans thought and they were certainly not under the influence of the Soviets (who were mainly supplying Chiang arms to fight off the Japanese). Accommodation between the communists and the Kuomintang had to be actively sought. In 1945 Lattimore thought such a compromise had sufficient historical force to be a real possibility. And, denying that Chiang was losing control or that he was a fascist, he put his hopes mainly on Chiang as a 'coalition statesman of genius'. (Not until 1948 did Lattimore lose faith in Chiang's ability to govern.)[7] The goal of American policy should therefore be to encourage the emergence of a coalition. But such external influence had to be exercised non-imperialistically, in the full recognition that for the Chinese 'their country and their society are theirs to build'.

I think it important to outline these ideas because they form the vital ideological context in which to interpret subsequent events. Lattimore was emphatic that Marxist thought could no longer be viewed as mere subversive propaganda but had to be accepted as an attractive and competitive political ideology throughout much of Asia. He rejected 'the absurd simplification' that revolutionary movements were a mere projection of Soviet influence and roundly castigated America's expert opinion on the Far East as 'so incompetent that usually when the majority of experts agree, they are wrong'. He sought, in his own writings and lectures, to shape a public opinion ready to hold expert opinion to account, prepared to lend its weight and influence to the grand geopolitical questions that the future of Asia posed.

The first hint of the drama to come came as early as 1946. In the wake of the Amerasia case and the celebrated picnic at Lattimore's home in Ruxton to which George Carter was invited (probably to facilitate Bowman's pet project of a school of geography in close liaison with the Page School), a press item from the *News Post* suggested that Lattimore was mixed up with communist front organizations and a 'principal writer' for the IPR, itself 'a veritable mill of Communist propaganda'.[8] The article was almost certainly inspired by Kohlberg, an ultra-right businessman

SAYS CARTER TOLD VIEWS TO MCCARTHY

Gave Senator Statement About Papers He Saw, Witness States

By PHILIP POTTER

[Washington Bureau of The Sun]

Washington, March 19 — Owen Lattimore, writer on Far Eastern affairs, today denounced as "an informer" a Johns Hopkins University colleague who allegedly gave Senator McCarthy (R., Wis.) a statement about a picnic he attended at Lattimore's home in 1945.

The man so characterized by Lattimore before the Senate Internal Security subcommittee investigating alleged Communistic influence on American foreign policy is George F. Carter, professor of geography at Johns Hopkins and chairman of the Isaiah Bowman School of Geography.

Carter, reached at his Baltimore residence, said, "I have no comment—no comment at all" on Lattimore's statement.

[7] See *Situation in Asia* and the text of Lattimore's 1949 memo to the State Department (published in the *Baltimore Sun*, 4 April 1950) on US policy in the Far East.
[8] *Baltimore News Post*, 7 June 1946.

at war with Lattimore within the IPR (see Newman). Lattimore was moved to issue an explicit denial. But the incident was the tip of an iceberg already in motion. The FBI interviewed Bowman in 1946 and by the end of his presidency, in 1948, Bowman appears to have turned against him, advising his successor, Detlev Bronk, to get rid of Lattimore.[9] It is hard to know whether Bowman's moves were for political reasons, due to financial difficulties within the Page School of International Affairs, or part of Bowman's 'pet project' to consolidate and protect the newly formed Department of Geography, of which George Carter, already Lattimore's chief antagonist, was the chair.

The storm finally broke over Lattimore on Monday, 13 March 1950, when Senator Joseph McCarthy, forced to substantiate charges made earlier ('I have here in my

INVESTIGATION MAJORITY CLEARS ALL ACCUSED OF CHARGES OF COMMUNISM

hand a list of 205 – a list of names that were known to the Secretary of State as being members of the Communist Party and nevertheless are still working and shaping policy . . .') fingered Lattimore in a Senate speech as 'one of four pro-communist advisors to the Department of State,' adding for the private delectation of newsmen that he thought Lattimore was without doubt 'the top Russian espionage agent in this country'. By 22 March, McCarthy was prepared to go public with that allegation, adding that his whole case would 'stand or fall' on the Lattimore evidence alone. McCarthy may have gotten hold of Lattimore's name from Kohlberg, though both Lattimore and Newman suggest that Carter played a key role (to my knowledge Carter has never denied the allegation).

The matter was serious enough to warrant investigation. This the Tydings Committee of the Senate provided, reporting on 17 July 1950 that no case existed against Lattimore and that McCarthy's charges were a 'fraud and a hoax' on the American people, 'the most nefarious campaign of half truths and untruth in the history of the republic'.[10] To this negative judgement, Lattimore added his own. *Ordeal by Slander*, published in July 1950, gave a blow-by-blow account of the Lattimores' experience

[9] Personal interview, independently corroborated by Neil Smith in an interview with Jean Gottman (then on the Hopkins faculty and fired by Bowman at about the same time).
[10] US Senate, Subcommittee on Foreign Relations, *State Department Employee Loyalty Investigations*. 81st Congress, 2nd Session. Washington, DC: Government Printing Office, 1950.

**Denies Spy Accusation
Urges Senator Resign**

Wirephoto on Page 2

By Frank R. Kent, Jr., and Bradford Jacobs
(Evening Sun Washington Bureau)

Washington, April 6 —Describing Senator McCarthy (R., Wis.) as the unwit-
g dupe of a Chinese Nationalist lobby, Owen Lattimore today before a Senate Foreign
ations subcommittee denied that he is or ever has been a Communist.

The Baltimorean, director of the Walter Hines Page School of International Rela-
ns, at the Johns Hopkins University, has been accused by McCarthy of being the top
ussian espionage agent in this country.

(part was written by Eleanor). Compulsive reading, it became the focus of
a vigorous counterattack, in the name of constitutional freedoms and the
Bill of Rights, against the gathering darkness of McCarthyism.

McCarthy was outraged. He termed the Tydings report an 'evil fraud'
and sent his cohorts into Maryland as
part of a vicious and successful campaign (that reverberated in State politics many years thereafter) to unseat
Senator Tydings (a rather conserva-

**MCCARTHY HAS
ANOTHER SAY**

tive Democrat) in the fall of 1950, a year that saw control of the Senate
pass to the Republicans. And McCarthy likewise indicated he was in no
mood to let up on Lattimore. He reiterated the charges against him in a
book and sent his henchman all over Maryland to dig up any dirt they
could (he charged in July 1950, for example, that Lattimore had been
'paid off' by the communists through a particularly profitable property
transaction on Cape Cod).[11] Early in 1951, McCarthy's agents had the
break they were looking for. They located, and illegally seized, the IPR
files in a barn in Massachusetts.

The Senate Internal Security Subcommittee, now under Republican
control and chaired by Senator Pat McCarran, a close confidante of
McCarthy, sifted through this new mine of evidence stretching back to
the early 1930s and orchestrated a most extraordinary inquisition against
Lattimore and many others associated with the IPR.[12] The hearings of the
McCarran Subcommittee ran from July 1951 until June 1952 and cover
nearly 6,000 pages of printed record. The final report relieved Mao and
millions of Chinese for any responsibility for their revolution and concluded that 'but for the machinations of the small group that controlled
(the IPR), China would be free, a bulwark against the Red hordes.' It also

[11] *The Baltimore Evening Sun*, 28 July 1950.
[12] US Senate, Subcommittee of the Committee of the Judiciary to Investigate the
Administration of the Internal Security Act and other Internal Security Laws, *The
Institute of Pacific Relations.* 82nd Congress, 1st and 2nd Sessions.

recommended the indictment of Lattimore for perjury, charging that he had been 'a conscious articulate instrument of the Soviet conspiracy' since the 1930s.

Lattimore began his testimony on 26 February 1952 and stepped down on 21 March, after twelve days, and more than forty hours, of continuous grilling. He was not permitted to take advice of counsel during the sessions nor could he easily refer to notes and documents. He had to work from memory before staff lawyers Robert Morris (who turns up again in 1980 as chair of a National Committee to Restore Internal Security that held hearings on that subject in the Senate, courtesy of Senator Hatch)[13] and J. G. Sourwine, who had at their fingertips every letter and memo that Lattimore had ever written. Furthermore he had been preceded by a wealth of antagonistic witnesses (Wittfogel, Colegrove, and the like) and informers (like Budenz and Utley) who, under the privileged protection of a Senate hearing (cross-examination is not allowed), had been prompted, or badgered, to say anything derogatory they could about Lattimore's opinions and associations.

Lattimore tried to take the offensive by reading from a prepared statement that echoed exactly the line taken in *Ordeal by Slander*. It took him three hours to get through the first eight sentences. The transcript of his sessions makes for extraordinary reading. In the end his full statement was read into the record to make way for extensive questioning that allowed Morris and Sourwine to spring many a factual trap against him. As he wearily complained on the seventh day,

> a number of times in these hearings the names of people have been mentioned whom I totally failed to recall, and later on some memorandum or other document is brought out which indicates that I did meet them. This is part of the whole procedure which I should very respectfully like to criticize.

McCarran sternly replied, 'You are not here for the purpose of criticizing: you are here for the purpose of testifying under oath.' It was a raw and lopsided contest of Lattimore's recollections versus the documented record of exactly where he had been, who he had talked to, and what had been said, over a period of nearly twenty years. A lesser person would probably have cracked under the strain (to this day he has an intense facial twitch that first appeared during the hearings). But Lattimore often countered with humorous banter. Here he is in action against Senator Smith, who was bent on showing that the fall of China followed upon the

[13] Judis, 1981.

removal of known anti-communists from their positions in the Department of State:

> *Mr Lattimore* Is your argument, Senator, a *post hoc, ergo propter hoc?*
> *Senator Smith* I believe you said you did not want to indulge in legal or technical language, so I am asking you in plain language if, after these men were removed, it is not a fact that there have been great advances by communism in the Far East?
> *Mr Lattimore* Yes. Of course the advances of communism since the death of Julius Caesar have been even greater.

And here he is again in action against Wittfogel, who had claimed that Lattimore knew he was a communist because when he (Wittfogel) had denied the suggestion to an American journalist in 1935, Lattimore had turned to him and smiled. Said Lattimore:

> If I smiled at all it was certainly a non-communist smile. Now I would be willing to believe that Communists have an arsenal of secret signals, but I would never suppose that it included anything as good-natured as a smile . . . If I am wrong, and if a smile is a secret Red signal, I confess that I used to smile a good deal. In the pre-McCarthy days I used to think that life was lots of fun.

The conflict between Wittfogel and Lattimore during the hearings is one of many extraordinary features. The former's testimony was key to one of the later perjury charges. But Wittfogel went to quite extraordinary lengths to depict Lattimore as a conscious exponent of the Stalinist line. This was particularly hammered home in relation to Lattimore's use of the terms 'feudal' and 'feudal survival' to describe Chinese society. There had been an extensive debate over that concept within the communist International since 1922, a debate that Stalin settled uncompromisingly in the 1930s. Wittfogel thereafter viewed the concept of feudalism as a litmus test for Stalinism. In the hearings, he put it as follows:

> By concentrating the whole problem on what they call 'feudal' land owner-ship, the Communists focus the energy of the peasants to the property question, and divert it from a ruling bureaucracy which they have had in the past and which the Communists intend to establish everywhere in the future.

He went on to point out that while many might use the term innocently, it had great political significance when someone of Lattimore's sophisti-cation used it. Lattimore thought the whole argument absurd: 'I am sorry that I did not know that the Communists had a patent on the term and that to use it was as dangerous as to smile.' What lay behind the break between Lattimore and Wittfogel (in 1947) and the latter's pointed denunciations requires more careful consideration than I can give here.[14]

Lattimore's deportment before the Committee was as controversial as his testimony. By gutsy confrontation, he seems to have kept his own sense of integrity under difficult and demanding circumstances (Committee members often harped on his lack of formal education, for example). But this allowed McCarran to pillory him as an 'insolent, over-bearing, arrogant and disdainful' witness who had delivered a calculated affront to the Senate and, thereby, to the people and institutions of the US. He was also forced into mistakes (though very few under the circumstances). And he forfeited a lot of support in the press, influential sections of which were now prepared to accept the overall conclusions of the McCarran Hearings. In this sense Lattimore lost the contest because large sections of the American public now accepted what to him must have been a most extraordinary conclusion: that the success of the Chinese revolution depended upon the machinations of the small group of people that controlled the IPR. And there was a broader loss that the hearings merely confirmed, for Lattimore was forced into innumerable anti–communist and anti–Marxist statements which, though probably consistent with his own beliefs, helped etch ever deeper in the American consciousness the idea that all communists and Marxists put themselves outside of constitutional protections because they are necessarily part of a subversive foreign conspiracy. Thus was Lattimore forced to tacitly admit in America a proposition he avidly denied in Asia.

The indictment for perjury came on 16 December 1952.[15] The first and most important count charged that Lattimore had lied in denying that he had ever been a promoter of communism and communist interests. The other six counts derived from contradictions in this testimony (several of which came from the period 1934–41). The Johns Hopkins University immediately granted Lattimore leave of absence with pay. George Boas, head of the Philosophy department, headed up a $40,000 defense fund (and assembled and published at his own expense testimony to Lattimore's outstanding qualities as a scholar).[16] And the Washington

---

[14] Wittfogel's version of what happened can be found in Ulmen 1978: 267–94.

[15] The following account is pieced together out of the Lattimore file of press clippings in the Enoch Pratt Library (mainly from the morning and evening versions of The Baltimore Sun).

[16] Boas and Wheeler 1953. The letter that Boas sent out began as follows: 'The recent charges against our colleague, Owen Lattimore, include that of not being a reputable scholar. It is admitted that he is a clever journalist, but since he has no university degree and no title of "Professor", he is accused of occupying a position on the Faculty under false pretenses.' Thirty-seven responses from scholars in the field were published in Lattimore's support.

MCCARRAN UNIT DEMANDS JURY PROBE FOR PERJURY; PROFESSOR ISSUES DENIAL

Federal Grand Jury Indicts Lattimore

firm of Arnold and Porter (who had acted as Lattimore's counsel all along) took on the case with all their renowned legal skill.

On 2 May 1953, Judge Youngdahl of the US District Court in Washington threw out four of the government's seven perjury charges on the ground that they were unconstitutional and expressed serious doubts

LATTIMORE ENTERS PLEA OF NOT GUILTY

Trial On Perjury Indictment Is Set For Early March

By PHILIP POTTER
[Washington Bureau of The Sun]

Washington, Dec. 19—Owen Lattimore today pleaded innocent to a seven-count perjury indictment in Federal District Court.

The Johns Hopkins faculty member spoke the words "not guilty" in a loud, firm voice, which contrasted sharply with the muttered pleas of more than a score of defendants arraigned ahead of him

New Lattimore Count Dismissed, Called Too Obscure

Washington, Jan. 18 (P)—Judge Luther W. Youngdahl today threw out the Government's key charges in the Owen Lattimore perjury case.

Youngdahl ruled the accusation that the controversial Far Eastern affairs specialist swore falsely in denying he ever had been a follower of the Communist line or a promoter of Red interests, was too vague to define an offense or outline a field for defense.

In different words Youngdahl attributed to the new charges the same faults he found in the Government's first accusation that Lattimore perjured himself when he denied he had been a Communist sympathizer or a promoter of Red causes.

whether the remaining three charges could pass 'the test of materiality'. Youngdahl was particularly hard on the first count, depicting it as 'nebulous and indefinite' and contrary to the First and Sixth Amendments. He went on to argue that we could not afford to weaken the Bill of Rights even in the cause of fighting against tyranny. The government appealed to the US Court of Appeals which, after the customary delays for filing briefs, heard the case in January 1954 and issued its opinion on 8 July. By an eight-to-one majority it upheld the dismissal of the main count, but reinstated two very minor counts by a five-to-four majority. The government could either appeal to the Supreme Court or reconsider its position.

It decided on the latter course and on 7 October reindicted Lattimore on two fresh perjury counts. The first charged that 'he had knowingly and intentionally followed the Communist line' and cited 130 instances from his writings and statements to prove it. The technique was bizarre. Statements of policy from the Soviet or US Communist parties were paralleled against anything that sounded even remotely similar in Lattimore's prolific writings. The defense had fun with that and produced a list of its own statements sympathetic to communist positions by Eisenhower, Churchill and *Time* magazine.[17] The second count charged that Lattimore had sworn falsely when he denied he had been a 'promoter of communist interests in his activities'. The Justice Department then challenged the fitness of Justice Youngdahl to try the case by reason of prejudice in favor of Lattimore. The judge responded on 23 October: a blistering attack on what he saw as a scandalous attempt to intimidate an independent judiciary. The following January he threw out the two new counts of perjury on the ground that they were too nebulous and vague, involved matters of opinion and were in violation of the First and Sixth Amendments. In April, the government, hoping to resurrect the five-to-four majority on the minor counts won previously, went back to the Court of Appeals. But a replacement on the bench for one of that majority would have to disqualify himself because he had earlier been actively involved in preparing the Justice Department's case – the new person was none other than Warren Burger, now Chief Justice of the United States. Faced with a four-four split at

JUDGE CALLS BIAS CHARGE SCANDALOUS

Aims Severe Criticism At U.S. Attorney Who Filed Accusation

Washington, Oct. 23 *(P)*—Federal Judge Luther W. Youngdahl bluntly rejected today a Government demand that he disqualify himself in the Owen Lattimore perjury case.

best, and with the excesses of McCarthyism on the wane, the government finally decided to drop all charges against Lattimore on 30 June 1955.

For nearly five and a half years Lattimore had been enveloped in a poisonous cloud of suspicion. He was initially swept up in a social movement

---

[17] United States Court of Appeals, District of Columbia, *Brief of Appellee* (USA versus Lattimore).

that sought to crush everyone of his ilk, and thereafter caught in the jaws of a governmental process that either could not, or would not, let go until all possible avenues of harrassment had been exhausted, even though the FBI files that Professor Newman has so patiently combed show that they knew all along that there was not a shred of evidence against him. It was appalling, as Lattimore himself put it only halfway through his grand ordeal, 'that innocence should have to be so long defended against such vengeful harrassment . . .'[18]

How the University reacted throughout is of some interest. It had the best of both worlds.[19] It appeared to support academic freedom by giving Lattimore paid leave of absence to fight his case and returning him to the status of lecturer in the department of History when the case was dropped. At least two members of the Board of Trustees resigned over the incident (one of whom put his money instead into the fountains that flow in downtown Baltimore's Hopkins Plaza, an ironic memorial as it were). On the other hand, the Page School was disbanded, the flow of graduate students dried up, and Lattimore was marginalized under threat because he had neither tenure nor formal academic credentials (the attempt to remove him was overtly

Owen Lattimore Again Lecturing At Johns Hopkins

Owen Lattimore has returned to active status as lecturer at the Johns Hopkins University, a university spokesman said yesterday in answer to an inquiry.

Mr. Lattimore was placed on leave of absence, with pay, in December, 1952, by Dr. Detlev W. Bronk, at that time president of the university. This action followed the Federal indictment of Mr. Lattimore for perjury.

The Government formally dropped the prosecution June 30 of this year. The return of Mr. Lattimore to active status at the university was automatic, the spokesman said.

based not on his politics but on his lack of a PhD which explains why George Boas went to such extraordinary lengths to gather academic testimony to Lattimore's outstanding achievements as a scholar). It was only through the vigorous support of colleagues like Boas and Painter (who chaired the History department) that Lattimore survived. And, of course, the Geography department was closed to him until M. G. Wolman replaced Carter as chair in 1958 and invited Lattimore to lecture within the

[18] *Baltimore Sun*, 17 December 1952.
[19] I have pieced together these ideas from conversations with many who were at Johns Hopkins during these years.

department. All of this explains why to this day Lattimore scrupulously refrains from any public criticism of the university but in private makes no bones about his resentments.

The broader academic and intellectual effects are harder to pin down. Lattimore's case was but one of many that hit scholars in the China field. In this arena Lattimore may have won his own private battle but he lost the war when it came to influencing US public opinion and hence US policy towards China and the Far East. It was not until after the painful relearning of the Vietnam War, and Nixon's final playing of 'the China Card' in the early 1970s, that any of that ground could be recovered. But cases of the Lattimore sort also demonstrate how dangerous it is to cultivate perspectives on geopolitical questions that deviate from certain narrow conceptions of national interest or offend some dominant political line. Hardly surprisingly, geopolitics dropped out of Geography and political geography became a dull backwater after the McCarthy years. Geographers felt safer behind the 'positivist shield' of a supposedly neutral scientific method. All of which adds up to a decided abrogation of social responsibility, an understandable but ignoble currying of professional safety and security in a highly politicized world. The epistemological break in our own discipline known as 'the quantitative revolution' must therefore be seen at least in part as a *political* break, of which the marginalization of such a towering and controversial figure as Lattimore is but a potent symbol. Few graduate students in Geography would now know who he is, let alone read his work. Ironically, too, George Carter, one of his denouncers, long ago received the Outstanding Contributor Award of the Association of American Geographers, while Lattimore's contributions remain scandalously unacknowledged.

There remains the question of Lattimore's intellectual politics. To this day he steadfastly resists the idea that he is a 'Marxist' (let alone a communist sympathizer per se).[20] But it is not hard to see how others might so view him in a world of simplistic judgements. While he profoundly proclaims that he has never read Marx, he is wont to say that he learned his theory of commodity by following them on the ground rather than through the pages of Marx's *Capital*.[21] In the intellectual atmosphere of the 1930s, when Marxism was very much in vogue even amongst those, like Sydney Hook, who subsequently became the most ardent of cold war

---

[20] Lattimore examines this issue implicitly in his preface to *Studies in Frontier History* and also went into it at length in personal conversations.

[21] He repeated to me what he had said to a reporter in 1942: 'I developed an interest in economics from an association with commodities themselves and not from books. As a result, I think in terms of concrete rather than theoretical problems.' *Baltimore Sun*, 12 April 1942.

warriors, Lattimore was at the height of his powers and at the center of a whole network of scholars who most certainly had read Marx and absorbed his message well. Lattimore was clearly not uninfluenced by that thinking but seems to have accepted only those aspects of it that gelled with his own experience. From this standpoint he had to be seen as a real down-to-earth historical materialist. When I put that idea to him he replied: 'Of course. Any damn-fool historian in his right mind has to be an historical materialist.' He also openly confesses to a partiality to deal with 'peoples rather than governments' and to use class analysis as the basis for much of his work.[22] But he steadfastly denies that Marxism has any patent on such positions (any more, in his response to Wittfogel, could he concede that Stalinists had a patent on the word 'feudal'). His sympathy with peoples rather than governments also leads him naturally into positions of sympathetic understanding for popular struggles, including those waged in the name of anti-imperialism. It is understandable that someone who proclaims the virtues of historical materialism and class analysis, and who takes strongly anti-imperialist positions, should be viewed by many as a Marxist. That he prefers the label of 'radical conservative' also makes sense. But then did not Marx also say, 'I am not a Marxist'? Perhaps a shot of the sort of 'radical conservatism' that Lattimore espouses is what Marxism needs from time to time to preserve it from its more dogmatic predilections. Certainly, Lattimore's perpetual concern to keep as close as possible to the people he sought to understand has a most healthy ring to it.

Lattimore remained at Hopkins until 1963 when he accepted an offer from Leeds University to head up a newly founded Chinese Studies program. There he experienced the renewed excitement of once more being surrounded by young scholars anxious to learn and to know. He continued to write and research (though American publishers, in a lingering testimony to the power of McCarthyism, did not touch his books again until the late 1960s). Lattimore retired from Leeds University in 1969.

When he left Hopkins in 1963 he gave a farewell lecture to a packed audience. Everyone expected him to talk about McCarthyism. He talked about society and culture in Mongolia. I suspect that is where his heart was all along.[23]

---

[22] See his 'preface' to *Studies in Frontier History*.
[23] Owen Lattimore died on May 31st, 1989. Professor Newman's biography was published in 1992.

CHAPTER 6

On the history and present condition of geography: an historical materialist manifesto

First published in The Professional Geographer, *1984.*

The present condition of geography and proposals for its transformation must be firmly grounded in an understanding of history. The roles and functions of geographical knowledge, together with the structures of that knowledge, have changed over time in relation to, and in response to, shifting societal configurations and needs. The history of our discipline cannot be understood independently of the history of the society in which the practices of geography are embedded. The rise of merchant, and later industrial and finance forms of capitalism in the West, paralleled as it was by increasing spatial integration of the world economy under western politico–economic hegemony, demanded and depended upon the crystal-lization of new forms of geographical knowledge within an increasingly fragmented professional and academic division of labor. The difficulties and alternatives geographers now face are likewise rooted in conflictual processes of societal transformation. Proposals for the transformation or stabilization of our discipline are, whether we like it or not, positions taken in relation to grander processes of social change. Awareness of that basic fact must inform debate over where our discipline is going and how it is to be restructured to meet contemporary challenges and needs.

On the history of geography and society

Geographical knowledge records, analyzes and stores information about the spatial distribution and organization of those conditions (both naturally occurring and humanly created) that provide the material basis for the reproduction of social life. At the same time, it promotes conscious aware-ness of how such conditions are subject to continuous transformation through human action.

The form and content of such knowledge depends upon the social context. All societies, classes, and social groups possess a distinctive

108

'geographical lore', a working knowledge of their territory, of the spatial configuration of use values relevant to them, and of how they may intervene to shape the use values to their own purposes. This 'lore', acquired through experience, is codified and socially transmitted as part of a conceptual apparatus with which individuals and groups cope with the world. It may take the form of a loosely-defined spatial and environmental imagery or of a formal body of knowledge – geography – in which all members of society or a privileged elite receive instruction. This knowledge can be used in the struggle to liberate peoples from 'natural' disasters and constraints and from external oppression. It can be used in the quest to dominate nature and other peoples and to construct an alternative geography of social life through the shaping of physical and social environments for social ends.

The form and content of geographical knowledge cannot be understood independently of the social basis for the production and use of that knowledge. Pre-capitalist societies, for example, produced highly sophisticated geographical understandings but often of a particular and localized sort, radically different from geography as we know it (Hallowell 1955; Levi-Strauss 1966). The trading empires of Greece, Rome, Islam and China all produced elaborate geographies of the world as they knew it (Herodotus 1954; Ibn Khaldûn 1958; Needham 1954; Strabo 1903–06). These geographies typically mirrored the movement of commodities, the migrations of peoples, the paths of conquest, and the exigencies of administration of empire.

The transformation from feudalism to capitalism in western Europe entailed a revolution in the structures of geographic thought and practice. Geographical traditions inherited from the Greeks and Romans, or absorbed from China and above all Islam, were appropriated and transformed in the light of a distinctively western Europe experience. Exchange of commodities, colonial conquest and settlement formed the initial basis, but as capitalism evolved, so the geographical movement of capital and labor power became the pivot upon which the construction of new geographical knowledge turned. Six aspects of geographical practice stand out in the bourgeois era:

First, concern for accuracy of navigation and the definition of territorial rights (both private and collective) meant that mapping and cadastral survey became basic tools of the geographer's art (Brown 1949; Skelton 1958). In the imperialist era, for example, the cartographic basis was laid for the imposition of capitalist forms of such rights in areas of the world (Africa, the Americas, Australia and much of Asia) that had previously lacked them. Such activity laid the basis for exclusive class-based privileges and rights to the appropriation of the fruits of both nature and labor within well-defined spaces. On the other hand, it also opened up the possibility for the rational organization of space and nature for the universal welfare of humankind.

Second, the creation of the world market meant 'the exploration of the earth in all directions' in order to discover 'new, useful qualities of things' and the promotion of 'universal exchange of the products of all alien climates and lands' (Marx 1973: 409). Working in the tradition of natural philosophy, geographers such as Alexander von Humboldt (1849–52) and Carl Ritter (1822–59) set out to construct a systematic description of the earth's surface as the repository of use values, as the dynamic field within which the natural processes that could be harnessed for human action had their being. The accurate description of physical and biotic environments has remained central to geography ever since.

Third, close observation of geographical variations in ways of life, forms of economy and social reproduction has also been integral to the geographer's practice. This tradition degenerated (particularly in the commercial geography of the late nineteenth century) into the mere compilation of 'human resources' open to profitable exploitation through unequal or forced exchange, the imposition of wage labor through primitive accumulation, the redistribution of labor supplies through forced migration, and the sophisticated manipulation of indigenous economies and political power structures to extract surpluses. Geographical practices were deeply affected by participation in the management of Empire, colonial administration, and the exploration of commercial opportunities (Capel 1981). The exploitation of nature under capitalism evidently often went hand in hand with the exploitation of peoples. On the other hand, the construction of such knowledge in the spirit of liberty and respect for others, as for example in the remarkable work of Elisée Reclus (1982), opened up the possibility for the creation of alternative forms of geographical practice, tied to principles of mutual respect and advantage rather than to the politics of exploitation.

Fourth, the division of the world into spheres of influence by the main capitalist powers at the end of the nineteenth century raised serious geopolitical issues. The struggle for control over access to raw materials, labor supplies, and markets was struggle for command over territory. Geographers like Friedrich Ratzel (1923) and Sir Halford Mackinder (1962) confronted the question of the political ordering of space and its consequences head on, but did so from the standpoint of survival, control, and domination. They sought to define useful geographical strategies in the context of political, economic and military struggles between the major capitalist powers, or against peoples resisting the incursions of empire or neocolonial domination. This line of work reached its nadir with Karl Haushoffer, the German geopolitician who actively supported and helped shape Nazi expansionist strategies (Dorpalen 1942). But geopolitical thinking continues to be fundamental within the contemporary era, particularly in the pentagons of military power and amongst those concerned with foreign policy. By force of historical circumstance, all national liberation movements must define themselves geopolitically if they are to succeed.

Fifth, concern with the use of 'natural and human resources' and spatial distributions (of population, industry, transport facilities, ecological complexes, and the like) led geographers to consider the question of 'rational' configurations of both. This aspect of geographical practice, which emerged strongly with the early geological, soil, and land-use surveys, has increased markedly in recent years as the capitalist state has been forced to intervene more actively in human affairs (Dear and Scott 1981). Positive knowledge of actual distributions (the collection, coding and presentation of information) and normative theories of location and optimization have proved useful in environmental management and urban and regional planning. To a large degree, these techniques entailed acceptance of a distinctively capitalist definition of rationality, connected to the accumulation of capital and the social control of labor power. But such a mode of thought also opened up the possibility for planning the efficient utilization of environments and space according to alternative or multiple definitions of rationality.

Sixth, geographical thought in the bourgeois era has always preserved a strong ideological content. As science, it treats natural and social phenomena as things, subject to manipulation, management, and exploitation. As art, it often projects and articulates individual and collective hopes and fears as much as it depicts material conditions and social relations with the historical veracity they deserve. For example, geographical literature often dwells upon the bizarre and quaint at the expense of dealing openly with the legitimate aspirations of peoples. Although it aspires to universal understanding of the diversity of life on earth, it often cultivates parochialist, ethnocentric perspectives on that diversity. It can be an active vehicle for the transmission of doctrines of racial, cultural, sexual, or national superiority. Ideas of 'geographical' or 'manifest' destiny, of 'natural' geographical rights (for example, US control over the Panama Canal), of the 'white man's burden' and the civilising mission of the bourgeoisie or of American democracy, are liberally scattered through geographical texts and deeply embedded in popular geographical lore (Buchanan 1974; Davis 1978; Weinberg 1963). Cold war rhetoric, fears of 'orientialism,' and the like, are likewise pervasive (Said 1979). Furthermore, the 'facts' of geography, presented often as facts of nature, can be used to justify imperialism, neocolonial domination, and expansionism. Geographical information also can be presented in such a way as to prey upon fears and feed hostility (the abuse of cartography is of particular note in this regard). But there is a brighter side to all of this. The geographical literature can express hopes and aspirations as well as fears, can seek universal understandings based on mutual respect and concern, and can articulate the basis for human cooperation in a world marked by human diversity. It can become the vehicle to express utopian visions and practical plans for the creation of alternative geographies (Kropotkin 1898; Reclus 1982).

The rise of geography as an academic discipline within a professional division of labor

Academic geographers sought to combine experience gained from these diverse practices into a coherent discipline within an academic division of labor that crystallized towards the end of the nineteenth century. They have not been altogether successful in this project. To begin with, they often remained eclectic generalists posing grand questions on such topics as environmental determinism, the social relation to nature, the role of geography in history, and so on) in an academia increasingly dominated by professional analytic expertise. Also, rejecting historical materialism as a basic frame of reference, they lacked methods to achieve synthesis and overcome the innumerable dualisms within their subject, between, for example, physical and human geography, regional specialization and systematic studies of global variation, unique and generic perspectives, quantitative and qualitative understandings. The dominant institutions within the discipline (such as the Royal Geographical Society) were more concerned with the practices of discovery and subordination of nature and the techniques of management of empire than they were with the creation of a coherent academic discipline (Capel 1981). Academic geography, as a consequence, posed grand questions but all too frequently trivialized the answers.

In the face of external pressures and internal disarray, geography has tended to fragment in recent years and seek salvation in a far narrower professionalization of its parts. But the more successful it has been in this direction, the more its method has coalesced into a monolithic and dogmatic positivism and the more easily the parts could be absorbed into some cognate analytic discipline (physical geographers into geology, location theorists into economics, spatial choice theorists into psychology, and so on). Geographers thereby lost their *raison d'être* as synthesizers of knowledge in its spatial aspect. The more specialized they became, also, the more they distanced themselves from the processes of construction of popular geographical knowledge. What was once an important preserve for the geographer fell into the hands of popular magazines and the producers of commercial travelogues and brochures, television films, news, and documentaries. The failure to help build appropriate popular understandings to deal with a world undergoing rapid geographical integration was a startling abrogation of responsibility.

Caught between lack of academic identity and profundity on the one hand and a weak popular base on the other, academic geography failed to build a position of power, prestige, and respectability within the academic division of labor. Its survival increasingly depended upon cultivation of very specialized techniques (such as remote sensing) or the production of specialized knowledges for powerful special interests. Big government, the corporation, and the military provided a series of niches into which geographers might conveniently crawl. The academic evolution of the discipline is now threatened by total submission to the dictates of powerful special interests.

Yet to be conscious of the facts of geography has always meant to exercise responsibility with respect to them (Buttimer 1974; Hérodote 1975; Peet 1977; Sayer 1981; Stoddart 1981). How that responsibility is expressed depends upon the social context and the individual and collective consciousness of geographers. Some, in the name of academic freedom and objectivity, have sought to raise the study of geography up onto some universal plane of knowledge, to create a positivist science above the influence of any mundane special interest. Others sought to confront the relation between power and knowledge directly, to create antidotes to what they see as one-sided geographical understandings and so become advocates for the legitimate aspirations of indigenous peoples or oppressed groups. Still others have struggled to help build an historical materialist science of human history in its geographical aspect, to create a knowledge that would help subject peoples, classes, and groups gain closer control over, and the power to shape, their own history.

The failing credibility of positivism in the late 1960s opened the way to attempts to create a more directly radical or Marxist tradition. Geographers were faced with a peculiar mix of advantages and disadvantages. Old-style geography – global, synthetic, and dealing with ways of life and social reproduction in different natural and social milieus – lent itself easily to historical materialist approaches, but was dominated by establishment thinkers attached to the ideology of empire or actively engaged in the service of national interests. A radical element lurked within this rather stuffy tradition. Reclus (1982) and Kropotkin (1898) brought anarchism and geography together to express their common social concerns in the late nineteenth century. More recently, writers like

Owen Lattimore (1962) and Keith Buchanan (1970) tried to portray the world not from the standpoint of the superpowers, but from that of indigenous peoples (*From China Looking Outwards* is a typical Lattimore title). The active repression of such thinkers, particularly during the cold war and McCarthyism (Newman), led many progressive geographers thereafter to express their social concerns behind the supposed neutrality of 'the positivist shield'. The main line of battle in the late 1960s, therefore, was over whether social concerns could be adequately expressed from behind the positivist shield or whether that shield was indeed as neutral as at first sight it appeared.

The radical and Marxist thrust in geography in the late 1960s concentrated on a critique of ideology and practice within the positivism that then reigned supreme (Quaini 1982). It sought to penetrate the positivist shield and uncover the hidden assumptions and class biases that lurked therein. It increasingly viewed positivism as a manifestation of bourgeois managerial consciousness given over at worst to manipulation and control of people as objects and at best capable of expressing a paternalistic benevolence. It attacked the role of geographers in imperialist endeavors, in urban and regional planning procedures designed to facilitate social control and capital accumulation. It called into question the racism, sexism, enthnocentrism, and plain political prejudice communicated in many geographical texts.

But the critics also had to create geographical thought and practice in a new image. Marxism (Peet 1977; Quaini 1982), anarchism (Breitbart 1979; Breitbart 1981; Quaini 1982), advocacy (Corey 1972), 'geographical expeditions' (Bunge 1977), and humanism (Ley and Samuels 1978) became some of the rallying points for those seeking alternatives. Each had then to identify and preserve those facets of geography relevant to their project. The more mundane techniques, such as mapping, information coding, and resource inventory analysis, appeared recuperable if not unavoidable to any reconstitution of geographic practice. The problem was to shake them free from their purely positivist presentation and integrate them into some other framework. Bourgeois geographers had also long sought to understand how different peoples fashion their physical and social landscapes as a reflection of their own needs and aspirations. They had also shown that different social groups (children, the aged, social classes, whole cultures) possess distinctive and often incomparable forms of geographical knowledge, depending upon their experience, position, and traditions. These ideas also seem recuperable as the basis for fresh geographical practice. Historical and cultural geographers, insofar as they had paid attention to the processes of spatial integration, regional transformation, and changing geographical configurations through time, provided relevant raw materials.

From the initial fumblings and searchings, a new agenda for geography emerged, rooted deep in tradition yet original and breathtaking in scope, exhilarating if often frustrating in its practice. The study of the active construction and transformation of material environments (both physical and social) together with critical reflection on the production and use of geographical knowledge within the context of that activity, could become the center of concern. The focus is on the process of *becoming* through which people (and geographers) transform themselves through transforming both their natural and social milieus. For the humanists, this process of becoming could be viewed religiously or secularly through the philosophical lenses of Heidegger and Husserl. The Marxists had to look no further than Marx's characterization of human labor as a process through which human beings, in acting on the external world and changing it, at the same time changed their own natures (Marx 1967: 177). Anarchists could appeal to Reclus who argued that 'humankind is nature becoming conscious of and taking responsibility for itself' (1982: vol. 1, p. 106). Those actively engaged in advocacy could feel they were integral to processes of social transformation.

While the commonality of the new agenda was frequently masked by bitter backbiting amongst the participants, there could be no question as to the common core of concern. But there were deeper problems that inhibited its execution and threatened it with early extinction, crushed under the overwhelming critical silence of a positivist reaction. The problems are in part external to the profession, the product of a societal condition that does not favor experimentation, innovation, and intellectual debate but which seeks to discipline unruly academics to more narrowly based immediate and practical concerns as defined by powerful special interests. But the problems are also internal. Advocates for community cannot justify a stance of 'community right or wrong' if one community's gain is another's loss any more than environmentalists can reasonably proceed oblivious of employment consequences. Humanists, if they are to avoid the trap of narcissistic radical subjectivism, need a more powerful theory than agency and structure to grapple with macro-problems of money power, inflation, and unemployment. Anarchists, while sensitized to ecological and communitarian concerns, lack the social theory to understand the dynamics of capitalism in relation to state power. Marxists come armed with a powerful theory but find it hard to cope with ecological issues or with a subject matter in which highly differentiated activities of individuals and social groups within the particularities of space and place are of paramount concern.

What is lacking is a clear context, a theoretical frame of reference, a language which can simultaneously capture global processes restructuring social, economic and political life in the contemporary era and the specifics of what is happening to individuals, groups, classes, and communities at particular places at certain times. Those who broke out from behind the safety of the positivist shield ruptured the political silence within geography and allowed conscience and consciousness freer play. But they spoke with many voices, generated a veritable cacophony of competing messages, and failed to define a common language to voice common concerns.

Between the safety of a positivist silence and the risk of nihilistic disintegration lies the passage to a revitalized geography, an intellectual discipline that can play a vital, creative, and progressive role in shaping the social transformations that beset us. How to negotiate that passage is *our* dilemma of *this* time.

The present condition of geography

Geography is too important to be left to geographers. But it is far too important to be left to generals, politicians, and corporate chiefs. Notions of 'applied' and 'relevant' geography pose questions of objectives and interests served. The selling of ourselves and the geography *we* make to the corporation is to participate directly in making *their* kind of geography, a human landscape riven with social inequality and seething geopolitical tensions. The selling of ourselves to government is a more ambiguous enterprise, lost in the swamp of some mythic 'public interest' in a world of chronic power imbalances and competing claims. The disenfranchised (and that includes most of us when it comes to interest rates, nuclear strategy, covert operations, and geopolitical strategizing) must be heard through the kind of geography we make, no matter how unpopular that voice within the corridors of power or with those who control our purse strings. There is more to geography than the production of knowledge and personnel to be sold as commodities to the highest bidder.

The geography we make must be a peoples' geography, not based on pious universalisms, ideals, and good intents, but a more mundane enterprise that reflects earthly interests, and claims, that confronts ideologies and prejudice as they really are, that faithfully mirrors the complex weave of competition, struggle, and cooperation within the shifting social and physical landscapes of the twentieth century. The world must be depicted, analyzed, and understood not as we would like it to be but as it really is, the material manifestation of human hopes and fears mediated by powerful and conflicting processes of social reproduction.

Such a peoples' geography must have a popular base, be threaded into the fabric of daily life with deep taproots into the well-springs of popular consciousness. But it must also open channels of communication, undermine parochialist world views, and confront or subvert the power of dominant classes or the state. It must penetrate the barriers to common understandings by identifying the material base to common interests. Where such a material base does not exist, it must frankly recognize and articulate conflict of equal and competing rights that flows therefrom. To the degree that conflicting rights are resolved through tests of strength between contending parties, so the intellectual force within our discipline is a powerful weapon and must be consciously deployed as such, even at the expense of internalizing conflicting notions of right within the discipline itself. The geographical studies we make are necessarily a part of that complex of conflictual social processes which give birth to new geographical landscapes.

Geographers cannot remain neutral. But they can strive towards scientific rigor, integrity and honesty. The difference between the two commitments must be understood. There are many windows from which to view the same world, but scientific integrity demands that we faithfully record and analyze what we see from any one of them. The view from China looking outwards or from the lower classes looking up is very different from that from the Pentagon or Wall Street. But each view can be represented in a common frame of discourse, subject to evaluation as to internal integrity and credibility. Only in this way can the myriad masks of false conflict be stripped away and the real structure of competing rights and claims be exposed. Only in this way too, can we insure that the geography we make is used and not abused in the struggles of our time.

The intellectual task in geography, therefore, is the construction of a common language, of common frames of reference and theoretical understandings, within which conflicting rights and claims can be properly represented. Positivism undermines its own virtues of objective materialism by spurious claims to neutrality. Historical materialism, though appropriate, is too frequently held captive within the rigidities of some political orthodoxy that renders windows on the world opaque and substitutes subjectively conceived political fantasy for hard-nosed objective materialism. Under such conditions the construction of a common discourse for describing and theorizing becomes a tough task.

But the very nature of the intellectual baggage accumulated these past years makes the geographers' contributions potentially crucial. For example, the insertion of concepts of space, place, locale, and milieu into any social theory has a numbing effect upon that theory's central propositions. Microeconomists working with perfect competition find only spatial monopoly, macroeconomists find as many economies as there are central banks and a peculiar flux of exchange relations between them, and

Marxists looking to class relations find neighborhoods, communities, and nations. Marx, Marshall, Weber, and Durkheim all have this in common: they prioritize time over space and, where they treat the latter at all, tend to view it unproblematically as the site or context for historical action. Whenever social theorists of whatever stripe actively interrogate the meaning of geographical categories and relations, they are forced either to make so many ad hoc adjustments to their theory that it splinters into incoherency, or else to abandon their theory in favor of some language derived from pure geometry. The insertion of spatial concepts into social theory has not yet been successfully accomplished. Yet social theory that ignores the materialities of actual geographical configurations, relations, and processes lacks validity (Giddens 1981; Gregory and Urry 1984).

The temptation then exists to abandon theory, retreat into the supposed particularities of place and moment, resort to naive empiricism, and produce as many ad hoc theories as there are instances. All prospects for communication then break down save those preserved by the conventions of common language. The ambiguities of the latter masquerade as theory, and theory itself is lost in a swamp of ambiguous meanings. Ambiguity may be preferable to rigid and uncompromising orthodoxy, but it is no basis for science. Retreat from explicit theory is retreat from the challenge to make conscious and creative interventions in the construction of future geographies. The junction between geography and social theory, therefore, is one of the crucial flash-points for the crystallization of new conceptions of the world and new possibilities for active intervention.

The political implications of a resolution of such real and highly charged intellectual dilemmas between geography and social theory are legion. Consider, for example, the clash between anarchist and Marxist perspectives both politically and within the history of geography. Reclus and Kropotkin, geographers both, were impressed by the remarkable diversity of life, culture, community, and environment revealed by their geographical studies. They respected that diversity and sought to preserve it through a political project that linked the peoples of the earth into some vast federation of autonomous self-governing communities. This entailed a highly decentralized and profoundly geographical vision of how an alternative society should look. It has helped fuel a political tradition concerned with worker self-management, community control, ecological sensitivity, and respect for the individual. Is it accidental that the radical urge in nineteenth-century geography was expressed through anarchism rather than through Marxism? The sensitivity to issues of place, ecology, milieu, and geographical particularities still makes the anarchist vision appealing. Yet it is seriously flawed by the absence of any powerful theory of the dynamics of capitalism. Reclus (1982) in his last work recognized that the intriguing geographical variety for which he had such respect was even then being swept away, crushed under the homogenizing heel of the circulation and accumulation of capital. The universality of that experience demands a global political response born out of more powerful universal understandings of the dynamics of capitalism than Reclus constructed. His political vision and his intellectual contribution are undermined by this crucial absence.

Marx, for his part, occupies the pinnacle of social theoretic power at the expense of excluding geographical variation as an 'unnecessary complication' (1967: vol. 2, p. 470; see also Chapter 13). From that highpoint he can proclaim a politics of universal class struggle founded on universal proletarian consciousness and solidarity. To be sure, Marx frequently admits of the significance of space and place within both his theory and his practice (the opposition between English and Irish working-class interests parallels oppositions in his theoretical work between town and country, inner and outer transformations, and the like). But none of this is thoroughly integrated into theoretical formulations that are powerful with respect to time but weak with respect to space. His political vision and theoretical contribution founder on his failure to build a systematic and distinctively geographical dimension into his thought. This was the 'error' that Lenin and the theorists of imperialism sought to rectify. They opened up the possibility of an alternative rhetoric within the Marxist tradition in which centers exploit peripheries, the first world subjugates the third, and capitalist powers compete for domination of protected space (markets, labor power, raw materials). People in one place exploit and struggle against those in another place. Ad hoc concessions to spatial structure provoke redefinitions of exploitation that coexist uneasily with Marx's view of a capitalist dynamic powered by the exploitation of one class by another. The theoretical foundations of Marxism-Leninism are thereby rendered ambiguous, sparking savage disputes over the right to national self-determination, the national question, the prospects for socialism in one country, the significance of geographical decentralization in political practice (Davis 1978; Harvey 1983; Lenin 1963; Luxemburg 1976).

There is more to the split between anarchists and Marxists (or divisions within the latter camp) than their respective approaches to geographical questions. But the Marxists, while proclaiming in principle the significance of geographical uneven development, have had a hard time integrating space or evolving a sensitivity to place and milieu within otherwise powerful social theories. The anarchist literature abounds with such sensitivity but founders on lack of theoretical and political coherence. All of which provokes the intriguing though somewhat idle thought: what would our political and intellectual world be like if Marx had been a better geographer and the anarchists better social theorists? That rhetorical question underlines the contemporary political importance of a theoretical project dedicated to the unification of geographical sensitivities and understandings with the power of general social theories formulated in the tradition of historical materialism. Such a theoretical project is more than just a tough academic exercise. It is fundamental to our thinking on the prospects for the transition to socialism.

An historical materialist manifesto

The tasks before us can now be more clearly defined. We must:

1. Build a popular geography, free from prejudice but reflective of real conflicts and contradictions, capable also of opening new channels for communication and common understanding.
2. Create an applied peoples' geography, unbeholden to narrow or powerful special interests, but broadly democratic in its conception.
3. Accept a dual methodological commitment to scientific integrity and non-neutrality.
4. Integrate geographical sensitivities into general social theories emanating from the historical materialist tradition.
5. Define a political project that sees the transition from capitalism to socialism in historico–geographical terms.

We have the power through our collective efforts as geographers to help make our own history and geography. That we cannot do so under historical and geographical circumstances of our own choosing is self-evident. In part our role is to explore the limits imposed by the dead-weight of an actually-existing geography and an already-achieved history. But we must define, also, a radical guiding vision: one that explores the realms of freedom beyond material necessity, that opens the way to the creation of new forms of society in which common people have the power to create their own geography and history in the image of liberty and mutual respect of opposed interests. The only other course, if my analysis of the trajectory of contemporary capitalism is correct (Harvey 1983), is to sustain a present geography founded on class oppression, state domination, unnecessary material deprivation, war, and human denial.

CHAPTER 7

Capitalism:
the factory of fragmentation

First published in New Perspectives Quarterly, *1992.*

The drive for capital accumulation is the central motif in the narrative of historical-geographical transformation of the western world in recent times and seems set to engulf the whole world into the twenty-first century. For the past 300 years it has been the fundamental force at work in reshaping the world's politics, economy and environment. This process of using money to make more money is not the only process at work, of course, but it is hard to make any sense of social changes these past 300 years without looking closely at it.

Contemporary historical materialism attempts to isolate the fundamental processes of capital accumulation that generate social, economic and political change and, through a careful study of them, get some understanding of the whys and hows of those changes. The focus is on *processes*, rather than on things and events. It is a bit like watching a potter at work on a wheel: the process may be simple to describe, but the outcomes can be infinitely varied in shape and size.

However, to say there is a simple process at work is not to say that everything ends up looking exactly the same, that events are easily predictable or that everything can be explained by reference to it alone. The drive for capital accumulation has helped create cities as diverse as Los Angeles, Edmonton, Atlanta and Boston, and transformed out of almost all recognition (though in quite different ways) ancient cities like Athens, Rome, Paris and London. It has likewise led to a restless search for new product lines, new technologies, new lifestyles, new ways to move around, new places to colonize – an infinite variety of stratagems that reflect a boundless human ingenuity for coming up with new ways to make a profit. Capitalism has, in short, always thrived on the production of difference.

Yet the rules that govern the game of capital accumulation are relatively simple and knowable. Capitalism is always about growth, no matter what the ecological, social or geopolitical consequences (indeed, we define 'crisis' as low growth); it is always about technological and lifestyle changes ('progress' is inevitable); and it is always conflictual (class and other forms of struggle abound).

Above all, capitalism generates a lot of insecurity: it is always unstable and crisis-prone. The history of capitalist crisis formation and resolution is, I maintain, fundamental to understanding our history. Understanding the rules of capital accumulation helps us understand why our history and our geography take the forms they do.

The worship of fragments

In *The Condition of Postmodernity*, I tried to put this style of thinking to work in explaining recent changes in economy and culture in the advanced capitalist world. I noticed that postmodern thought tended to deny anything systematic or general in history, and to jumble together images and thoughts as if criteria of coherence did not matter: it emphasized separation, fragmentation, ephemerality, difference and what is often now called 'otherness' (a strange term that is mainly used to indicate that I have no right to speak *for* or even, perhaps, *about* others or that when I do speak about them I 'construct' them in my own image).

Furthermore, some postmodern theorists argued that the world was not knowable because there was no sure way of establishing truth and that even pretending to know or, worse still, holding to some version of 'universal truth' lay at the root of gulags, holocausts and other social disasters. The best that we could hope for, they said, was to let things flourish in their multiple and different ways, look for alliances where possible, but always to avoid peddling supposed universal solutions or pretending there were general, knowable truths. This sort of thinking carried over into architecture, the arts, popular culture, new lifestyles and gender politics.

Now, there is much that is refreshing about all of this, particularly the emphasis upon heterogeneity, diversity, multiple overlapping concerns of gender, class, ecology, and so on. But I just could not see why the sort of heterogeneity that postmodernism celebrates was in any way inconsistent with thinking the world was knowable through an appreciation, of, for example, processes of capital accumulation, which not only thrive upon but actively produce social difference and heterogeneity.

The postmodern phoenix

Since this shift in cultural sensibility paralleled some quite radical changes in the organization of capitalism after the capitalist crisis in 1973–5, it even seemed plausible to argue that postmodernism itself was a product of the process of capital accumulation.

After 1973, for example, we find that working-class politics went on the defensive as unemployment and job insecurity rose, economic growth slackened, real wages stagnated, and all sorts of substitutes for real productive activity took over to compensate for wave after wave of deindustrialization. Merger manias, credit binges and all the other excesses of the 1980s, which we are now paying for, were the only vital activity at a time of gradual dismantling of the welfare state and the rise of *laissez-faire* and very conservative politics. Strong appeal to individualism, greed and entrepreneurial spirit characterized the Reagan-Thatcher years. Furthermore, the crisis of 1973 set in motion a frantic search for new products, new technologies, new lifestyles and new cultural gimmicks that could turn a profit. And these years also saw a radical reorganization of international power relations, with Europe and Japan challenging a dominant US power in economic and financial markets.

This general shift from old-style capital accumulation to a new style, I call the shift from Fordism (mass assembly line, mass political organization and welfare-state interventions) to flexible accumulation (the pursuit of niche markets, decentralization coupled with spatial dispersal of production, withdrawal of the nation-state from interventionist policies coupled with deregulation and privatization). It seemed to me quite plausible to argue, therefore, that capitalism, in undergoing this transition, had produced the conditions for the rise of postmodern ways of thinking and operating.

Time-space compression

But it is always dangerous to treat simultaneity as causation, so I set about looking for some sort of link between the two trends. The link I believed worked best was the one between time and space. Capital accumulation has always been about speed-up (consider the history of technological innovations in production processes, marketing, money exchanges) and revolutions in transport and communications (the railroad and telegraph, radio and automobile, jet transport and telecommunications), which have the effect of reducing spatial barriers.

The experience of time and space has periodically been radically transformed. We see a particularly strong example of this kind of radical transformation since around 1970: the impact of telecommunications, jet cargo transport, containerization of road, rail and ocean transport, the development of futures markets, electronic banking and computerized production systems. We have recently been going through a strong phase of what I call 'time-space compression': the world suddenly feels much smaller, and the time-horizons over which we can think about social action become much shorter.

Our sense of who we are, where we belong and what our obligations encompass – in short, our *identity* – is profoundly affected by our sense of location in space and time. In other words, we broadly locate our identity in terms of space (I belong *here*) and time (this is *my biography*, *my history*). Crises of identity (Where is my place in this world? What future can I have?) arise out of strong phases of time-space compression. Moreover, I think it plausible to argue that the most recent phase has so shaken up our sense of who and what we are that there had to be some kind of crisis of representation in general, a crisis that is manifest in the contemporary world primarily by postmodern ways of thinking.

Embracing ephemerality as a desired quality in cultural production, for example, matches the rapid shifts in fashion and production designs and techniques that evolved as part of the response to the crisis of accumulation that developed after 1973.

Interestingly, when we look back on other phases of rapid time-space compression – the period after 1848 in Europe, the period just before and during the First World War, for example – we find similar phases of rapid change in the arts and in cultural activities. From this I conclude that it is possible to arrive at a *general* interpretation of the rise of postmodernism and its relation to the new experience of space and time that new forms of capital accumulation have produced.

But, again, I want to enter a caveat: this is not to say that everything is simply deterministic. I repeat, capitalism thrives on and produces heterogeneity and difference, though only within certain bounds.

Niche markets

There is nothing about postmodernism in general that inhibits the further development of capital accumulation. Indeed, the postmodern turn has proved a perfect vehicle for the development of new fields and forms of profit-making.

Fragmentation and ephemerality, for example, open up abundant opportunities to explore quick-changing niche markets for new products. But this does not mean that there has been any radical inversion of the historical materialist view of reality, an inversion where culture, not economics, has become the driving force of history. I think such a view misinterprets rather than misrepresents what is happening.

Marx held that production of any sort requires the prior exercise of the human imagination; it is always about the mobilization of human desires, purposes and intentions to a given end. The problem under industrial capitalism is that most people are denied access to this process: a select few do the imagining and designing, make all the decisions and set up technologies that regulate the worker's actions, so that for the mass of the population the full play of human creativity is denied.

That is a profoundly alienating situation, and much of history recounts attempts to respond to this alienation. The rich and the privileged, themselves not enamored of industrialism, countered alienation by developing a distinctive field of *culture* – think of romanticism and the cultivation of aesthetic pleasures and values – as a kind of protected zone for creative activities outside of the crass materialism of industrial capitalism.

Workers likewise developed their own creative pleasures when they could: hunting, gardening, tinkering with cars. These activities, which went under the general name of 'culture,' high or low, were not so much superstructural as compensatory for what industrial capitalism denied to the mass of the people in the workplace.

Over time, those compensatory pleasures have gradually become absorbed into the processes of capital accumulation and turned into new spheres for making profit. As industrial capitalism became less and less profitable, at least in the US and Britain, so these new spheres of profit-making became much more important, particularly after 1945 and even more so after the crisis of 1973–5.

So, there is a sense in which culture no longer trails other forms of economic activity but has moved into the vanguard, not as a protected zone of non-economic activity, however, but as an arena of fierce competition for profit-making. The accumulation of market niches, of diverse preferences and the promotion of new heterogeneous lifestyles, all occur within the orbit of capital accumulation.

The latter, furthermore, has had the effect of breaking down distinctions between high and low culture – it commercializes aesthetics – at the same time as it has thrived, as it always does, on the production of diversity, heterogeneity and difference. What we generally think of as 'culture' has become a primary field of entrepreneurial and capitalistic activity.

Through the postmodern door

The picture I have so far painted probably looks very pessimistic, with capital accumulation, market materialism and entrepreneurial greed ruling the roost. So let me look now at the opportunities and dangers that attach to this postmodern condition.

I notice, first of all, that capitalism has not solved its crisis tendencies and that capital accumulation, economic growth and sustained development into the foreseeable future are, if anything, more remote now than they were twenty years ago. When the fundamental irrationality of capitalism becomes plainer for all to see – as in the present depression on both sides of the Atlantic – the conditions are set up in which some kind of new direction has to be taken (if only throwing the ruling party out of power).

Secondly, the frantic promotion of cultural heterogeneity and difference over the past twenty years has opened up all kinds of new spaces for the exploration of different lifestyles, different preferences and a more generalized debate about human potentialities and the sources of their frustration. This is the positive side of what much of postmodernism stands for: it produces openings for a critique of dominant values, including those that directly attach to the rules of capital accumulation, and therefore all kinds of opportunities for radical politics. The corollary is that contemporary radical politics has as much to do with culture as with traditional problems of class struggle in production.

But here we encounter as many dangers as opportunities. The crisis of identity provoked by time-space compression can lead to the acceptance of exclusionary religious doctrines (the promise of eternity in a world of rapid change) or exclusionary territorial practices (maintaining the security and position of the home, the locality, the nation against external and international pressures). The rise of fascist and exclusionary sentiments across Europe and the progress of the Buchanan campaign in the US provide good examples. The refusal to accept that there are some basic processes at work and that knowable truths can be established can all too easily lead to head-in-the-sand politics ('I will pursue my particular political interest and to hell with all the rest').

The fetishism of the image at the expense of any concern for the social reality of daily life can divert our gaze, our politics, our sensitivities away from the material world of experience and into the seemingly endless and intricate webs of representations. And while it is true that the 'personal is political', we do not have to look much further than the present presidential campaign to see how that principle can be abused. Above all, the promotion of cultural activities as a primary field of capital accumulation promotes a commodified and prepackaged form of aesthetics at the expense of concerns for ethics, social justice, fairness, and the local and international issues of exploitation of both nature and human nature.

So postmodernism opens a door to radical politics but for the most part has refused to pass through it. To pass to a thoroughly radical critique of contemporary capitalism, which is plainly languishing not only economically but culturally and spiritually, requires that we grapple with the central processes of capital accumulation that are so radical in their implications for our lives. Capitalism has transformed the face of the earth at an accelerating pace these past 200 years. It cannot possibly continue on that trajectory for another 200 years. Someone, somewhere, has to think about what kind of social system should replace it. There seems no alternative except to build some kind of socialist politics that will have as its central motif the question: what could life be about if capital accumulation no longer dominated? That question deserves the close attention of everyone.

CHAPTER 8

A view from Federal Hill

First published in
The Baltimore Book: New Views on Urban History, *1992.*

An' they hide their faces,
An' they hide their eyes,
'Cos the city's dyin'
An' they don' know why;
Oh!Baltimore!!
Man it's hard, jus livin'
Jus livin'.

Randy Newman

A city center, it has been said, is a great book of time and history. The view of Baltimore from Federal Hill (*site 1* on Figure 8.1) is an impressive introduction to that book and conveys a powerful image of what the city is about. But we have to learn to read all the signs of the landscape.

Certain things stand out in a city. A medieval European city immediately signals that religion and aristocracy were the chief sources of power by the way cathedrals and castles dominate. The United States struggled long and hard to get rid of aristocratic privilege, but Baltimore's downtown skyline says that a financial aristocracy is alive and well. As you look down on the city from Federal Hill, banks and financial institutions tower over everything else, proclaiming in glass, brick, and concrete that they hold the reins of power.

The Federal Building (*site 2*), buried in the midst of all these financial institutions, signals a system of governance that is, as Mark Twain once put it, 'the best that money can buy'. City Hall (*site 3*), attractive and classical though it may be, is neither centrally located nor conspicuous enough to suggest it has more than a marginal role to play in determining the city's fate. As for churches, they can be seen only when you look across the densely packed rowhouses of ethnic and working-class East Baltimore. God, it seems, has meaning for the working class; mammon is fully in control downtown.

The other image that stands out is the importance of water, of Chesapeake Bay, which formed Baltimore's commercial lifeline to the

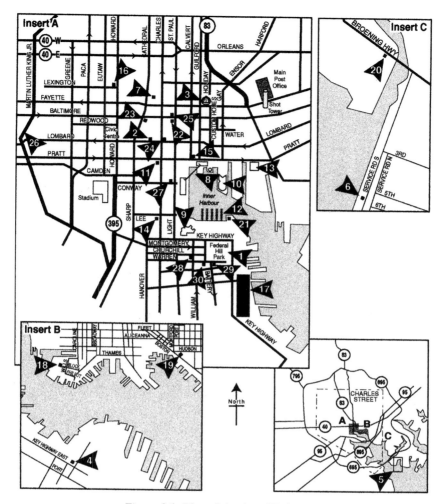

Figure 8.1 *Plan of the city of Baltimore*

Italic type indicates original building site or function
Bold italic indicates original building is not extant

1. Federal Hill
2. The Federal Building
3. City Hall
4. Domino Sugar plant
5. Bethlehem Steel plant, Sparrows Point
6. Dundalk Marine Terminal
7. Charles Center
8. Pavilions at Harborplace
9. Maryland Science Center
10. National Aquarium
11. Convention Center
12. Marina at Inner Harbor
13. Power Plant at Inner Harbor
14. Harbor Court
15. Gallery at Inner Harbor
16. *Hutzler's Palace*
17. *Bethlehem Steel shipyards*
18. ***Allied Chemical plant***
19. American Can Company
20. Western Electric plant
21. Rusty Scupper at Inner Harbor
22. Maryland National Bank
23. Mercantile Safe Deposit and Trust Company
24. First National Bank
25. *Merritt Commericial Savings and Loan,* now Citibank
26. Martin Luther King Boulevard
27. Hyatt-Regency Hotel
28. 820 Churchill Place, condominuim
29. *Shofer Furniture Warehouse*; called The Paper Mill by developers; now Federal Park Condominium
30. *Southern High School,* now Harbor View

Figure 8.2 Baltimore Harbor in the 1930s. The Peale Museum, Baltimore.

Figure 8.3 Downtown and the Inner Harbor, from Federal Hill, 1966. The Maryland National Bank Building dominates the center, illustrating the power and authority of financial institutions compared to those of politics, symbolized by the less conspicuous Federal Building on the left and the dome of City Hall on the right. The Peale Museum, Baltimore.

Figure 8.4 *Downtown and the Inner Harbor from Federal Hill, 1988. The Federal Building and City Hall are engulfed by the mass of buildings given over to financial functions and the pavilions of Harborplace, a center of leisure and consumerism sprawled around the Inner Harbor edge.*

world and became the nexus for much of its now declining manufacturing industry. Signs of those connections abound: the Domino Sugar plant (*site 4*), grain elevators, the chemical plant, and oil tanks that line the edge of the bay, as it opens out from Federal Hill toward the Bethlehem Steel plant at Sparrows Point (*site 5*), and the Dundalk Marine Terminal (*site 6*), still one of the most important ports of entry on the East Coast of the US.

Nor is it hard to imagine that the Inner Harbor, now important as a tourist attraction and leisure park, was once the main port of entry to the city. Indeed, those functions were preserved there until shortly after the Second World War.

Though the view from Federal Hill tells us much about the city, it cannot tell us how what we see came into being. How was Baltimore built? Who decided that it should be a tourist mecca rather than an industrial city? Why do the buildings look the way they do, and to what traditions are they monuments?

Charles Center

Most of the downtown skyline has been in place since 1970 or so, though a transitional period dates to the mid 1950s. By then, the boom in production and trade that had powered Baltimore's economy during the Second World War had begun to fade. Strong currents of suburbanization, both of industry and of population, particularly the more affluent whites, the immigration of poor rural African Americans from the South, and the shift of port functions to deep water down the bay left Baltimore's downtown and inner city in a parlous state.

The formation of the Greater Baltimore Committee (GBC), an association of local business leaders, in 1956 marked a turning point. The committee recognized that downtown deterioration threatened the future of business in the city and that it was politically dangerous for any ruling elite to abandon the symbolic and political center of the metropolitan

Figure 8.5   The Mies van der Rohe building, One Charles Center, centerpiece of Baltimore's first attempts at downtown revitalization, 1988

region to an underclass of impoverished blacks and marginalized whites.

The committee developed a plan and then pressured city government into pursuing a downtown urban renewal project that would revive property development and corporate power in the downtown core. Federal urban renewal funds were available, Mayor Thomas D'Alesandro was persuaded, and the Greater Baltimore Committee/Charles Center Management Corporation was formed to promote and organize the renewal. This was the first of a succession of quasi-public agencies, dominated by corporate and business interests and outside any democratic control, that were to shape downtown renewal efforts over the next twenty years.

After nearly $40 million in public expenditures, which attracted a further $145 million of private investment, Charles Center (*site 7* on Figure 8.1) was essentially completed by the late 1960s. Modernist in design (its Mies van der Rohe building is considered a classic), Charles Center houses office workers and financial or governmental institutions in somewhat arid modern buildings punctuated by bleak open public spaces.

The city, it was argued, would receive two main benefits from such development: The increase in employment would help the city's economy, and the increase in the tax base would provide the city with more resources to meet the needs of its poor. Unfortunately, from the beginning, Charles Center was conceived and built as a property development scheme of direct benefit to corporate and finance capital. The city as a whole received very little benefit from it. Much of the new downtown employment, particularly in skilled and well-paying jobs, went to residents of the suburbs. The jobs created for city residents were either in temporary construction or low-paying services.

Moreover, Charles Center was so heavily subsidized that it was a drain on, rather than a benefit to, the city's tax base. This was particularly true before an upward revaluation in 1975, a year after it was revealed that tax assessments in Charles Center were lower than they had been before redevelopment.

The Inner Harbor

With the completion of Charles Center in the late 1960s, downtown realtors and business leaders turned their attention to the Inner Harbor. Plans were laid to extend development to the decaying waterfront of derelict piers and crumbling warehouses, marks of Baltimore's once significant water trade now rendered obsolete by the trucks that rolled across the expanding network of federally subsidized highways.

There were few takers for developing this zone until the early 1970s.

Figure 8.6 Tour boats, the converted Power Plant, and Scarlett Place, captured in a view from the pavilions of the Inner Harbor redevelopment.

And it took a basic shift in orientation and philosophy to bring about this new and most recent phase of construction.

Baltimore, like many cities in the 1960s, was racked by race riots and civil strife. Concentrated in the abandoned and decayed inner cities, this breakdown in civil order focused on racial discrimination in job and housing markets, unemployment, and the disempowerment and impoverishment of much of the city's African-American population. Investment in the inner city seemed neither safe nor profitable. The urban spectacles that drew the crowds downtown were race riots, anti-war demonstrations, and all manner of countercultural events.

The riots and burnings that gutted areas of Gay Street and North Avenue in the wake of the assassination of Dr Martin Luther King, Jr, in April 1968, left six people dead, some 5,000 arrested, massive property

damage, and streets patrolled by the military. In 1970, a day-long skirmish
between youths and the police at the city's flower mart – an annual event
promoted by Baltimore's elite since 1911 – indicated that anger was
common to disempowered blacks and discontented white youths.

On 25 October 1973, a group of women representing Baltimore's elite
placed a plaque at the Washington Monument to commemorate the end
of the flower mart. At the time it seemed a fitting symbol of the lack of
confidence and social malaise that inhibited any investment in the city's
future. The business climate in downtown Baltimore could not have been
less propitious.

It was precisely in this context that many in the city sought for some
way to restore a sense of civic pride, some way to bring the city together
as a working community, some way to overcome the siege mentality with
which investors and the citizenry viewed the inner city and its downtown
spaces. The coalition that was to form was much broader than the Greater
Baltimore Committee. It included church and civil rights leaders, distressed
that the riots generated as much self-inflicted pain as social redress for
those doing the rioting; academics and professionals, including downtown lawyers, suddenly made aware of the wretched living conditions of

*Figure 8.7 The postmodernism of Scarlett Place, with its preserved nineteenth-cen-
tury seed warehouse on its left-hand corner and its attempt to simulate a
Mediterranean-style hilltop village, contrasts with the austere modernism of Mies van
der Rohe.*

Figure 8.8 Baltimore riots in 1968, following the assassination of Dr Martin Luther King, Jr, resulted in significant property damage in several of the city's neighborhoods. The riots dramatically enacted the deep racial tensions of the city and represented the first time since the railroad riots of 1877 that the National Guard was summoned to Baltimore to enforce state power. The Baltimore Sun, 1968.

the majority of the city's population; city officials who had long striven to build a better sense of community; and downtown business leaders who saw their investments threatened.

In this climate, the idea of a city fair that would build on neighborhood traditions but would celebrate a common purpose began to take shape. In 1970, when the first fair was held, the fear of violence was great. But 340,000 people came during the weekend of the fair in peaceable fashion, proving that disparate neighborhoods and communities could come together around a common project.

'A city reborn through a fair of neighbors,' trumpeted Baltimore's newspapers. A Department of Housing and Urban Development report in 1981 recommended the fair to other urban governments in these terms: 'Spawned by the necessity to arrest the fear and disuse of downtown areas caused by the civic unrest of the late 1960s, the Baltimore City Fair was originated by individuals in city government who seized upon the idea of a country fair in the city as a way to promote urban redevelopment.'

By 1973, the fair was attended by nearly 2 million people. It had abandoned its location in the secure heart of Charles Center and moved to the edge of the Inner Harbor. In so doing, it suggested an entirely different set of uses to which that site could be put. The city fair proved that large numbers of people could be attracted downtown without having a riot. It also helped Baltimore rediscover the ancient Roman formula of bread and circuses as a means of masking social problems and controlling discontent.

The story of the Inner Harbor's construction is one of a steady erosion of the aims of the coalition that set it in motion and its capture by the narrower forces of commercialism, property development, and financial power. Two events had particular significance. The first was the election of a strong-willed and authoritarian mayor, William Donald Schaefer, in 1971.

Schaefer had grown up in Baltimore's Democratic party machine politics, and he was everything a machine politician should be. He believed strongly in a partnership of business and private enterprise for furthering the city's development and in an elaborate and often ruthless politics of social control over the city's neighborhoods. To offend the mayor was to risk retribution; to play along with him meant patronage and access to city services.

The second event was the recession of 1973–5, which brought a massive wave of plant closures and deindustrialization to the Baltimore region. Unemployment surged. The prognosis for the city's economic future was bleak. In 1973, after President Nixon announced that the urban crisis was over, Baltimore faced the beginning of the end of large-scale federal programs to assist cities with their problems.

Budgetary cutbacks in the Reagan years were the highwater mark of federal government withdrawal from its commitment to help the nation's cities. The recession of 1981–3 – along with sharper foreign competition from Japan, Western Europe, and a host of newly industrializing countries – added to the city's difficulties. The list of plant closures and lay-offs grew daily more threatening. A new international division of labor was coming into being, with manufacturing plants moved to cheap-labor locations overseas and basic US industries like steel falling behind Japan and South Korea in world markets. Baltimore now had to find its way in a hostile and highly competitive world.

Table 8.1: Where the jobs have gone

The recession of 1980–3 brought a powerful wave of job losses to the Baltimore region, as illustrated in this list of cutbacks in manufacturing and retail establishments from 1980 through 1985. The list is adapted from a chart appearing in the *Baltimore Sun* on 21 March 1985. Companies marked with an asterisk have shut down operations completely.

Company	Type of business	Number of jobs lost
* Acme Markets	Grocery chain	1,200
* Airco Welding	Cored wire	150
* Allied Chemical	Chromium	145
Bethlehem Steel	Steel	7,000
* Bethlehem Steel Shipyard	Ship repair	1,500
* Brager-Gutman	Retail stores	180
* Cooks United	Discount stores	220
Esskay	Meat packing	240
General Electric	Electrical products	550
General Motors	Auto parts and distribution	247
* Korvette's	Department stores	350
* Maryland Glass	Glass	325
* Maryland Shipping & Drydock	Ship repair	1,500
Max Rubins	Apparel	225
* Misty Harbour Raincoat	Rainwear	210
* Pantry Pride	Grocery chain	4,000
* Plus Discount Stores	Discount stores	150
* Two Guys	Discount stores	500
Vectra	Fiber and yarn	600
* Western Electric	Electrical products	3,500

The turn to tourism, the creation of an image of Baltimore as a sophisticated place to live, the razzle-dazzle of downtown, and the commercial 'hype' of Harborplace (*site 8* on Figure 8.1) have to be seen as Mayor Schaefer's (and the GBC's) distinctive solution to that problem. With the crowds pouring in, it was a short step to commercializing the city fair, first by adding all manner of ethnic festivals, concerts, and spectacular events – for example, the visit of the 'tall ships' during the 1976 bicentennial celebration – to draw even more people downtown.

Then, having proved the existence of a market, the next step was to institutionalize a permanent commercial circus through the construction of Harborplace, the Maryland Science Center (*site 9*), the National Aquarium (*site 10*), the Convention Center (*site 11*), a marina (*site 12*), and innumerable hotels, shopping malls, and pleasure citadels of all kinds. The strategy did not even have to be consciously thought out, it was such an obvious thing to try.

This thrust had the additional virtue of projecting a new persona for the city. The 'armpit of the East' had been the out-of-town image of Baltimore in the 1960s. But by transforming the entertainment spectacle into a permanent image, it became possible to use it to lure in developer capital, financial services, and entertainment industries, all big growth sectors in the US economy during the 1970s and 1980s.

The imaging of Baltimore itself became important. The mayor, the media, and civic leaders set out on a binge of civic boosterism that would brook no criticism. When excessive cancer rates were reported in a neighborhood long exposed to chemical wastes, the mayor criticized those who did the reporting because they had sullied the city's image. When an impoverished population took advantage of a heavy snowstorm in 1978 to loot city stores, the mayor accused them of creating unemployment because they had damaged the city's image. So pervasive did the campaign become that when someone dreamed up the catchy slogan 'Think pink', the mayor had downtown sidewalks painted pink.

Image building of this sort had definite rewards. The mayor, designated the best mayor in America by *Esquire* in 1984, appeared more and more to be the savior of a city, a magician who had made Renaissance City emerge phoenixlike out of the ashes of the civil strife of the 1960s. Twice featured in *Time* magazine, Baltimore's Inner Harbor began to gain national and even international recognition as an example of urban revitalization. In November 1987, even the UK's *Sunday Times* newspaper bought the idea, lock, stock, and barrel:

> Baltimore, despite soaring unemployment, boldly turned its derelict harbor into a playground. Tourists meant shopping, catering and transport, this in turn meant construction, distribution, manufacturing – leading to more jobs, more residents, more activity. The decay of old Baltimore slowed, halted, then turned back. The harbor area is now among America's top tourist draws and urban unemployment is falling fast.

If people could live on images alone, Baltimore's populace would have been rich indeed.

After fifteen years as mayor, Schaefer was elected governor in 1986. Only then could another tale of Baltimore be freely told. *Baltimore 2000*, a report commissioned by the Goldseker Foundation in 1987, summed up Baltimore this way:

Over the last twenty-five years, Baltimore has lost a fifth of its population, more than half of its white population, and a hard to enumerate but very large proportion of its middle class, white and black. It has lost more than ten per cent of its jobs since 1970, and those that remain are increasingly held by commuters. By 1985, the city's median household income was just over half that of surrounding counties and the needs of its poor for services were far more than the city's eroded tax base could support.

There was plenty of 'rot beneath the glitter', as one consultant to the report put it. The depth of that rot can perhaps best be illustrated by the rapid rise in the city's status to that of fifth-worst-off city in the nation, according to a 1984 congressional estimate. The city was ranked next to last among the nation's fifteen largest cities in the proportion of twenty- to twenty-four-year olds who had completed high school, in part reflecting the more than 15 per cent decline in municipal spending on education between 1974 and 1982.

Impoverishment in inner-city neighborhoods increased. 'Of the officially designated neighborhoods in the city,' wrote Marc Levine in an article in *Urban Affairs*, '210 (75.8 per cent) experienced increases in the percentage of their residents living below the poverty line between 1970 and 1980,' while almost 90 per cent of the city's predominantly African-American neighborhoods saw their poverty rates rise. A *Baltimore Sun* survey of the Gay Street neighborhood, scene of some of the worst rioting in 1968, showed little change in conditions of impoverishment between 1966 and 1988. Yet the city's expenditures on social services for the poor fell by an astounding 45 per cent in real terms over the 1974–82 period.

These facts cannot be seen from Federal Hill, but they belie the image of affluence and fun that the Inner Harbor conveys. Nor can we see the more-than-40,000 families that wait patiently for access to public housing and the many others suffering from housing deprivation.

We cannot see the 45 per cent of the population over age sixteen who either do not or cannot enter the job market, the desperate plight of female-headed households, the record number of teenage pregnancies, the severe problems of infant mortality that put some neighborhoods on a par with Mexico or Venezuela, the problems of rats, high cancer rates, and a resurgence of tuberculosis and lead poisoning. The conditions of grinding poverty in the city do not in any way appear to have been assuaged by all that massive downtown redevelopment.

This failure of the downtown redevelopment to make any substantial dent in the city's social and economic problems is all the more shocking when the vast public subsidy is taken into account. According to a US Civil Rights Commission report of 1983, the first phase of the Inner

Table 8.2: Gay Street: Baltimore then and now

The statistics that follow appeared in the *Baltimore Sun* on 4 April 1988. They were compiled from surveys commissioned by the Baltimore Urban Renewal and Housing Agency in 1966 and by the *Baltimore Sun* in April 1988, and are reprinted with permission.

	1966	1988
Economic percentages		
Adult unemployment rate	7	19
Households receiving welfare	28	30
Households with incomes under $10,000 (1988 dollars)	41	47
Households with incomes over $20,000 (1988 dollars)	16	18
Adults who are high school graduates	10	49
Households in which at least one person owns a car	23	36
Percentage employed as laborers	43	8
Percentage doing clerical work	1	30
Household and family structure		
Median household size	2.9	1.9
Percentage of adults retired	13	30
Percentage of population under 18 years of age	45	34
Percentage of households with children that contain a male adult	56	43
Percentage of one-person households	16	31
Percentage of households with five or more people	30	12
The neighborhood		
Most commonly cited 'good' aspect	people	people
Most common complaint	housing	drugs, crime
Percentage of residents who are renters	85	78
Percentage of adults who have lived in neighborhood 10 or more years	48	60
Percentage who think neighborhood is getting better	N.A.	14

Harbor development (costing $270 million) was 90 per cent funded from the public treasury 'either in infrastructure, business subsidies, or loans/grants'. Yet the management of the project remained entirely in corporate hands.

Where did the benefits of all this public investment go? There is no easy answer to that question, but some tentative conclusions can be

Figure 8.9 Strolling along the dockside at Harborplace.

drawn. First, most of the development so far has been hugely profitable to those who undertook it, with a few signal exceptions, such as the conversion of the old Baltimore Gas and Electric Company power plant into the Six Flags Power Plant entertainment center (*site 13* on Figure 8.1).

Second, though not as seriously undertaxed as in the early 1970s, present tax flows barely match public expenditures on the Inner Harbor. Indeed, a recent internal study suggested that Baltimore spends $17 million a year more on servicing the downtown and Inner Harbor than it gets back in tax revenues.

Third, the Renaissance has indeed brought jobs to the city, but most are low-paying jobs (janitors, hotel staff, service workers). Those who hold well-paid managerial jobs, such as the six directors of T. Rowe Price (a dynamic Baltimore money fund that grew rapidly in the 1980s) each of whom gets more than $600,000 a year, tend to live in the suburbs. Some middle-level managers stay downtown and create a demand for gentrified housing and condominiums.

Fourth, and perhaps most problematic, the redevelopment has certainly brought money into the city through a rapid growth of the convention and tourist trades. But there is no guarantee that the money stays in Baltimore. Much of it flows out again, either as profits to firms or payments for goods from Europe, Hong Kong, South Korea, Japan, England, or elsewhere. Spending money at Benetton or Laura Ashley does not stimulate the Baltimore economy. Evidence is hard to find, but the Inner Harbor may function simply as a harbor, a transaction point for money flowing from and to the rest of the world.

Baltimore's urban elite have struggled to make a new city. Powerless to prevent deindustrialization and recession, they have tried to create a profitable growth machine that has focused on tourism, leisure, and conspicuous consumption as an antidote to falling profits and urban decline. In limited ways, the strategy has worked, though mainly for them. By putting Baltimore on the map and by creating a prideful image of place and community, they have to some degree secured the political compliance of the majority. This can be measured by Mayor Schaefer's re-election victories of 1979 and 1983, in which community activists lost heavily to machine politicians.

The close public-private partnership forged between City Hall and dominant corporate power helped turn Baltimore into an entrepreneurial city that fared rather better in a highly competitive world than some of its rivals, cities like Detroit, Newark, Cleveland, or even Pittsburgh. Yet such victories may prove pyrrhic. Excess investment in shopping malls, entertainment facilities, high-priced condos, office space, convention centers, and sports stadiums throughout urban America spells trouble for some cities, and Baltimore may or may not be one.

The failure of the Six Flags Power Plant amusement park in the Inner Harbor and the difficulties encountered selling high-priced condos in Harbor Court (*site 14*) are warning signals. And there are signs that the city is robbing Peter to pay Paul in the downtown commercial redevelopment stakes. James Rouse's Gallery (*site 15*), a three-story shopping mall at Harborplace, is a success, but Hutzler's Palace, (*site 16*) four blocks away on Howard Street, has had to close its doors.

Several festival marketplaces in other cities (Norfolk, Toledo, Flint, and even New York's South Street Seaport) are awash in red ink. Houston, Dallas, Atlanta, and Denver experienced overinvestment in hotels and office space in the 1970s, with catastrophic effects on the financial health of local banks and savings and loans. There is no reason to think that Baltimore is immune. There are already signs that the tourist trade is leveling off (according to Baltimore Office of Promotion and Tourism data), while employment in financial services took some hard knocks in the wake of the stock market crash of October 1987.

Furthermore, a serious social danger attaches to creating an island of affluence and power in the midst of a sea of impoverishment, disempowerment, and decay. Like the city fair, the Inner Harbor functions as a sophisticated mask. It invites us to participate in a spectacle, to enjoy a festive circus that celebrates the coming-together of people and commodities. Like any mask, it can beguile and distract in engaging ways, but at some point we want to know what lies behind it. If the mask cracks or is violently torn off, the terrible face of Baltimore's impoverishment may appear.

The lost treasures of Chesapeake Bay

Turn your back on all the downtown glitter and look down the long reach of the Chesapeake Bay, and you will see another, far less glamorous Baltimore. The landscape reflects the change from manufacturing to service industry and the growing influence of foreign capital in the Baltimore economy.

Along with this has come harder times for Baltimore workers. Total employment in the metropolitan region has remained fairly constant since 1970, but the average wage has declined substantially. Employment has shifted radically from blue-collar jobs (many in relatively high-paying unionized industries sprawling around the edge of the bay) to white-collar occupations (many in low-paying and insecure service jobs, often held by women, and concentrated downtown). Where family incomes have risen, it is nearly always because more women have entered the workforce.

For example, at the foot of Federal Hill, on the eastern side, you can see the abandoned Bethlehem Steel Corporation shipyard (*site 17*), once a thriving centerpiece of Baltimore industry that employed some 1,500 blue-collar workers, many of whom lived in South Baltimore. The yard was closed in 1983, put out of business by foreign competition, particularly from the Far East, and world recession, in spite of wage concessions and give-backs by the workers.

To the chagrin of even South Baltimore gentrifiers, the yard was bought by a developer who proposed to convert the site into a marina, a repair yard for pleasure boats, a large office and commercial complex, and more than 1,500 expensive condos with two twenty-nine-storey towers that would block views of the harbor. The Coalition of Peninsula Organizations protested loudly and won some concessions, but they lost the battle for the site. A zoning change from industrial uses to residential and commercial uses was approved in 1985.

But the developers went bankrupt, and the project's most recent $100 million incarnation, Harbor Keys, is funded by a consortium of investors from Singapore, Malaysia, Hong Kong, and Australia, all brokered by the Bangkok Bank of Thailand, which financed the purchase of the site. This means that jobs lost in the region through competition from the Far East allow capital abroad to return to dominate Baltimore.

The closure of the Allied Chemical plant (*site 18*) directly across the Inner Harbor is another sign of lost industrial power. A gray-striped eyesore, it was the last barrier to continuous condos and conversions on the northern side of the harbor from downtown through Fells Point to Canton. Developers would like to build condos here too if a way can be found to get the poisonous chromates out of the soil underneath.

The list of plant closures and industrial loss grows longer as we look down the harbor's edge: the American Can Company at Fells Point (*site 19*), the Western Electric plant on Broening Highway (*site 20*) that eliminated 3,500 jobs in 1984, and the host of abandoned warehouses and rotting piers that testify to Baltimore's decline from a once-powerful port and manufacturing city.

Even with the costly modernization projects recently undertaken with taxpayer dollars, the port of Baltimore is barely competitive as a major seaport on the East Coast. But the price has been tighter labor contracts and rapidly falling employment for Baltimore workers. The death of a union picket in 1985 in a struggle to stop the use of non-union labor may signal a return to a bitter era of labor relations. The International Longshoreman's Association, once a powerful voice in Baltimore's labor movement, now has to balance a struggle to improve wages and working conditions against the kind of concessions demanded to keep Baltimore competitive with Norfolk, Charleston, and other ports. The difficulty of dredging the Bay and disposing of the soil, the long journey up the Bay, and the canalization of the Mississippi-Tennessee river system also threaten the viability of the port.

We should be careful not to romanticize the lost era of powerful industry and commerce and the strong traditions and labor culture it nurtured. Many of the traditional industries (including the port before containerization) were onerous and dangerous. The division in the labor force between relatively affluent white male workers and the less-skilled, less-powerful women and African Americans was always a barrier to efforts to improve the lot of working people. Moreover, the economy was heavily involved in the exploitation of Third World resources and labor and was largely dependent on defense contracts. The Domino Sugar plant reminds us, for example, of the strong connection between Baltimore and Havana that had Baltimore businessmen rooting for Fulgencio Batista and against Fidel Castro precisely because of the cheap sugar produced by wretchedly paid Cuban sugarcane cutters. The Cuban revolution forced a major shift in Baltimore's trade. Interestingly enough, Domino Sugar has recently been sold to a British company (Tate and Lyle), illustrating once again how vulnerable Baltimore's industry is to international forces.

The Rusty Scupper, a restaurant at the foot of Federal Hill (*site 21*), is another reminder of the negative aspects of international trade. Permission to build the Rusty Scupper was held up by local protests

Figure 8.10 The Rusty Scupper Restaurant provides a foreground to Scarlett Place.

because the developer was a subsidiary of Nestlé, a Swiss corporation accused in the early 1980s of the deaths of thousands of babies in the Third World by marketing its infant formula as a substitute for breast feeding. The Rusty Scupper opened only after Nestlé agreed to change its practices in the face of widespread international protest.

Military contracts have always been an important source of employment in the Baltimore region. Steel and shipbuilding were heavily favored in the Second World War, but military expenditures in recent years have focused on more high-tech materials for which Baltimore was not so well positioned. It was the loss of military contracts that put the final nail in the coffin of the Bethlehem shipyard.

Some of the region's most thriving firms, such as Martin Marietta and Westinghouse, depend heavily on defense contracts. This fact of Baltimore's existence has not changed, despite efforts by local peace activists to focus attention on the waste of such expenditures relative to the social needs of the city. It should be possible, they argue, to convert industries producing instruments of death and destruction into activities that serve more human and benign social purposes.

For most visitors, Baltimore's dependence on military production and its connection with the exploitation of Third World labor are the least visible aspects of a view from Federal Hill. Stroll along the Inner Harbor or climb to the top of Federal Hill, and you are more likely to notice how pretty the sight is and to appreciate it as a place of entertainment and diversion. But whether or not you care to consider it, the landscape of the center city is a great book of time and history, proclaiming in glass, brick, and concrete who holds the reins of power.

Baltimore's banks

Banks and financial institutions dominate the downtown skyline. There are no huge corporate headquarters in Baltimore of the sort we would encounter in Pittsburgh or Cincinnati because Baltimore is a branch manufacturing city run by financiers rather than industrialists. Only one of the Fortune 500 largest manufacturing companies has its headquarters here. It has been like that since the turn of the century when many local industrialists sold out to the trusts and cartels that were forming at the time.

Table 8.3: Bank control of Baltimore

Until the recent deregulation, a few Baltimore banks controlled much of the economic activity in the Baltimore region. The concentration of banking power, documented for 1968 in the congressional Wright-Patman Report, showed the statistics that follow.

Bank	Percentage of bank trust assets in the region[a]	Number of companies on the boards of which bank directors had positions	Number of companies in which bank held more than 5 per cent of the stock
Mercantile Safe Deposit and Trust Company	63.38	196	213
Equitable Trust Company	13.85	137	53
Maryland National Bank	13.51	86	13
First National Bank of Maryland	5.28	213	56
Union Trust Company	3.66	74	55

[a] Bank trust assets are investment funds, such as pension money, administered by the bank on behalf of others.

Bethlehem Steel, General Motors, Westinghouse, and the now-departed Western Electric play a small role in what happens in Baltimore, because corporations with branch plants are usually less concerned about the effects of plant closure and less involved with local education, cultural facilities, and the like. In contrast, local banks and financial institutions are much more interested in property development than they are in employment and education.

Historically, the governance of Baltimore has been heavily influenced by a small group of local banks. As recently as 1968, a congressional report depicted Baltimore as one of the most monopolistically-organized cities in the US with respect to its financial structure. The Maryland National Bank (*site 22*), together with the Mercantile Safe Deposit and Trust Company (*site 23*) and the First National Bank (*site 24*), decided what the city was to be about and who was to run it.

A list of directors of these institutions reads like a who's who of Baltimore's elite. It includes members of the media and educational institutions (for example, The Johns Hopkins University), and leaders of the city's cultural and business life. It was in the Maryland National Bank building that Mayor Schaefer waited for the result of his final mayoral election.

Consider, then, the two buildings that form the pinnacle of the downtown skyline. A venerable old Baltimore institution, the Maryland National Bank lies at the center; beside it rises, like an upraised finger, the newer and more formidable-looking building owned by Citibank (*site 25*), a recent interloper from out of state.

A prime funder of real estate speculation and blockbusting in the 1960s, Maryland National persisted in its lack of concern for mortgage financing for low- to moderate-income inner-city neighborhoods, a recent study showed. The effect has been to promote the deterioration of housing conditions for the less well-off, and so prepare the way for more urban development and gentrification. At the same time, Maryland National was using the deposits of Baltimore residents to invest in South Africa. Only after an intense campaign by an activist group called the Maryland Alliance for Responsible Investment did Maryland National agree to pull out of South Africa and promise $50 million, over a five-year period, for financing inner-city housing.

The Citibank Building was begun in the early 1980s as the prestigious headquarters of the Merritt Commercial Savings and Loan Association, a fast-growing financial institution that rivaled another savings and loan, Old Court, in competing for deposits and in undertaking spectacular ventures. Caught in the shifting sands of interest-rate fluctuations and recession, Maryland's state-insured savings and loan industry came crashing down in the late spring of 1985 because Merritt, Old Court, and several other savings and loan institutions had engaged in shady deals and made preposterous unsecured loans.

The Maryland Deposit Insurance Corporation, which is supposed to guarantee customer deposits, went bankrupt, forcing the state government into a crisis that took two years to resolve. Depositors could not gain access to their funds, and the flamboyant philanthropist Jeffrey Levitt,

Figure 8.11   Citibank Building, formerly the Merritt Tower, site of the ill-fated Merritt Commercial Savings and Loan that went into state conservatorship during the savings and loan crisis of 1985.

head of Old Court, ended up in jail for bilking his savings and loan of millions. Both Merritt and Old Court were put into state conservatorship. Merritt was eventually sold to Citibank, an out-of-state bank that appropriated Merritt's building for its own use and gained access to a Baltimore market long monopolized by local banks.

This opening-up of Baltimore's financial market to out-of-state banks marked the end of local control. In 1983, for example, the largest bank in Ireland acquired a stake in First Maryland, the second-largest bank in the state with assets of $6.1 billion, and plans a total takeover. Breaking the local financial monopoly opens Baltimore even further to the chill winds of competition for money capital, which these days flashes around the world in the twinkling of an eye. Thus, the local economy becomes much more vulnerable to the whims and insecurities of international finance.

Redevelopment in the Inner Harbor

Maryland Science Center

Four sites in Baltimore's Inner Harbor may help explain what the redevelopment process has been about. The Maryland Science Center, which opened in 1976, was one of the first buildings planned there. It looks like a fortress. It has no entrance facing the community or even the street. The building was designed in the wake of the 1968 riots, at a time when a substantial African-American population inhabited the close-by community of Sharp-Leadenhall. The fortress design is deliberate; it is designed to keep out social unrest and minimize property damage.

The Maryland Science Center functions as a kind of strategic outpost, now rendered largely irrelevant by the gentrification of South Baltimore, at the south end of the Inner Harbor. Another example of strategic building is Martin Luther King Boulevard (*site 26*). Besides relieving traffic congestion downtown, it creates an easily patrolled line of defense between the mass of downtown buildings and the low-income and largely African-American communities of West Baltimore.

Figure 8.12 The Maryland Science Center, designed like a fortress, without windows, guards the southern approach to the Inner Harbor from potential rioting by neighborhood residents.

Figure 8.13 The western edge of the Inner Harbor with the Maryland Science Center on the left, Harbor Court towers rising behind, and the Hyatt-Regency Hotel to the right.

Hyatt-Regency Hotel

The shimmering, glass-fronted Hyatt-Regency Hotel (*site 27*), costing $35 million, was almost entirely financed by a $10-million federal subsidy in the form of an Urban Development Action Grant, plus loans that the city secured. The owners, the Pritzker and Hyatt interests, put up only $500,000. They took no risks and ended up with a $35 million hotel. Holiday Inn and other hotel chains in the city protested that the arrangement was unfair. Since Urban Development Action Grants were originally set up to help cities deal with problems of urban distress, their diversion into this project was justified in terms of employment and tax-base benefits. But the benefit and subsidy to the developers was enormous compared to the numbers of relatively low-paid service jobs created and a tax benefit that barely kept pace with public costs.

Harbor Court

Harbor Court is another example of a public subsidy for private gain. In 1984, the city transferred one of the prime pieces of development property on the East Coast to David Murdock, a California developer. Mayor Schaefer agreed (over oysters at Lexington Market, some say) to sell the land at a net loss of $500,000 in return for a promise that Murdock would

Figure 8.14 The Harbor Court tower block, 1988.

help redevelop the predominantly black retailing district around Lexington Market.

Murdock took the property, erected an $85 million building, then sold condos in it for up to $1 million each, in a city with a huge waiting-list for public housing. In 1986, Murdock withdrew from the Lexington Mall development proposal with no penalty, leaving the city with nothing except a supposedly improved tax base and an ugly tower to block the view.

Harborplace Pavilions

The pavilions of Harborplace, built by James Rouse, were the subject of considerable controversy. Rouse was originally offered the option to convert the long-abandoned Baltimore Gas and Electric Company power plant at the harbor's edge into a pavilion, but he refused. He wanted to use the public land at the strategic corner of the Inner Harbor. The city agreed but was legally obligated to put the proposed transfer of the land to a public referendum. Opposition was strong, especially from South Baltimore residents, who felt the waterfront and their access to it would be lost to private control. They also feared the impact of the development on this traditional working-class community, which had long lived and worked in South Baltimore. But the referendum passed.

Rouse's project opened in the summer of 1980 and was an instant commercial and popular success. It became the crown jewel of Baltimore's Renaissance, supposedly drawing in more visitors than Disney World.

The rate of return per square foot of rental space is reputed to be one of the highest in the US, yet the tax benefit to the city is relatively modest, given the public expenditures required. Much of the attraction and charm of Harborplace comes from the people who mass there and provide the spectacle – the same crowd that pays for the overpriced goods and services that generate such fabulous financial returns for private and corporate business.

South Baltimore

The Inner Harbor has had a substantial impact on areas immediately surrounding it, such as the streets in South Baltimore behind Federal Hill. A glance down the western side of the hill shows a solidly gentrified community with its sundecks, newly cleaned brick exteriors, and shuttered windows. The ubiquitous coach lamps, a symbol of the new urban gentry that lives here, march street by street into South Baltimore.

An African-American neighborhood church has been converted into condos (*site 28*), as was the old Shofer Warehouse (*site 29*). Developers call it the Paper Mill – a reflection of its earlier use – perhaps to make the price of $300,000 (and up) per unit a bit more palatable. Housing prices have shot up from the $10,000 level common in the early 1970s to well over $100,000 for a refurbished rowhouse in the late 1980s.

The effect has been to increase local tax assessments and property tax burdens (from $300 to $2,000 a year, in some cases), pushing poor people out and making way for speculators and developers. The displacement of local residents sparked local resistance, and the Coalition of Peninsula Organizations led the way in trying to rescue the neighborhood, in spite of (then) Mayor Schaefer's opposition. But with no more employment in the shipyards, South Baltimore has become vulnerable to the inflow of young professionals seeking a safe neighborhood close to downtown office jobs.

Figure 8.15 *The old Southern High School converted into the condominiums of Harbor View with the help of the 'shadow government' of city trustees.*

Figure 8.16 *A neighborhood church, once a place of worship for African–Americans in South Baltimore, now converted to condominiums.*

Figure 8.17 Shofer Warehouse, once an old paper mill, has been converted to condominiums.

City Hall's part in that transition can be seen most clearly in its role in the conversion of the old Southern High School (*site 30*), on the southwest corner of Federal Hill Park, into condos. Now called Harbor View, this building is one of many memorials to an undemocratic system of city governance that has allowed the City Council's control over city expenditures to be superseded by the formation of what became known as a shadow government.

In the 1970s, Mayor Schaefer designated two trustees to administer all federal loans and grants to private developers. As repayments from developers came in, the trustees built up a $200-million development bank, entirely under the mayor's control, that could be used as a revolving fund to promote further private development. One such project was the conversion of Southern High School by the Jolly Company, which acted as developer and builder. The company put up no money of its own, but borrowed everything it needed from the trustee fund. When the company failed to make interest payments on the loan they had taken to purchase the site, the city foreclosed. But the company continued with the lucrative business of building conversion at a price it had set as a developer.

The city ultimately managed to sell off the condos without too much loss, but Jolly the builder profited most handsomely from the conversion at public expense from an operation with no risk. In fact, none of the agencies responsible for the downtown and Inner Harbor redevelopment

Figure 8.18 Federal Park condos, after conversion from the warehouse.

was accountable to anyone but itself, even though they were awash with public funds. In 1980, a reporter from the *Baltimore Sun*, C. Frazier Smith, exposed the whole structure of shadow governance, including several quasi-public agencies (the Charles Center Management Corporation was the first) that controlled public funds for largely private purposes.

The issue, most agreed, was not corruption of the ordinary sort but circumvention of the democratic processes of government and of public accountability for the use of public money. The mayor argued, with some justification, that the trustee system was the only way he could bypass the conservatism of Baltimore banks. He wound up the trustee system as banking became more open and competitive in the mid-1980s, but other quasi-public corporations, still unaccountable to the City Council, have not been touched and have remained the vital center of Baltimore's so-called public-private partnership.

Figure 8.19 *The coach lamps of gentrifiers, marching into South Baltimore behind Federal Hill, leaving the formstone houses as a sign of long-standing residency.*

CHAPTER 9

Militant particularism and global ambition: the conceptual politics of place, space, and environment in the work of Raymond Williams

First published in Social Text, *1995.*

Local militancy and the politics of a research project

In 1988, shortly after taking up a position in Oxford, I became involved in a research project concerning the fate of the Rover car plant in that city. Oxford, particularly for outsiders, is usually imagined as a city of dreaming spires and university grandeur, but as late as 1973 the car plant at Cowley in East Oxford employed some 27,000 workers, compared to fewer than 3,000 in the employ of the university. The insertion of the Morris Motors car plant into the mediaeval social fabric of the city early in the century had had enormous effects upon the political and economic life of the place, paralleling almost exactly the three-stage path to socialist consciousness set out in *The Communist Manifesto*. Workers had steadily been massed together over the years in and around the car plant and its ancillary installations; they had become conscious of their own interests and built institutions (primarily the unions) to defend and promote those interests. During the 1930s and again in the 1960s and early 1970s, the car plant was the focus of some of the most virulent class struggles over the future of industrial relations in Britain. The workers' movement simultaneously created a powerful political instrument in the form of a local Labour Party that ultimately assumed continuous control of the local council after 1980. But by 1988 rationalizations and cutbacks had reduced the workforce to around 10,000, and by 1993 it was down to less than 5,000 (as opposed to the 7,000 or so then in the employ of the university). The threat of total closure of the car plant was never far away.

A book on the Cowley story, edited by Teresa Hayter and myself and entitled *The Factory and the City: The Story of the Cowley Auto Workers in Oxford*, was published late in 1993. It originated in research work conducted in support of a campaign against closure that began in 1988, when British Aerospace (BAe) acquired the Rover car company in a sweetheart privatization deal from the Thatcher government. Partial closure and rationalization at the plant was immediately announced and the prospect of total closure loomed. Land values in Oxford were high and BAe, with the property boom in full flood, acquired a property development company specializing in the creation of business parks (Arlington Securities) in 1989. The fear was that work would be transferred to Longbridge (Birmingham) or, worse still, to a greenfield non–union site in Swindon (where Honda was already involved in co–production arrangements with Rover), releasing the Oxford land for lucrative redevelopment that would offer almost no prospects for employment to a community of several thousand people that had evolved over many years to serve the car plant.

An initial meeting to discuss a campaign against closure drew representatives from many sectors. It was there agreed to set up a research group to provide information on what was happening and what the effects of any moves by BAe might be on the workforce and on the Oxford economy. The Oxford Motor Industry Research Project (OMIRP) was formed and I agreed to chair it. Shortly thereafter, the union leadership in the plant withdrew its support for both the campaign and the research, and most of the Labour members on the city council followed suit. The research was then left to a small group of dissident shop stewards within the plant and a group of independent researchers, some of whom were based at Oxford Polytechnic (now Oxford Brookes University) and Oxford University.

For personal reasons, I was not active in the campaign nor did I engage much with the initial research. I did help to publicize the results and to mobilize resources for the research project which the union leadership and the majority of the local Labour Party actively tried to stop – they did not want anything to 'rock the boat' in the course of their 'delicate negotiations' with BAe over the future of the plant and the future of the site. Fortuitously, OMIRP produced a pamphlet, *Cowley Works*, at the very moment when BAe announced another wave of rationalizations that would cut the workforce in half and release half of the land for redevelopment. The history of the plant together with the story of the struggle to launch a campaign and the dynamics of the subsequent rundown are well described in the book.

Teresa Hayter, the coordinator of OMIRP, received a three–year research fellowship at St Peter's College in 1989 to pull together a book about the history of Cowley, the failed campaign, and the political problems of mobilizing resistance to the arbitrary actions of corporate capital. The book entailed the formation of a broad-based group (including academics as well as political activists), with each member producing chapters on topics with which they were most familiar. Each chapter was read by others and comments went back and forth until a final version was arrived at. I agreed, partly for purposes of making the book more attractive to prospective publishers, to be a co-editor of the book with Teresa Hayter. This meant that in addition to the one chapter I co-authored, I spent quite a lot of time, along with Hayter, editing, commissioning new segments to ensure full coverage, and generally trying to keep the book as a whole in view while attending to the parts.

The book is a fascinating document. It brings together radically different positionalities, varying from an unnamed shop steward and others who had worked in the plant or had been long-term residents of East Oxford, to academics, planners, and independent leftists. The language differs drastically from one chapter to another. The activist voice emanating from the plant experience (what I will later, following Raymond Williams, call 'militant particularism') contrasts sharply with the more abstract judgments of the academics, for example, while the perspective from the community often reads rather differently from the perspective of the production line. The heterogeneity of voices and of styles is a particular strength of the book.

It was evident early, however, that the many contributors had quite different political perspectives and interpretations. Initially, these were negotiated, in the sense that everyone trod warily through a minefield of differences in order to get to the other side with a completed book. The difficulties arose with the conclusion. I originally proposed that we should write two conclusions, one by Hayter and one by myself, so that readers might get a better handle on the political differences and be left to judge for themselves. This, however, was rejected. And so I undertook to draft a conclusion based on the various ideas that had been put forward by several members of the group. That draft conclusion succeeded in exploding almost every mine that had been negotiated in the writing of the book. Matters became extremely tense, difficult, and sometimes hostile between Hayter and me, with the group to some degree polarized around us.

In the midst of these intense arguments, I recall a lunch in St Peter's College at which Hayter challenged me to define my loyalties. She was very clear about hers. They lay with the militant shop stewards in the plant, who were not only staying on and laboring under the most appalling conditions but daily struggling to win back control from a reactionary union leadership so as to build a better basis for socialism. By contrast, she saw me as a free-floating Marxist intellectual who had no particular loyalties to anyone. So where did my loyalties lie?

It was a stunning question and I have had to think about it a great deal since. At the time I recall arguing that while loyalty to those still employed in the plant (and perhaps to the Socialist Outlook Trotskyists who formed the core of the opposition but whose views were in a minority) was important, there were many more people in East Oxford who had been laid off or who had no prospects for employment (for example, alienated and discontented young people, some of whom had taken to joyriding and thus had left criminalization and police oppression for the whole community in their wake) who deserved equal time. All along, I noted, Hayter had treated my concerns for the politics of community as a parallel force to the politics of the workplace with scepticism. I further thought that some consideration had to be given to the future of socialism in Oxford under conditions in which the working-class solidarities that had been built around the plant were plainly weakening and even threatened with elimination. This meant the search for some broader coalition of forces, both to support the workers in the plant and to perpetuate the socialist cause. I also thought it would be disloyal in a general sense not to put a critical distance between us and what had happened in order to better understand why the campaign had failed to take off. Hayter refused to countenance anything that sounded even remotely critical of the strategy of the group that had tried to mobilize sentiments behind a campaign. Likewise, she rejected any perspective that did not accept as its basis the critical struggle for power on the shop floor of the plant.

But all sorts of other issues divided us. Deteriorating work conditions in the plant, for example, made it hard to argue unequivocally for the long-term preservation of what were in effect 'shit jobs', even though it was plainly imperative to defend such jobs in the short run because there were no reasonable alternatives. The issue here was not to subordinate short-term actions to long-term pipe-dreams, but to point to how difficult it is to move on a long-term trajectory when short-term exigencies demand something quite different. I was also concerned about the incredible overcapacity in the automobile industry in Britain, and in Europe in general. Something had to give somewhere, and we had to find some way to protect workers' interests in general without falling into the reactionary politics of the 'new realism' then paralyzing official union politics. But across what space should that generality be calculated? Britain? Europe? The world? I found myself arguing for at least a European-wide perspective on adjustments in automobile production capacity, but found it hard to justify stopping at that scale when pressed. There were also important ecological issues to be considered deriving not only from the plant itself (the paint shop was a notorious pollution source), but also from the nature of the product. Making Rover cars for the ultra rich and so contributing to ecological degradation hardly seemed a worthy long-term socialist objective. The ecological issue ought not to be ducked, I felt, even though it was plain that the bourgeois North Oxford-heritage interests would likely use it to get rid of the car plant altogether if given the chance. Again, the problem of time horizon and class interests needed to be explicitly debated rather than buried. Furthermore, while I would in no way defend the appalling behavior of BAe, I did think it relevant to point out that the company had lost about a third of its stock-market value in the first few months of 1992 and that its hopes for a killing on the property market had been seriously diminished in the property crash of 1990. This posed questions of new forms of public or community control over corporate activity (and in this case the turn to property speculation as an alternative to production) that would not repeat the bitter history of nationalization (such as the disastrous rationalizations and reordering of job structures already suffered by Rover, when it was British Leyland in the 1970s, for example).

I felt that it would be disloyal to the conception of socialism as some real alternative not to talk about *all* of these issues in the conclusion to the book. Not, I hasten to add, with the idea that they could be resolved, but because they defined an open terrain of discussion and debate that seemed to me at least to flow out of the materials assembled in the book. Such a conclusion would keep options open and so help readers to consider active choices across a broad terrain of possibilities while paying proper attention to the complexities and difficulties. But Hayter felt, even when she partially agreed on the long-term significance of such ideas, that raising matters of this sort would dilute the immediate struggle to keep jobs in Cowley and prevent their transfer to a greenfield non-union site in Swindon. The issues which I wanted to raise could be attended to, she held, only when the workforce and the progressive stewards had regained their strength and power in the workplace.

I was operating, it became plain, at a different level and with different kinds of abstraction. But the impetus for the campaign, the research, and the book did not come from me. It arose out of the extraordinary strength and power of a tradition of union militancy emanating from the plant. This tradition had its own version of internationalism and presumptions to universal truth, although a case could be made that its capture and ossification by a rather narrow Trotskyist rhetoric was as much a part of the problem as the more fundamental conflict between Hayter's and my perspective. But it would be wrong to depict the argument solely in sectarian terms. For the issue of a purely plant-based versus a more encompassing politics was always there. I could not abandon my own loyalty to the belief that the politics of a supposedly unproblematic extension outwards from the plant of a prospective model of total social transformation is fundamentally flawed. The view that what is right and good from the standpoint of the militant shop stewards at Cowley is right and good for the city and, by extension, for society at large is far too simplistic. Other levels and kinds of abstraction have to be deployed if socialism is to break out of its local bonds and become a viable alternative to capitalism as a working mode of production and social relations. But there is something equally problematic about imposing a politics guided by abstractions upon people who have given their lives and labor over many years in a particular way in a particular place.

So what level and what kinds of abstraction should be deployed? And what might it mean to be loyal to abstractions rather than to actual people? Beneath those questions lie others. What is it that constitutes a privileged claim to knowledge and how can we judge, understand, adjudicate, and perhaps negotiate different knowledges constructed at very different levels of abstraction under radically different material conditions?

Raymond Williams and the politics of abstraction

These were questions that preoccupied Raymond Williams, erupting again and again in his work; though, for reasons that will shortly become apparent, they are far better articulated in his novels than in his cultural theorizing. My purpose here is not, I should make plain, to hold up Williams as some paragon of virtue on these matters. Indeed, I accept the criticism that the nearer he steers in his theorizing to what might be called 'cultural holism' – the view that culture must be understood as a 'whole way of life' and that social practices have to be construed as 'indissoluble elements of a continuous social material process' – the closer he comes to an organicist view of the social order – a 'community' characterized by a certain 'structure of feeling' as a 'total way of life' – that cannot help but be exclusionary with respect to outsiders and in some respects oppressive for insiders, too. The critical interventions of Said (1989) and of Gilroy (1987) strongly point to the difficulty with respect to outsiders, the latter accusing Williams of complicity with a metropolitan colonialism and imperialism by virtue of his situatedness within the 'structures of feeling' that were associated with working-class support for the British Empire. A purely organicist view makes it equally difficult to examine multiple forces of oppression and domination within a cultural configuration. Williams, it is generally acknowledged, is nowhere near sensitive enough on the gender issue, for example (though, again, he felt he handled such questions much more firmly in his novels than in his theorizing).

Roman's (1993) sympathetic and constructive critique of some of the pitfalls into which Williams sometimes seems to fall exposes some of the dangers as well as opportunities that Williams creates from both a feminist and a more racially sensitive perspective. There is no doubt, either, that Williams's reluctance to let go of 'lived experience' leads him to accept, as Hall (1989: 62) has remarked, a rather 'empiricist notion of experience' as if there were nothing problematic about taking daily experience as a direct basis for theory construction. Williams's reticence in this regard has even led some critics to conclude, erroneously I believe, that Williams made no real theoretical contributions at all, save giving Gramsci's notions of hegemony a new and somewhat more nuanced lease on life (Snedeker 1993) . Yet there is a certain paradox at work here, for it is also true that Williams's influence, in spite of all his supposed defects, 'remains powerful in contemporary cultural studies, with their emphasis on the counterhegemonies of feminist, Third World, and working-class movements' (Snedeker 1993: 113).

I shall not here try either to defend or to offer a systematic critique of Williams's controversial stances on politics and culture (see the edited collections by Eagleton, 1989, and Dworkin and Roman, 1993, for extended discussions). But there are two crucial points concerning his work which may explain why so many of his most trenchant critics so often find themselves returning to his formulations. The first concerns the dialectical way in which his concepts get formulated. Consider, for example, the following passage:

> In most description and analysis, culture and society are expressed in an habitual past tense. The strongest barrier to the recognition of human cultural activity is this immediate and regular conversion of experience into finished products. What is defensible as a procedure in conscious history, where on certain assumptions many actions can be definitively taken as having ended, is habitually projected, not only into the always moving substance of the past, but into contemporary life, in which relationships, institutions and formations in which we are still actively involved are converted, by this procedural mode, into formed wholes rather than forming and formative processes. Analysis is then centred on relations between these produced institutions, formations, and experiences, so that now, as in that produced past, only the fixed explicit forms exist, and living presence is always, by definition, receding.
>
> (Williams 1977: 128–9)

Williams is not immune from the tendency to produce alienated conceptions that instantiate 'formed wholes' as dominants over 'forming and formative processes'. But certainly in this passage he declares a strong preference for dialectical readings that prioritize the understanding of processes over things, so that any organicist notion of community, for example, is necessarily tempered by the knowledge of the complicated flows and processes that sustain it. Williams here charts a terrain of theoretical possibilities in which the reduction of relations between people into relations between concepts can be continuously challenged, while our understanding of relationships, institutions, and forms can be brought alive by focusing attention on the processes that work to produce, sustain, or dissolve them.

The second point is that the manner of 'embeddedness' (as contemporary sociologists [such as Granovetter 1985] refer to it) of political action in what anthropologists like to term 'intimate culture' (Lomnitz-Adler 1991) is simultaneously empowering and problematic. But it also follows that the abstractions to which we appeal cannot be understood independently of whatever it is that political and theoretical activity is embedded in and whatever it is that social life is being intimate about. A study of some of Williams's formulations can here be extremely helpful, since he both uses and systematically questions the notion of embeddedness and intimate culture throughout his work. In what follows, I shall pay particular attention to the way Williams treats environment, space, and place as framing concepts that help define what these ideas might mean.

The novel as environmental history

Press your fingers close on this lichened sandstone. With this stone and this grass, with this red earth, this place was received and made and remade. Its generations are distinct but all suddenly present.

(Williams 1990: 2)

So ends the opening statement of both volumes of Williams's last and unfinished novel, *People of the Black Mountains*. The story begins in 23,000 BC and passes across periods of vast environmental and social change. The second story, for example, is set at the edge of the great ice sheet that surrounded the Black Mountains at the maximum point of glaciation in 16,000 BC. Subsequent episodes take up the advent of settled agriculture, writing, and other key moments of transformation of both the physical and social environment through human action. The earlier reconstructions draw heavily on archaeological, paleological, and environmental history (the list of sources furnished at the end of the second volume is extensive indeed), while the later periods lean much more heavily on the works of economic, social, and cultural historians, making this a fictional account deeply rooted in those material realities identified through research across a wide range of disciplines. In episode after episode, the people who have traversed and struggled in that place are imagined into life.

So why was one of Britain's most eminent socialist thinkers, in the very last fictional work he undertook, writing the social and environmental history of the Black Mountains? One partial answer to that question presumably lies in Williams's insistence that social beings can never escape their embeddedness in the world of nature and that no conception of political action could, in the final analysis, afford abstractions that did not encompass that fact. 'Nature' was, thus, a key word for Williams (1983: 219) – perhaps 'the most complex word in the language' since the idea of it 'contains, though often unnoticed, an extraordinary amount of human history. . . both complicated and changing, as other ideas and experiences change' (Williams 1980: 67). An enquiry into environmental history as well as into changing conceptions of nature therefore provided a privileged and powerful way to enquire into and understand social and cultural change. Williams construes the social and environmental dialectically, as different faces of the same coin. Close attention to the environmental side was, however, bound to throw into relief certain features that might otherwise be missed. His materialism and critical realism always see to it that work (or what he elsewhere calls 'livelihood') – broadly understood as a simultaneously life-giving and culturally creative activity – is the fundamental process through which our relation to and understanding of the world of nature gets constituted. 'Once we begin to speak of men mixing their labor with the earth, we are in a whole world of new relations between man and nature and to separate natural history from social history becomes extremely problematic' (Williams 1980: 76). Such a dialectical and transformative view of how specific social relations connect to new ways of mixing labor with the land is not unique to Williams. It echoes, for example, the views of Marx and Engels that 'as long as men exist, the history of nature and the history of men are mutually conditioned' because by 'acting on the external world and changing it, (we) at the same time change (our) own nature' (Marx 1967: 173). William Cronon (1983: 13–14), the doyen of the contemporary movement to create a distinctively environmental form of history, makes a similar argument:

> An ecological history begins by assuming a dynamic and changing relationship between environment and culture, one as apt to produce contradictions as continuities. Moreover, it assumes that the interactions of the two are dialectical. Environment may initially shape the range of choices available to a people at a given moment but then culture reshapes environment responding to those choices. The reshaped environment presents a new set

of possibilities for cultural reproduction, thus setting up a new cycle of mutual determination. Changes in the way people create and re-create their livelihood must be analysed in terms of changes not only in their *social relations* but in their *ecological* ones as well.

But the environmental history of the Black Mountains is not something that evolves purely in place. The novel is also a story of wave after wave of migratory influences and colonizations that situate the history of the Black Mountains in a matrix of spatiality, constituted by the flows and movements pulsing across Europe and beyond. The distinctiveness, or what Williams affectionately calls the 'sweetness of the place', gets constructed through the working out in that place of interventions and influences from outside. The three themes of place, space, and environment are tightly interwoven in this particular novel as inseparable elements in complex processes of social and environmental transformation.

But why choose the novel as a vehicle to explore these themes? Why not write straight environmental history, or rest content with the abundant source materials upon which Williams draws? I think there are two reasons. The first is explicitly laid out again and again in the novels as key characters reflect on the nature of the knowledge and understandings they hold. In *People of the Black Mountains* (vol. 1, 10–12) we find Glyn – the person through whom the voices and tales of the past become historically present – looking for his uncle lost on the mountains, reflecting thus on the vast literatures assembled by different disciplines about the place:

> Yet the kinds of scrutiny that were built into these disciplines had their own weaknesses . . . They would reduce what they were studying to an internal procedure; in the worst cases to material for an enclosed career. If lives and places were being seriously sought, a powerful attachment to lives and to places was entirely demanded. The polystyrene model and its textual and theoretical equivalents remained different from the substance they reconstructed and simulated . . . At his books and maps in the library, or in the house in the valley, there was a common history which could be translated anywhere, in a community of evidence and rational inquiry. Yet he had only to move on the mountains for a different kind of mind to assert itself: stubbornly native and local, yet reaching beyond to a wider common flow, where touch and breadth replaced record and analysis; not history as narrative but stories as lives.

This is a very familiar theme in all of Williams's novels (and it is interesting to note how it presages the move within history of a shift from narrative to story form). In *Border Country*, we similarly encounter Matthew Price, like Williams, a Cambridge-educated son of a railway signalman from a rural Welsh community, but now fictionally placed as a university lecturer in economic history in London. His work on population movements in Wales in the nineteenth century has hit an impasse. The data are all there but something is missing:

> The techniques I have learned have the solidity and precision of ice cubes, while a given temperature is maintained. But it is a temperature I can't really maintain: the door of the box keeps flying open. It's hardly a population movement from Glynmawr to London, but it's a change of substance, as it must have been for them, when they left their villages. And the ways of measuring this are not only outside my discipline. They are somewhere else altogether, that I can *feel but not handle, touch but not grasp*.
>
> (Williams, 1988a: 10; emphasis mine)

The implication is clear enough and applies with great force to Williams's own work. Concerned as he always was with the lived lives of people, the novel form allows him to represent the daily qualities of those lives in ways that could not be handled or grasped by other means. So while on the one hand Williams insists that his novels should not be treated as separate from his cultural theorizing, he also freely admits that he found some themes far easier to explore in his novels than in his theoretical work (Williams 1989a: 319).

But I think there is another reason behind the choice of the novel form. He wants always to emphasize the ways in which personal and particular choices made under given conditions are the very essence of historical-geographical change. The novel is not subject to closure in the same way that more analytic forms of thinking are. There are always choices and possibilities, perpetually unresolved tensions and differences, subtle shifts in structures of feeling, all of which stand to alter the terms of debate and political action, even under the most difficult and dire of conditions. It was for precisely this reason that Williams admired Brecht's theater. Brecht, he says, discovered:

> ways of enacting genuine alternatives: not so much as in traditional drama, through the embodiment of alternatives in opposing characters, but by their embodiment in one person, who lives through this way and then that and invites us to draw our own conclusions.
>
> (Williams 1961: 157)

This means, he goes on to point out, that 'there is no imposed resolution – the tension is there to the end, and we are invited to consider it.' All of Williams's central characters live that tension. The stories of the people of the Black Mountains are precisely about that also. Politically, this allows Williams to remind us of the way in which these people, by virtue of the choices they made and the ways they lived their lives, are 'all historically present'. His aim is empowerment in the present through celebrating the strength and capacity to survive. But it is not only that:

> The crisis which came to me on the death of my father, who was a socialist and a railway worker – I haven't been able to explain this to people properly, perhaps I explained it partly in my novel *Border Country* – was the sense of a kind of defeat for *an idea of value*. Maybe this was an unreasonable response. All right, he died, he died too early, but men and women die. But it was very difficult not to see him as a victim at the end. I suppose it was this kind of experience which sent me back to the historical novel I'm now writing, *People of the Black Mountains*, about the movements of history over a very long period, in and through a particular place in Wales. And this history is a record of . . . defeat, invasion, victimization, oppression. When one sees what was done to the people who are physically my ancestors, one feels it to be almost incredible . . . The defeats have occurred over and over again, and what my novel is then trying to explore is simply the condition of anything surviving at all. It's not a matter of the simple patriotic answer: we're Welsh, and still here. It's the infinite resilience, even deviousness, with which people have managed to persist in profoundly unfavourable conditions, and *the striking diversity of beliefs in which they've expressed their autonomy. A sense of value which has won its way through different kinds of oppression of different forms . . . an ingrained and indestructible yet also changing embodiment of the possibilities of common life.*
>
> (Williams 1989a: 321–2; emphasis mine)

The embeddedness that Williams here wants to celebrate is the ability of human beings, as *social* beings, to perpetuate and nurture in their daily lives and cultural practices the *possibility* of that sense of value that seeks a commonality to social life even in the midst of a striking heterogeneity of beliefs. But the maintenance of such a sense of value depends crucially upon a certain kind of interpersonal relating that typically occurs in particular places.

The dialectics of space and place

So what were people building in the Black Mountains? It was place that was being 'received and made and remade'. But what did 'place' mean to Williams? It is not one of his key words (though 'community', which is generally given a place-bound connotation in his work, is). Nevertheless:

> A new theory of socialism must now centrally involve *place*. Remember the argument was that the proletariat had no country, the factor which

differentiated it from the property-owning classes. But place has been shown to be a crucial element in the bonding process – more so perhaps for the working class than the capital-owning classes – by the explosion of the international economy and the destructive effects of deindustrialization upon old communities. When capital has moved on, the importance of place is more clearly revealed.

(Williams 1989a: 242)

The embeddedness of working-class political action is, according to this account, primarily in 'place'. In his novels, however, the meaning of place becomes particularly clear, since it is almost as if the processes of place creation and dissolution – again a very dialectical conception as compared to the formed entity of an actual place – become active agents in the action. But the constitution of place cannot be abstracted from the shifting patterns of space relations. This is well established in *People of the Black Mountains* and was, of course, the guiding principle that allowed Williams to construct the incredibly rich literary analysis deployed in *The Country and the City*. But this material relation is rendered even more vivid in the strike episode in *Border Country*: political consciousness in a rural Welsh village community, traversed by a railway line along which goods and information flow, gets transformed by virtue of its relation to the miners' strike in South Wales, only in the end to be sold out by decisions taken in London. In an essay on the general strike of 1926, Williams (1989a: 1065–6) makes clear how the episode in *Border Country* was shaped only after long conversations with his father. He then reflects on the structure of the problem as follows:

These men at that country station were industrial workers, trade unionists, in a small group within a primarily rural and agricultural economy. All of them, like my father, still had close connections with that agricultural life ... At the same time, by the very fact of the railway, with the trains passing through, from the cities, from the factories, from the ports, from the collieries, and by the fact of the telephone and the telegraph, which was especially important for the signalmen, who through it had a community with other signalmen over a wide social network, talking beyond their work with men they might never actually meet but whom they knew very well through voice, opinion and story, they were part of a modern industrial working class.

The point of the strike episode is to show how something special is achieved, in this case a realization of class consciousness and an understanding of the *possibility* (and this word is always lurking in the margins of all of Williams's discussions) of a real alternative. But this possibility is arrived at precisely through the internalization within that particular place and community of impulses originating from outside. How those external impulses were transformed and internalized as a very local 'structure of feeling' is a crucial part of the story. Something very special occurred in the fictional Glynmawr (the strike, he narrates, had raised the prospects of common improvement 'to an extraordinary practical vividness' [Williams 1988a: 153]) and in the actual Pandy, giving a meaning to socialism that was of a peculiarly high order, thus making the tragedy of its sellout from afar particularly devastating.

But there is a counterflow at work here. After the collapse of the strike, one of its dynamic leaders, Morgan Prosser, takes to doing business deals until he ultimately becomes the biggest businessman in the valley, only in the end to be bought out by corporate capital. Says Morgan:

> this place is finished, as it was. What matters from now on is not the fields, not the mountains, but the road. There'll be no village, as a place on its own. There'll just be a name you pass through, houses along the road. And that's where you'll be living, mind. On a roadside.
>
> (Williams 1988a: 242)

While Morgan always professes his willingness to give up his business ways if another genuine alternative for common betterment can be found, he pushes home relentlessly the view that the only choice is either to 'settle' in place and take what comes or to internalize whatever can be had from the external forces at play and use them to particular, personal, or place-bound advantage.

In *The Fight for Manod*, this local place-bound internalization of capitalistic values becomes even more apparent. Says Peter Owen, the radical sociologist coopted to look at what a new town built in the rural backwater of Manod in Wales might mean, 'the actual history is back there in the bloody centre: the Birmingham-Dusseldorf axis, with offices in London, Brussels, Paris, Rome.' 'What always breaks us up is this money from outside,' complains a local resident Gwen (Williams 1988b: 140). As the tale of secret land company procurements comes to light, we see how a faceless capitalism exercises a deeply-corrupting influence on everyone:

> The companies. And then the distance, the everyday obviousness of the distance, between that lane in Manod, all the immediate problems of Gwen and Ivor and Trevor and Gethin and the others: the distance from them to this register of companies, but at the same time the relations are so solid, so registered. The transactions reach right down to them. Not just as a force from outside but as a force they've engaged with, are now part of. Yet still a force that cares nothing about them, that's just driving its own way.
>
> (Williams 1988b: 153)

What follows for Matthew is the bitter realization that: 'to follow what seem our own interests, as these farmers are doing in Manod, isn't against [this process] but is part of it; is its local reproduction' (Williams 1988b: 153). All of this poses acute problems of political identity depending upon the spatial range across which political thought and action is construed as possible:

> 'This is Tom Meurig,' Peter said. 'He lives in Llanidloes or in Europe, I can't remember which.' Tom Meurig laughed . . . 'He can't make up his mind,' Peter said, 'whether to proclaim an immediate federation of the Celtic Peoples, with honorary membership for the Basques, or whether simply to take over Europe, with this new communal socialism they've been dreaming up in the hills.' 'Either of those,' Meurig said, 'or the third possibility: getting one of our people on to the District Council.'
> (Williams 1988b: 133)

The humor of that exchange conceals an incredible tension. It turns out that the internalization of these external forces in Manod depends crucially upon a farmer on the District Council having privileged knowledge of plans being hatched elsewhere. The relevant place and range of political action (as well as action in the novel) cannot get resolved outside of a particularly dialectical way of defining loyalties to place across space. And within such loyalties we will always find a peculiar tension between resistance and complicity

Williams tries to incorporate 'place' more directly into socialist theorizing. The key phrase here is what Williams calls 'militant particularism'. I want to pay particular attention to this idea since it captures something very important about both the history and prospects for socialism, at least insofar as Williams saw them. Williams (1989a: 249, 115) reflects as follows:

> The unique and extraordinary character of working-class self-organization has been that it has tried to connect particular struggles to a general struggle in one quite special way. It has set out, as a movement, to make real what is at first sight the extraordinary claim that the defence and advancement of certain particular interests, properly brought together, are in fact the general interest.

Ideals forged out of the affirmative experience of solidarities in one place get generalized and universalized as a working model of a new form of society that will benefit all of humanity. This is what Williams means by 'militant particularism' and he sees it as deeply ingrained in the history of progressive socialism in Britain as well as 'a most significant part of the history of Wales'. It is not hard to generalize the point, even though Williams himself was reluctant to let go of the particularities and specificities of actual places as the fundamental basis for his thinking. The French revolutionaries, after all, proclaimed doctrines of 'the rights of man'; the international workers movement proclaimed the global transition to socialism for the benefit of all; the Civil Rights movement in the United States articulated a politics of universal racial justice; certain wings of contemporary feminism and the ecology movement project their militant particularism as the basis for a wide-ranging social reconstruction that will benefit, if not save, us all.

Williams appears to suggest that many, if not all, forms of political engagement have their grounding in a militant particularism based in particular structures of feeling of the sort I encountered in Cowley. But the difficulty is:

> That because it had begun as local and affirmative, assuming an unproblematic extension from its own local and community experience to a much more general movement, it was always insufficiently aware of the quite systematic obstacles which stood in the way.
>
> (Williams 1989a: 115)

Such obstacles could only be understood through abstractions capable of confronting processes not accessible to direct local experience. And here is the rub. The move from tangible solidarities understood as patterns of social life organized in affective and knowable communities to a more abstract set of conceptions that would have universal purchase involves a move from one level of abstraction – attached to place – to another level of abstractions capable of reaching out across space. And in that move, something was bound to be lost. 'In came,' Williams ruefully notes, 'necessarily, the politics of negation, the politics of differentiation, the politics of abstract analysis. And these, whether we liked them or not, were now necessary even to understand what was happening.' Even the language changes, shifting from words like 'our community' and 'our people' in the coalfields to 'the organised working class,' the 'proletariat,' and the 'masses' in the metropolis where the abstractions are most hotly debated (Williams 1989c: 293).

The shift from one conceptual world, from one level of abstraction to another, can threaten that sense of value and common purpose that grounds the militant particularism achieved in particular places:

> This was my saddest discovery: when I found that in myself . . . that most crucial form of imperialism had happened. That is to say, where parts of your mind are taken over by a system of ideas, a system of feelings, which

really do emanate from the power centre. Right back in your own mind, and right back inside the oppressed and deprived community there are reproduced elements of the thinking and the feeling of that dominating centre... If that negative politics is the only politics then it is the final victory of a mode of thought which seems to me the ultimate product of capitalist society. Whatever its political label it is a mode of thought which really has made relations between men into relations between things or relations between concepts.

(Williams 1989a: 117)

This tension between the different levels and kinds of abstractions to which individuals necessarily appeal in order to understand their relation to the world is particularly vivid in his novels, often internalized within the conflicting emotions of the protagonists. In *Border Country*, Matthew takes that name given by his father into the wider world, but in Glynmawr he is always known as Will, the name his mother wanted. The duality of that identity – who is he, Matthew or Will? – is perpetually at work throughout the novel. Caught in that duality it becomes almost impossible to find a language with which to speak:

He was trained to detachment: the language itself, consistently abstracting and generalizing, supported him in this. And the detachment was real in another way. He felt, in this house, both a child and a stranger. He could not speak as either; could not speak really as himself at all, but only in the terms that this pattern offered.

(Williams 1988a: 83)

The tension is registered even in the way in which a familiar landscape gets remembered:

It was one thing to carry its image in his mind, as he did, everywhere, never a day passing but he closed his eyes and saw it again, his only landscape. But it was different to stand and look at the reality. It was not less beautiful; every detail of the land came up with its old excitement. But it was not still, as the image had been. It was no longer a landscape or a view, but a valley that people were using. He realized as he watched, what had happened in going away. The valley as landscape had been taken, but its work forgotten. The visitor sees beauty: the inhabitant a *place* where he works and has friends. Far away, closing his eyes, he had been seeing this valley, but as a visitor sees it, as the guide book sees it: this valley, in which he had lived more than half his life.

(Williams 1988a: 75)

This distinction between a 'tourist gaze' and lived lives in place is vital to Williams. Lived lives and the sense of value that attaches thereto are embedded in an environment actively molded and achieved through work, play, and a wide array of cultural practices. There is a deep continuity here between the environmental ambience of *Border Country* and the more explicit environmental history of *People of the Black Mountains*. Only at the end of the former novel can Matthew/Will come together, perhaps to reconcile the different structures of feeling that arise through the mind that asserts itself walking on the mountains and the knowledge achieved through the 'polystyrene models and their theoretical equivalents': 'now it seems like the end of exile. Not going back, but the feeling of exile ending. For the distance is measured, and that is what matters. By measuring the distance, we come home' (Williams 1988a: 351).

Again and again, this same duality erupts in Williams's novels. The battle between different levels of abstraction, between distinctively understood particularities of places and the necessary abstractions required to take those understandings into a wider realm, the fight to transform militant particularism into something more substantial on the world stage of capitalism – all of these elements become central lines of contradiction and tension that power the story line of the novels. *Loyalties* turns crucially on such tensions. And in that novel we get a far more profound exploration of certain dilemmas than comes from any of the theoretical work.

A question of loyalties

The story of *Loyalties* begins with a meeting in 1936 between Welsh miners and Cambridge University students on a farmstead in Wales to work out common means to fight fascism in Spain. Out of that meeting comes a brief passionate liaison between a Welsh girl, Nesta, who has striking artistic talents, and Norman, a young Cambridge student from an upper-class background. The question of their distinctive places, both materially and in the structure of society, is raised immediately. She maintains that the place, Danycapel, has made her what she is; he graciously concedes that it must therefore be a good place but then urges her not to get stuck in it. She remains there for the rest of her life – the woman embedded in the particular place that has both nurtured her and which she continues to nurture – while he, the man, returns to a more cosmopolitan, internationalist, and seemingly rootless world of international political intrigue and scientific enquiry. Though the two never talk again after their brief initial encounters, the novel turns on the continuance of the tension between them, primarily in the figure of Gwyn, the son born out of wedlock between two class and gender positions – the one closely place-bound and the other ranging more widely across space – within a supposedly common politics defined largely through the Communist Party. Gwyn, like Matthew Price in *Border Country*, internalizes the tension: raised in that place where Nesta dwells, he eventually goes to Cambridge to study, in part at the insistence of Norman's sister, who performs a crucial link role nurturing a familial connection to Gwyn that Norman broadly ignores.

The place-bound politics arising out of the experience of class solidarities and gender relations in Wales is radically different from the more abstract conceptions held by academics and party leaders. The difference is not, it should be noted, between parochialism and universalism. The miner, Bert, who ultimately marries Gwyn's mother and becomes Gwyn's real father, fights in Spain alongside other workers and students. When the student, who was close to Norman at Cambridge, is killed in action, Bert acquires his binoculars (a symbolic terrain of vision?) only on his deathbed to hand them on to Gwyn. Bert also fights in the Second World War (billed as 'the ultimate war against fascism'), and suffers a hideous injury in Normandy that permanently disfigures his face – Bert forever carries the marks of his internationalist commitments on his body.

Norman, Gwyn's biological father, dwells in a different world and fashions loyalties to the Party and to the cause in a radically different way. Perhaps modelled on Burgess, Maclean, Philby, and Blunt (the Cambridge group who became Soviet agents during the 1930s), Norman is involved in passing on scientific knowledge to the Communist powers, suffering interrogation and perpetual mental pressures, acquiring internal mental scars as he anguishes over whether to sustain loyalties contracted in one era when they made sense, in a Cold War world where conscience might dictate another course of action. Williams does not, interestingly, condemn Norman, though Bert's bitter deathbed judgment is powerfully registered against these 'runaways from their class' – 'they used us . . . we know now we got to do it by ourselves.' Gwyn echoes this judgment: Norman and his ilk were the very worst 'because they involved in their betrayal what should have been the alternative: their own working-class party, their socialism.'

But Gwyn's final angry confrontation with Norman (see below) is paralleled by an extraordinary outburst directed against Gwyn by his mother Nesta. The occasion arises when she reveals to him two sketch portraits she has hidden away: one of the young Norman, fair-haired and ethereal, and the other of a now-deceased Bert, drawn after his return from the war, a portrait that 'was terrible beyond any likeness, as if the already damaged face was still being broken and pulled apart.' Gwyn is deeply moved but can only say how 'intensely beautiful' the latter portrait is:

She was staring at him angrily. Her face and body seemed twisted with sudden pain. He was bewildered because he had never seen her in even ordinary anger. She had been always so contained and quiet and pleasant, always younger than her age, self-possessed and slightly withdrawn.

'It is not beautiful!' she screamed, in a terrible high voice.

'Mam, please, I didn't mean that,' Gwyn struggled to say. 'Do you understand nothing?' she screamed. 'Do you know nothing? Have you learned nothing?'

'Mam, all I meant –'

'It is not beautiful!' she cried again. 'It's ugly. It's destroying! It's human flesh broken and pulped!'

'Yes. Yes in him. But the truth, that you saw the truth –'

'It's ugly, it's ugly!' she screamed now past all control.

(Williams 1989c: 347–8)

This violent clash of sensibilities, of 'structures of feeling' as Williams puts it, says it all. The problem here is not only the level of abstraction at which the world view of socialist politics gets constituted, but of the very different structures of feeling that can attach to those different levels of abstraction. Gwyn has acquired the distance to look upon the portrait of his father as a work of art, as an aesthetic event, as a thing of beauty precisely because it can capture and represent the awfulness of disfigurement with an elemental truth. But for Nesta it is not the representation that matters, but what is being represented; the sheer pain of that always remains fundamental and elemental.

The difficulties posed by the search for any kind of critical distance then come more clearly into focus. In *Border Country*, for example, Matthew/ Will takes to climbing the nearby mountain, the Kestrel, and admiring the view from on high. Looking at 'the patch' where he had been raised, he knew it

was not only a place, but people, yet from here it was as if no one lived there, no one had ever lived there, and yet, in its stillness, it was a memory of himself . . . The mountain had this power, to abstract and to clarify, but in the end he could not stay here: he must go back down where he lived.

(Williams 1988b: 293)

And then:

On the way down the shapes faded and the ordinary identities returned. The voice in his mind faded and the ordinary voice came back. Like old Blakely asking, digging his stick in the turf. What will you be reading, Will? Books, sir? No better not. History, sir. History from the Kestrel, where you sit and watch memory move, across the wide valley. That was the sense of it: to watch, to interpret, to try and get clear. Only the wind narrowing the eyes, and so much living in you, deciding what you will see and how you will see it. Never above, watching. You'll find what you're watching is yourself.

(Williams 1988b: 293)

But it is only partly the *level* of abstraction at which different representations operate that is vital here. For there is something else going on in these interchanges that derives from the kind of abstraction achievable given different ways of acquiring knowledge of the world. And here there is a definite polarization in Williams's argument. Ingold (1993: 41), in a rather different context, describes the opposition as that between a vision of the world as a sphere which encompasses us or as a globe upon which we can gaze:

> the local is not a more limited or narrowly focused apprehension than the global, it is one that rests on an altogether different mode of apprehension – one based on an active, perceptual engagement with components of the dwelt-in world, in the practical business of life, rather than on the detached, disinterested observation of a world apart. In the local perspective the world is a sphere . . . centred on a particular place. From this experiential centre, the attention of those who live there is drawn ever deeper into the world, in the quest for knowledge and understanding.

Both Bert and Nesta seem always to be reaching out from their centered place – Danycapel – whereas Norman always tries to understand the world in a more detached way en route to arriving at his political commitments. Gwyn internalizes both perspectives and is riven with conflicting thoughts and feelings. Yet, Williams seems to be saying, we cannot do without both kinds of abstraction any more than we can do without the conflicting modes of representation that necessarily attach thereto. Williams tries to define a complementary, even dialectical relation between the two visions, though I think it is evident on what side of that opposition he feels most comfortable. We should, he again and again insists, never forget the brute ugliness of the realities of lived experience for the oppressed. We should not aestheticize or theorize those lived realities out of existence as felt pains and passions. To do so is to diminish or even to lose the raw anger against injustice and exploitation that powers so much of the striving for social change. The formulaic view that 'truth is beauty,' for example, deserves to be treated with the wrath that Nesta metes out.

The question of loyalties is defined, then, both by the level and kind of abstraction through which political questions are formulated. As an affective and emotive political force, loyalties always attach to certain definite structures of feeling. The richest characters in all of Williams's novels are precisely those who internalize different and conflicting loyalties to radically different structures of feeling: Gwyn in *Loyalties* or Matthew Price in *Border Country* and Owen Price in *Second Generation*. And it is no accident here that Williams turns to the novel form to explore the conflicts and tensions. The Brechtian strategy is everywhere apparent and suggests not only that the tensions can never be resolved but that we should never expect them to be so. By perpetually keeping them open, we keep open a primary resource for the creative thinking and practices necessary to achieve progressive social change.

This is a telling formulation of a problem that many of us can surely recognize. I certainly recognize it not only as someone who, like Williams, went from an English state school to a Cambridge education, but also more immediately in the contested politics of the Cowley project. Where did my loyalties lie? Williams's warnings are salutary. The possibility of betrayal looms, in our heads as well as in our actions, as we move from one level of abstraction or from one kind of epistemology to another. The dissident shop stewards in the Cowley car plant probably said unkindly words about me of exactly the sort that Bert said of 'the class runaways' in *Loyalties*. Interestingly, Hayter (though herself even more of a 'class runaway' than me) inserted into the conclusion the very strong words of a shop steward in the plant: 'Betrayal is a process, not an individual act, and it is not always conscious.' While the comment was not directed at me, it could well have been in the light of our discussions.

But betrayal is a complex as well as a bitter term. Let me go back to the fictional account in *Loyalties* for a moment. Here is how Norman's close associate defends him to Gwyn:

> 'There are genuine acts of betrayal of groups to which one belongs. But you have only to look at the shifts of alliance and hostility, both the international shifts and within them the complex alliances and hostilities of classes, to know how dynamic this definable quantity becomes. There are traitors within a class to a nation, and within a nation to a class. People who live in times when these loyalties are stable are more fortunate than we were.'
>
> 'Not only in times. In places,' Gwyn said.
>
> (Williams 1989c: 317)

In any case, Norman was involved in scientific research that had a completely different domain of reference. This entailed

> a dynamic conflict within a highly specialized field. It was vital to prevent it, through imbalance, reaching that exceptionally dangerous stage in

which, by its own logic, it passed beyond nations and classes and beyond all the loyalties that any of us had known. Except, perhaps, in the end, a simple loyalty to the human species.

(Williams 1989c: 317–19)

Nothing of such moment was involved in the Cowley case, of course. Although there is one minor twist at the end of *Loyalties* that would make the connection. Norman, allowed to retire without disgrace, has bought in a wood to save it from development. In the face of Gwyn's accusation of class betrayal, of betrayal of 'the morality of shared existence' that underlies the militant particularism of a community like Danycapel, Norman argues:

> 'You abuse what you call my class but what you are really abusing is know-ledge and reason. By the way the society is, it is here, with us, that ideas are generated. So it has been with socialism: at once the good ideas and the errors. Yet we have begun to correct them, and this is all that can be done. In reason and in conscience our duty now is not to something called social-ism, it is to conserving and saving the earth. Yet nothing significant for either is generated among what you call your fellow countrymen. Indeed, that is, precisely, their deprivation. It is also their inadequacy, and then what are you asking of me. That I should be loyal to ignorance, to short-sightedness, to prejudice, because these exist in my fellow countrymen? That I should stay still and connive in the destruction of the earth because my fellow countrymen are taking part in it? And that I should do this because of some traditional scruple, that I am bound to inherit a common inadequacy, a common ignorance, because its bearers speak the same tongue, inhabit the same threatened island? What morality, really, do you propose in that?'

Gwyn's response is sharp enough: 'What you thought about communism, what you now think about nature, is no more than a projection of what suited you. The fact that for others each belief is substantial merely enabled you to deceive them' (Williams 1989c: 364).

The argument in *Loyalties* is not, of course, resolved. And I think Williams's point is to insist that it will never be. Loyalties contracted at one scale, in one place and in terms of a particular structure of feeling, cannot easily or simply be carried over without transformation or trans-lation into the kinds of loyalties required to make socialism a viable movement either elsewhere or in general. But in the act of translation something important necessarily gets lost, leaving behind a bitter residue of always unresolved tension.

Loyalties, identities, and political commitments

Accepting this leads to some uncomfortable political reflections. Let me depict them at their starkest. The socialist cause in Britain has always been powered by militant particularisms of the sort that Williams described in Wales, and that I encountered in Cowley. A good deal of historical evidence could, I believe, be assembled in support of that argument. A recent volume of essays on *Fighting Back in Appalachia* (Fisher 1993) documents the point brilliantly within the United States. But those militant particularisms – even when they can be brought together into a national movement, as they have been at various historical moments by the Labour Party in Britain – are in some senses profoundly conservative because they rest on the perpetuation of patterns of social relations and community solidarities – loyalties – achieved under a certain kind of oppressive and uncaring industrial order. While ownership may change (through nationalization, for example), the mines and assembly lines must be kept going, for these are the material bases for the ways of social relating and mechanisms of class solidarity embedded in particular places and communities. Socialist politics acquires its conservative edge because it cannot easily be about the radical transformation and overthrow of old modes of working and living – it must in the first instance be about keeping the coal mines open and the assembly lines moving at any cost (witness the tangled industrial policy of successive British Labour governments in the 1960s and 1970s). Should the struggle at Cowley be to keep the increasingly oppressive jobs in the car plant going, or to seek out different, better, healthier, more satisfying jobs in some quite different and more ecologically sensitive system of production? At a time of weakness and no alternatives, the Cowley struggle necessarily focused on the former objective, but I had the distinct impression that even in the long run and under the best of circumstances it would always be thus for those working on the shop floor, for those most strongly imbued with the militant particularism associated with working in the plant.

There is another way of putting this. Can the political and social identities forged under an oppressive industrial order of a certain sort operating in a certain place survive the collapse or radical transformation of that order? The immediate answer I shall proffer is 'no' (and again I think a good deal of evidence can be marshalled to support that conclusion). If that is the case, then perpetuation of those political identities and loyalties requires perpetuation of the oppressive conditions that gave rise to them. Working-class movements may then seek to perpetuate or return to the conditions of oppression that spawned them, in much the same way that those women who have acquired their sense of self under conditions of male violence return again and again to living with violent men.

That parallel is instructive here. It is, as many feminists have argued and many women have shown, possible to break the pattern, to come out of the dependency. Working-class movements can similarly retain a revolutionary impulse while taking on new political identities under transformed conditions of working and living. But it is a long hard process that needs a lot of careful work. Williams recognizes this difficulty explicitly in his discussion of the ecological issue:

> It is no use simply saying to South Wales miners that all around them is an ecological disaster. They already know. They live in it. They have lived in it for generations. They carry it with them in their lungs . . . But you cannot just say to people who have committed their lives and their communities to certain kinds of production that this has all got to be changed. You can't just say: come out of the harmful industries, come out of the dangerous industries, let us do something better. Everything will have to be done by negotiation, by equitable negotiation, and it will have to be taken steadily along the way.
>
> (Williams 1989a: 220)

The worry at the end of that road of negotiation is that socialist parties and governments will only succeed in undermining the social and political identities and loyalties that provide the seedbed of their own support (again, quite a bit of evidence can be marshalled for that proposition in Western Europe since the Second World War). Socialism, it could be argued, is always about the negation of the material conditions of its own political identity. But it so happens that capitalism has fortuitously taken a path these past twenty years towards the elimination of many of the militant particularisms that have traditionally grounded socialist politics – the mines have closed, the assembly lines cut back or shut down, the shipyards turned silent. We then either take the position that Hayter voiced to me – that the future of socialism in Oxford depended on the outcome of a struggle to get mass employment in car production back into Cowley (a view I could not accept) – or else we have to search for new combinations of both old and new forms of militant particularism to ground a rather different version of socialist politics. I see no option except to take the latter path, however difficult and problematic it may be. This does not entail abandoning class politics for those of the 'new social movements', but exploring different forms of alliances that can reconstitute and renew class politics. Put pragmatically, class politics in Oxford could survive the total closure of the Cowley plant, but only if it secures a new basis.

There is still another dimension to all this, which has to do with the question of spatial scale and temporal horizon. With respect to the former, Neil Smith (1992: 72–3) has recently remarked how we have done a very bad job of learning to negotiate between and link across different spatial scales of social theorizing and political action. He emphasizes what I see as a central confusion in contemporary constructions of socialism arising out of 'an extensive silence on the question of scale':

> The theory of geographical scale – more correctly the theory of the production of geographical scale – is grossly underdeveloped. In effect, there is no social theory of geographical scale, not to mention an historical materialist one. And yet it plays a crucial part in our whole geographical construction of material life. Was the brutal repression of Tiananmen Square a local event, a regional or national event, or was it an international event? We might reasonably assume that it was all four, which immediately reinforces the conclusion that social life operates in and constructs some sort of nested hierarchical space rather than a mosaic. How do we critically conceive of these various nested scales, how do we arbitrate and translate between them?

Capitalism as a social system has managed not only to negotiate but often to actively manipulate such dilemmas of scale in its forms of class struggle. This has been particularly true of its penchant for achieving uneven sectoral and geographical development so as to force a divisive competitiveness between places defined at different scales. But where does 'place' begin and end? And is there a scale beyond which 'militant particularism' becomes impossible to ground, let alone sustain? The problem for socialist politics is to find ways to answer such questions, not in any final sense, but precisely through defining modes of communication and translation between different kinds and levels of abstraction.

On conclusions

I conceded that Hayter write the conclusion to *The Factory and the City*. The book, after all, was largely the result of her efforts. The result reads very oddly. Broadly 'workerist' assertions that focus exclusively on the struggle to regain radical control in the plant are ameliorated here and there by questions about overcapacity, community involvement, and the environment. The effect is strange since it does not, I think, arrive at any sort of identifiable or productive internalized tension. I think this a pity. For there was an opportunity here not to seek closure of an argument but to use the materials in the book to reflect upon and learn from what had happened, to open up a terrain of discussion and debate. I cannot help contrasting our effort with the far more thoughtful conclusion – largely focusing upon the tension between class-based and plant-based Marxist perspectives on the one hand and neopopulist communitarian perspectives on the other – provided by Stephen Fisher in *Fighting Back in Appalachia*, an edited collection on incidents of struggle and conflict in Appalachia that has many parallels in terms of the multiple voices it incorporates. Our failure helps explain, I think, why Williams resorted to the novel form to explore certain dilemmas. The closure that we often seem compelled to search for in a piece of cultural or political economic research can more easily remain perpetually open for reflection in the novel form, even when, as happens to Matthew Price, some sort of reconciliation becomes possible once 'the distance is measured'. Dual conclusions to the Cowley book would have gone some way towards keeping issues open and the tensions alive, at the same time highlighting the question of different levels and kinds of abstractions.

In view of all this, I was quite startled to read Williams's novel *Second Generation*, sometime after the Cowley book was finished. This novel was published in 1964 and set in Oxford at around that time. It revolves around the tensions between a university-based socialism on the one hand and the contested politics within the car plant on the other. The opening paragraph sets the scene for the problem of socialist politics in a divided city:

> If you stand, today, in between Town Road, you can see either way: west to the spires and towers of the cathedral and colleges; east to the yards and sheds of the motor works. You see different worlds, but there is no frontier between them; there is only the movement and traffic of a single city.
>
> (Williams 1988c: 9)

Kate Owen, a local Labour Party organizer and wife of a union leader in the plant, is torn between loyalty to family and community and the sexual freedom that beckons from the other side of the class divide within a university-based socialism. Peter Owen, her son, is likewise caught in between. He is studying for his doctorate in industrial sociology at an Oxford college in the midst of violent shop-floor struggle that wears his father down. All the themes that Williams develops elsewhere concerning the kinds of knowledge that it is possible to acquire and hold are richly developed here, including the interplays of gender and class within the 'structures of feeling' that get incorporated into a socialist politics.

But, interestingly, many of the substantive issues that arose in the work on the Cowley project actually crop up, without resolution, in *Second Generation*. Had I read it before, rather than after, becoming associated with the Cowley research, I think my approach might well have been different. I would on the one hand have insisted much more strongly on the Brechtian strategy of keeping the conclusions open. But on the other I would have taken much more careful notice of Williams's (1989: 220) injunction that 'everything will have to be done by negotiation, equitable negotiation, and it will have to be taken steadily along the way.'

Evaluations and possibilities

The three words *space*, *place*, and *environment* encompass much of what geographers do. Their meaning has been contested within geography over the years in fierce debates (particularly in the radical journal *Antipode*) over, for example, how and why localities and places might be said to matter and how properly to view relations between place and space (see, for example, Agnew and Duncan 1989; Cooke 1989, 1990; Massey 1991; Pred 1984; Smith 1987; Swyngedouw 1989, 1992a). And in the course of this discussion, the question of level of abstraction and scale has again and again been raised (see Cox and Mair 1989; Cooke 1989; Duncan and Savage 1989; Horvath and Gibson 1984; Merrifield 1993; Swyngedouw 1992b; and Smith 1990, 1992). But geographers are not the only ones to deal in such matters. In recent years the meanings to be attributed to space, place, and nature have become a crucial matter of debate in social, cultural, and literary theory (see, for example, Carter, Donald and Squires, 1993), a debate in which geographers have certainly participated (see Bird et al. 1993; Gregory and Urry 1985; Keith and Pile 1993). These sorts of concerns and interests have been impelled in part by the question of the relations between what appears to be an emergent global capitalist culture on the one hand and the reassertion of all sorts of reactionary as well as potentially progressive 'militant particularisms' based in particular places on the other, coupled with a seemingly serious threat of global environmental degradation. But the concerns have also in part been produced by a burgeoning tradition of cultural studies that Raymond Williams helped to define, with its emphasis upon structures of feeling, values, embeddedness, difference, and the particularities of the counterhegemonic discourses and social relations oppositional groups construct.

Williams thought a great deal about questions of space, place, and environment and evidently worried as to how they might be brought into play both in his cultural theory and in his views on how socialism might be constructed. The social transformations of space, place, and environment are neither neutral nor innocent with respect to practices of domination and control. Indeed, they are fundamental framing decisions, replete with multiple possibilities, that govern the conditions (often oppressive) over how lives can be lived (see, in particular, the collection by Keith and Pile 1993, on this point). Such issues cannot, therefore, be left unaddressed in struggles for liberation. Furthermore, such struggles must internalize a certain reflexivity, if not an unresolvable tension, concerning both the levels and kinds of abstractions they must necessarily embrace as part and parcel of their working tools for practical action.

The fact that Williams's dealings and concerns over space, place, and environment are voiced primarily in his novels suggests, however, a certain hesitancy on his part, if not an outright difficulty in getting this tripartite conceptual apparatus into the very heart of cultural theory. The conclusion is not, however, that space, place, and environment cannot be incorporated into social and cultural theory, but that practices of theorizing have to be opened up to the possibilities and dilemmas that such an incorporation requires. By treating Williams at his word, and seeing his novels and his critical cultural theory as complementary aspects within a unified field of endeavor, we find him opening up a terrain of theorizing far more profound than many of the high theorists of contemporary culture who ignore such dimensions could ever hope to achieve. Theory cannot be construed as a pure achievement of abstraction; more importantly, theoretical practice must be constructed as a continuous dialectic of the militant particularism of lived lives on the one hand, and the struggle to achieve sufficient critical distance and detachment on the other. In this regard, the problematic that Williams defines is surely universal enough to bring its own rewards. The search for a critical materialist and thoroughly grounded, as opposed to a confined metaphorical and purely idealist, incorporation of place, space, and environment into cultural and social theory is on. And the stakes are high. The return of theory to the world of daily political practices across a variegated and hierarchically structured geographical space of social and ecological variation can then become both the aim and the reward of a particular kind of theoretical practice.

One of the most moving chapters in *Fighting Back in Appalachia* is entitled 'Singing Across Dark Spaces'. It is a personal account by Jim Sessions and Fran Ansley of the union/community takeover of Pittston's Moss 3 Plant in the bitter coal strike of 1989 in Appalachia, a takeover that proved crucial in resolving the strike on terms much more acceptable to the miners. Jim Sessions, who was on the inside of the plant during the takeover as 'an unaffiliated witness', and Fran Ansley, who remained on the outside, recorded their day-by-day experiences. Wrote the latter, after two days of the occupation, 'there are moments of transcendence that are capable of teaching us, of making us *feel* the possibilities that reside in us,

in the people around us, and in the groups of which we are or can be part'
(Fisher 1993: 217). Theorists can also learn to sing across the dark spaces
of increasingly violent and bitter social and cultural conflict. But only if
we open ourselves to the possibilities that Williams created.

CHAPTER 10

City and justice:
social movements in the city

*First delivered to the Conference on Model Cities in Singapore in April 1999
and published by the Urban Redevelopment Authority of Singapore as
conference proceedings,* Model Cities: Urban Best Practices *in 2000;
the revised version, published here, was presented to the Man and City:
Towards a more Human and Sustainable Development Conference
in Naples, September 2000.*

The history of cities and of thinking about cities has periodically been marked by intense interest in the transformative role of urban social movements and communal action. Such movements get variously interpreted, however, depending upon historical and geographical conditions. The Christian reformism culminating in the social control arguments of Robert Park and the Chicago School of Urban Sociology (evolved during the inter-war years in the United States and exported around the world in the post-war period as standard fare for urban sociologists) contrast, for example, with both the pluralist 'interest group' model of urban governance favored by Robert Dahl and the more radical and revolutionary interpretations arrived at (mainly in Europe and Latin America) during the 1960s and and 1970s (culminating in Castell's *magnum opus* on *The City and the Grassroots*).

In its most recent incarnations, interest has variously focused upon ideals of citizenship (Douglass and Friedmann 1998), on the role of religious and ethnic identities (communalism) or on a secular political communitarianism, in the evident belief that the real sources of urban change (no matter whether cast in a positive or negative light) lie (or ought to lie) in civil society rather than in the 'official' spheres of the state apparatus (see Sandercock 1998). In some instances and places, loss of confidence in the state apparatus and political parties has resulted in the coalescence of political thinking around ideals of local and people-based action as the main means to humanize, ameliorate, transform or in some instances even to revolutionize the qualities of urban life. A deep and abiding faith in this latter path to social change even underpins that most pervasive of all strategies for urban change, that of the so-called 'public-private partnership.'

It is not my intention here to conduct any intensive critical or comparative review of this extensive literature. But what does strike me as curious is the way academic, intellectual and political interpretations of grassroots activism have ebbed, flowed and diverged without any clear or obvious relationship to the actual activities themselves. While the intensity and forms of the latter do vary, the attention paid to them in urban theorizing varies according to some other logic. Only at moments of intense turbulence or disruption do the two currents tend to flow together. But even then, events like the unrest in Los Angeles, the uprisings in Jakarta, the intercommunal violence in India or Sri Lanka, the riots in the suburbs of Paris and Lyon or even the extraordinary events in Prague and Berlin that saw the end of the Cold War, all too often catch urban theorists by surprise. Or, conversely, close observers of urban life find themselves perpetually surprised by the odd forms and manifestations of localized politics in the settings they trouble to study in any detail (see, for example, Seabrook 1996).

A crude but nevertheless fertile starting-point to understand the roots of this disjunction lies in the ebb and flow of both the *sense of possibility* and the *desire* for change in political and intellectual circles (often expressed as utopian dreams of alternative city forms) on the one hand and the need to identify political agents – such as a proletariat or urban social movements – capable of realizing such dreams, on the other. The dialectical relation between these two currents of thought and action is of course important. The flood of student and activist believers into the neighborhoods of Chicago or Paris in the late 1960s and early 1970s undoubtedly played an important role in infusing local social movements with global political ambitions. The subsequent retreat of such movements into what idealists construed as a rather ignoble and self-serving localism (degenerating into what many would regard as a reactionary 'not-in-my-backyard' or even actively exclusionary communal politics) played an important role in the political disillusionment and abandonment of leftwing utopianism that followed. On the other hand, those concerned to mobilize power – that, say, of Hindu revivalism – have largely done so by organizing and orchestrating communal movements in particular urban settings, impelling a completely different drift to urban transformations to that which the secular left would regard as healthy.

But this last example hardly appears as a simple example of a welling-up of grassroots sentiments. There is a great deal of orchestration from above. And this then raises the question as to the efficacy of grassroots activities in and for themselves of changing anything other than conditions in their own backyards. Sceptics, armed with a good deal of historical and geographical comparative information, might reasonably conclude that left to themselves such movements amount to nothing more than minor perturbations on the deeper currents of socio-ecological change. Yet even the most sceptical analyst would be forced to wonder why it is that again and again social theorists and political practitioners turn to local grassroots movements as some sort of seedbed from which major social changes can arise. Is there, then, a more general way to understand the role of urban social movements that goes beyond mere episodic and particular constructions? I here explore a tentative theoretical framework to answer that question.

Militant particularism and the politics of collectivities

The thesis of militant particularism to which I am strongly attached (see Harvey 1996; 2000) holds that all politics (of no matter what sort and no matter whether it is local, urban, regional, national or global in focus) have their origins in the collective development of a particular political vision on the part of particular persons in particular places at particular times. I presume that an undercurrent of grassroots ferment is omnipresent in all places and localities, though its interests, objectives and organizational forms are typically fragmented, multiple and of varying intensity. The only interesting question under this formulation is how and when such militant particularisms become internally coherent enough and ultimately embedded in or metamorphosed into a broader politics.

Collective grassroots politics often flow, of course, in constrained and predictable channels. As such, they often pass unremarked precisely because they seem more to be about business as usual than about social change. In the US, for example, it is the homeowner associations, dedicated to protection of their property values, privileges and lifestyles, who dominate the urban/suburban scene (Davis 1990). The violence and anger that greets any threat to individualized property rights and values – be it from the state or even from agents of capital accumulation like developers – is a powerful political force. It permeates religious institutions (I suspect it grounds much of the work of the Christian Coalition in the US, for example) as well as much of what passes for active politics at the local, state and federal levels.

Such collective movements preclude rather than promote the search for alternatives (no matter how ecologically wise or socially just). They tend to preserve the existing system, even as they deepen its internal contradictions, ecologically, politically and economically. For example, suburban separatism in the US – based upon class and racial antagonisms – increases car dependency, generates greenhouse gasses, diminishes air quality and encourages the profligate use of land, fossil fuels and other agricultural and mineral resources. Militant particularism here functions as a seemingly immovable conservative force to guarantee the preservation of the existing order of things. Even when such a politics dresses up in democratic or radical clothing, its drift lies towards exclusionary and authoritarian practices. Etzioni (1997), a leading proponent of the new communitarianism in the US (a movement that largely presents itself as progressive and antagonistic to market values), actively supports the principle of closed, exclusionary and gated communities. Collective institutions can also end up merely improving the competitive strength of territories in the high-stakes game of the uneven geographical development of capitalism, as Putnam (1993) purports to show in the case of contrasting patterns of economic development in Italy. For the wealthy, therefore, 'community' often means securing and enhancing privileges already gained. For the marginalized, it all too often means 'controlling their own slum'. Inequalities multiply rather than diminish. What appears as a just procedure, produces unjust consequences (a manifestation of the old adage that 'there is nothing more unequal than the equal treatment of unequals').

I cite these cases because high levels of local activism often signal strong barriers to progressive and juster forms of social change. Reformers interested in even such mildly transgressive objectives such as 'smart growth' and resource conservation in the US then have to confront or circumvent strong community-based powers if they are to make even the mildest headway with the policies they advocate. The greater equalization of well-being through spatial reorganizations faces formidable localist barriers. But militant particularism is not *inherently* conserving and conservative. There are plenty of cases to show that this is not always or necessarily so. The militia or neo-fascist movements on the right (a fascinating form of insurgent politics), the movement towards religious communalism, the active forms of militant particularism that lead to inter-communal violence and ethnic cleansing, all illustrate how insurgent forms of politics can connect with grassroots movements. Though these instances might be characterized as reactionary, the left has its own pantheon of examples to cite as well (the Paris Commune, the storming of the Winter Palace in St Petersburg and so on). This evidence suggests that insurgent and transformative politics are constantly intermingling with local mobilizations. An understanding of how local solidarities and political cohesions are or can be constructed (particularly in today's unruly urban settings) is essential for thinking through how proposals for social change (particularly those emanating from ideological, political and intellectual circles) might become a reality.

All political movements have therefore to confront the issue of 'locality' and 'community' somehow. And in some instances, such as the turn to communitarianism (or even to a form of communalism inspired by religious or ecological beliefs), such concepts have become foundational rather than instrumental in the quest for alternative forms of social change. Articulating the place and gauging the significance of militant particularism – the coming-together of individuals in local patterns of solidarity – within a broader frame of politics becomes, as many observers have noted, a crucial task for urban theory and practice.

'Community' must, however, be viewed as a process of coming together not as a thing. It is therefore important to understand the processes that produce, sustain and dissolve the contingent patterns of solidarity that lie at the basis of this 'thing' we call 'a community'. But it is also important to recognize the 'thing-like' qualities of what gets created. The dialectic of the 'process-thing' relationship (see Harvey 1996, Chapter 2) is all too often ignored or forgotten in urban studies. Exactly how a structure of something called community gets precipitated out of the social process deserves careful attention. There is, for example, the tangible struggle to define its limits and range (sometimes even its distinctive territory) and its rules and conditionalities of membership and belonging (so crucial to identity formation). The social struggle to create and sustain its institutions (through social networks and collective powers such as the churches and other religious institutions, the unions, neighborhood organizations, local governments, and the like) is often bitterly waged. Such struggles simultaneously shape community, the sense of a proper way to live and the identities of those within its sphere of influence. It is precisely within such struggles that we must look for hints and possibilities of insurgent forms of change and the quest for social and environmental justice.

But the re-making and re-imagining of 'community' works in more general directions only if it connects to and becomes embedded in a more broadly-based politics that challenges the status quo in some way or other. The crystallization of a relatively permanent and coherent form of local organization, though not sufficient, is a necessary condition for broader kinds of political action. This means that systems of authority, consensus-formation and 'rules of belonging' must be set up and these inevitably become exclusionary in certain respects and even controlling of the social processes that grounded solidarities in the first instance. The dialectic between a free flow of processes involved in imagining and building something called 'community' and the stolid permanence of an institutionalized political presence lies at the contradictory heart of what militant particularism in general is about.

This points to a singular and important conclusion: although community 'in itself' has meaning as part of a broader politics, community 'for itself' almost invariably degenerates into regressive exclusions and fragmentations. The danger then exists that the institutionalized thing we call community will stifle the living processes that gave birth to it. Community organizations can become hollow at their core and liable to easy and almost instantaneous collapse when challenged or to easy manipulation by external political forces. If they are to function as meaningful agents of change, therefore, such movements must remain strongly nurtured by continuous processes of solidarity formation and reaffirmation. But one of the prime means whereby a community can remain alive to its constituents and resist the deadening effect of becoming 'for itself' is to be embedded in broader processes of social change. Militant particularist movements must either reach out across space and time to shape broader political-economic processes or, like the home-owner associations, become embedded in some more integrated and broad-based process of historical-geographical change. Militant particularism and local solidarities must be understood, therefore, as crucial *mediators* between individual persons and a more general politics. Their liveliness and influence depends crucially upon how they play that mediating role. Understanding their situatedness in this way, locates their importance in terms of relations established inwards to the individuals that comprise them and outwards to the broader world of political economy.

The dialectics of particularity and universality

Consider first the dialectical relation between grassroots movements and more general social processes. The critical problem for the vast existing array of localized and particularistic struggles is to transcend particularities and arrive at some conception of a more global if not universal politics. For oppositional movements (as opposed to those primarily dedicated to reinforcing the existing state of things) this means defining a general alternative to that social system which is the source of their difficulties. Grassroots movements only become interesting to the theorist or advocate of social change to the degree that they transcend such particularities. It is therefore important to understand how this transcendence can occur.

There is much to be learned in this regard from the study of the historical and geographical record of grassroots movements in general and urban social movements in particular as well as from the synthetic statements arrived at by someone like Raymond Williams (who first coined the phrase 'militant particularism' and did much to unravel its problematics) or Castells. But I am looking for a more general and theo-retical way to situate the problem.

Dialectics is here helpful. It teaches that universality always exists *in relation* to particularity: neither can be separated from the other even though they are distinctive moments within our conceptual operations and practical engagements. The notion of social justice, for example, acquires universality through a process of abstraction from particular instances and circumstances, but, once established as a generally accepted principle or norm, becomes particular again as it is actualized through particular actions in particular circumstances. But the orchestration of this process depends upon mediating institutions (those, for example, of language, law and custom within given territories or among specific social groups). These mediating institutions 'translate' between particularities and uni-versals and (like the US Supreme Court) become guardians of universal principles and arbiters of their application. They also become power centers in their own right. This is, very broadly, the structure set up under capitalism, with the state and all of its institutions being fundamental as 'executive committees' of capitalism's systemic interests. Capitalism is replete with mechanisms for converting from the particular (even per-sonal) to the universal and back again in a dynamic and iterative mode. Historically, of course, the primary mediator has been the nation state and all of its institutions, including those that manage the circulation of money. And, as I have already argued, community and grassroots movements also play such a mediating role.

But this line of analysis points to a singular conclusion. No social order can evade the question of universals. The contemporary 'radical' critique of universalism is sadly misplaced. It should focus instead on the specific institutions of power that translate between particularity and universality rather than attack universalism per se. Clearly, such institutions favor certain particularities (such as the rights of ownership of means of pro-duction) over others (such as the rights of the direct producers) and promote a specific kind of universal. But there is another difficulty. The movement from particularity to universality entails a 'translation' from the concrete to the abstract. Since a violence attaches to abstraction, a tension always exists between particularity and universality in politics. This can be viewed either as a creative tension or, more often, as a destructive and immobilizing force in which inflexible mediating institutions come to dominate over particularities in the name of some universal principle.

But there always exists a creative tension within the dialectic of partic-ularity-universality which is hard to repress, particularly under a social system like capitalism which demands change as a condition of its own survival. Mediating institutions, under such conditions, cannot afford to ossify. The optimal configuration that emerges is one of sufficient permanence of institutional and spatial forms (for example, urban governance and physical infrastructures) to provide security and continuity, coupled with a dynamic negotiation between particularities and universals so as to force mediating institutions and their associated spatial structures to be as open as possible. At times, capitalism has worked in such a way (consider how, for example, the law gets reinterpreted to confront new socio-economic conditions and how new spatial structures and spatialities have been constructed).

Any alternative, if it is to succeed, must follow capitalism's example in this regard. It must find ways to negotiate between the security conferred by fixed institutions and spatial forms on the one hand and the need to be open and flexible in relation to new socio-spatial possibilities on the other. That process demands that grassroots movements be an integral part of any process of negotiating future trajectories of development. Without them, universals will remain empty and remote at best and authoritarian impositions at worst. Letting the dialectic work between the grassroots and mediating authorities becomes a vital strategy for pursuing social change of any sort (including that required to keep a capitalist dynamic in motion). If grassroots movements did not exist, then higher-order power structures would have to invent, shape and implant them (as has often happened as political parties set up neighborhood organizations or as religious institutions colonised spaces through conversions and congregation building). The dialectic between particularity and universality is a shadowy stalking-horse for relations between different sources of power: local and more general. And it is often a biased relation that we are here contemplating in the sense that power is not necessarily evenly distributed at different scales. Grassroots politics become a focus of interest when they start to assume their own powers (by their own exertions or, as in the present circumstances, more by default) rather than simply deriving them instrumentally from some higher-order power such as the nation state.

Institutions and mediations

The formation of institutions that can mediate the dialectic between particularity and universality is, then, of crucial importance. Many of these institutions become centers of dominant discourse formation as well as centers for the exercise of power. In metropolitan areas, the offices of finance, the budget committees, the highways and transport committees, the public works departments, a wide range of non-governmental and civic organizations, as well as powerful individuals with particular interests, are all active in urban governance and operate in effect as mediators between particular localized interests and global social and political-economic relations.

Such institutions are often organized territorially and define a sphere of action at a particular spatial scale. The intermediate institutions typically take the militant particularism at work at the local grassroots level and use it or translate it, both theoretically and in terms of material action, to construct a workable spatial order facilitative of certain social processes operating at a quite different spatial scale (that, say, of the metropolis as a whole). In the process, they necessarily formulate universal principles (such as legally-binding zoning and land-use controls or, more informally, smart growth policies, philosophies of public/private partnership or urban entrepreneurialism) as guides to action. Decisions have to be made, of course, and arbitrary authority and power are invariably implicated in the process. Universal principles (of, say, urban planning and control and of neighborhood organization) can then be imposed from on high. If the organization at the grassroots is fragmented, badly articulated and partially instrumentalized by a higher power, then that higher power can easily prevail. But then the danger exists of the hollowing-out of local institutions by the gradual demise of processes of solidarity formation at their base.

No mediating institution is, however, free-floating or outside of the process-thing dialectic of the social process as a whole. What we then identify are layers of mediating institutions, often organized into some rough hierarchy, operating as transmission centers through which social processes unevenly flow. Metropolitan governments, operating in a complex relationship to grassroots movements, for example, may be forced into economic competition with each other for investments or for support from some higher authority (such as the nation state or international agencies like the World Bank or European Union). Metropolitan governance 'precipitates out' as a distinctive institutional layering characterised by corporatist forms of organization and entrepreneurial modes of behavior. It may then act predominantly as a mediator to impose upon grassroots movements a logic derived from, say, competitive globalization. That this has predominantly been the case in recent times does not mean, however, that metropolitan governance cannot also be organized as an oppositional rather than a compliant force in relation to, say, neoliberal market forces. It can function as a 'protector' of localities from the ravages of neoliberalism or, as in cases like Porto Alegre (see Abers 1998), become an active seedbed for some alternative at the grassroots scale.

Two conclusions then follow. First, the context in which to understand local social movements is set by a fluid but highly complex interaction between processes and institutions operating at a variety of quite different spatial scales (such as national, regional, metropolitan and local). If, as I believe to be the case, we have a very weak understanding of how relations and processes work across such different scales, then we have a very weak context in which to locate our understanding and interpretation of the dynamics of militant particularist movements. The danger then exists that the latter will either be fetishised as a form of political salvation or dismissed as totally irrelevant in relationship to powers and influences operating at an entirely different scale (for example, national or global). Secondly, since all universal principles are filtered through these multiple layers and scalars of institutionalized discourses, the dialectics of universality and particularity can become refracted, distorted or even thoroughly opaque. These two conclusions are hardly startling in themselves but what is surprising is how easily thinking about them gets lost in our analytical frames.

The formulation of universal principles – like social and environmental justice – is consequently fraught and frequently contested (as one might properly expect) but on grounds that are not well understood. This condition is frequently reflected in arguments within planning theory as well as within the extraordinary diversity of formulations available concerning the role of social movements in urban life. Again, I cannot hope to summarize, let alone resolve, such conflicts here. But there is one particular difficulty to which I do want to pay some attention. This concerns how multiple militant particularisms may be brought into some kind of constructive relation to each other.

Translations

The fragmented heterogeneity of grassroots movements requires a common language, a coherent politicised discourse, if it is to coalesce into a broader movement with more universal impact. Of course, it is in this domain, as Foucault has again and again pointed out, that discourses of power, attached to distinctive mediating institutions (such as the state apparatus or, more informally, within the worlds of education, religion, knowledge production and the media) typically play their often over-whelming disciplinary and authoritarian role. Hegemony becomes the focus of political struggle. Imposing conceptions of the world and thereby limiting the ability to construe alternatives is always a central task for dominant institutions of power (consider how far and how deeply the ideology of free-market individualism and liberalism has penetrated in recent times).

But if grassroots alliances are to emerge as an alternative political force (as they periodically do), then the problem of how to construct some sort of alternative hegemonic discourse out of multiple militant particularisms has to be confronted. The benevolent dictator who wishes to acquire a minimalist aura of legitimacy and consent must likewise negotiate a language through which to rule since, as Italo Calvino remarks, the only means of communication no emperor – no matter how powerful – can ever control is language itself. And it is at this point that the question of translation moves to the fore as a means to codify a common political agenda.

For James Boyd-White (1990: 257–64), translation means:

> confronting unbridgeable discontinuities between texts, between languages, and between people. As such it has an ethical as well as an intellectual dimension. It recognises the other – the composer of the original text – as a center of meaning apart from oneself. It requires one to discover both the value of the other's language and the limits of one's own. Good translation thus proceeds not by the motives of dominance and acquisition, but by respect. It is a word for a set of practices by which we learn to live with difference, with the fluidity of culture and with the instability of the self.
>
> We should not feel that respect for the other obliges us to erase ourselves, or our culture, as if all value lay out there and none here. As the traditions of the other are entitled to respect, despite their oddness to us, and sometimes despite their inhumanities, so too our own tradition is entitled to respect as well. Our task is to be distinctively ourselves in a world of others: to create a frame that includes both self and other, neither dominant, in an image of fundamental equality.

This has, of course, a distinctively utopian ring and it is not hard to problematize it, as Said did so brilliantly in *Orientalism* (1979), as the power of the translator (usually white male and bourgeois) to represent 'the other' in a manner that dominated subjects (orientals, blacks, women, and the like) are forced to internalize and accept. Rather more subtly, translation can alter political meanings and messages (sometimes even without knowing it) and thereby alter the whole dynamic of political beliefs and action. Benedict Anderson (1998) shows, for example, how the English rendering of the executed Philippine national poet Jose Rizal's work (originally written in Spanish in the late nineteenth century before the US occupation) destroys so much of the original meaning as to put an immense distance between the founding concept of national identity and its contemporary manifestations.

Such historical understandings themselves provide a hedge against the kinds of representational repressions and distortions that many feminist and postcolonial writers have recorded. Furthermore, as White points out, 'to attempt to "translate" is to experience a failure at once radical and felicitous: radical, for it throws into question our sense of ourselves, our languages, of others; felicitous, for it releases us momentarily from the prison of our own ways of thinking and being.' The act of translation offers a moment of liberatory as well as repressive possibility.

The importance of translation becomes even more obvious in the multicultural (and increasingly linguistically-fragmented) settings that now prevail in many of the world's largest metropolitan areas. For translation offers a way to create common understandings without erasing differences. And there are two compelling reasons to push in this direction. First, as Zeldin (1994: 16) remarks, we know a great deal about what divides people but nowhere near enough about what we have in common (the universals that bind us as a species). Secondly, without translation and the construction of some common language, collectivization of grassroots action becomes impossible. Armed with a common language that respects differences, grassroots movements can coalesce to re-imagine and reconstruct their social world. Translation is the hard work that has to be done in taking militant particularism and grassroots activism onto some broader terrain of struggle and mobilizing grassroots powers to some higher purpose.

But translation does not merely entail the exploration of the commonalities that lie within the diverse structures of feeling that characterize the materialities of social relations and social belonging. For language itself is a multilayered system in which powerful abstractions have their role to play. We have access, for example, to an important historical legacy of universal principles: liberty, freedom, justice, rewards for creative endeavours, responsiveness to needs, and the like. And part of what translation is about is giving tangible meaning to those abstractions (such as environmental or social justice, human rights, liberty and compassion) in particular settings and, by so doing, reaffirm the significance, power and meaning of such universal principles. Universal principles and truths are not free-standing; they do not, and cannot, stand outside of us as abstract and absolute principles that descend from some ether of morality to regulate human affairs for all times and places. Once again we see that the processes of translation and conversion depend upon institutionalized practices and mediating institutions (such as those of education, religion, the media, the law, governments, and the like). But also, and in the final analysis, no universal principle holds good that is not connected to individuals and persons who act as conscious bearers of such principles. And it is at this point that we find ourselves forced to reflect upon the processes that nourish militant particularism and grassroots movements in the first place.

The personal is political

Looked at through the other end of the analytical telescope, we see militant particularism and grassroots activism as a particular kind of collective expression of personal and individual needs, wants and desires. At this level we see a different kind of dialectic at work which helps us understand both its limitations and potentialities.

The beginning point is to understand how the personal is always political. Through changing our world we change ourselves. We cannot talk, therefore, about social change without at the same time being prepared, both mentally and physically, to change ourselves. Conversely, we cannot change ourselves without changing our world. That relation is not easy to negotiate. We encounter all manner of unintended consequences of our actions. And taking on struggles with some better-organized external power at some larger scale (like the state apparatus) is a daunting enough task to be discouraging in itself.

But there is a more subtle problem to be confronted. Foucault (1984), for example, worried that the 'fascism that reigns in our heads' is far more insidious than anything that gets constructed outside. And it is important to understand what he might have meant by that and how it relates to the powers and limitations of grassroots activism.

Consider, then, the question of 'the person' as the irreducible moment for the grounding of all politics and social action. That person is not some absolute and immutable entity fixed in concrete but in some respects a social being open to influence and control. A relational conception of the person, for example, puts emphasis upon our porosity in relation to the world. But this then poses the key question. Does the collectivity shape the person or does the person shape the collectivity? This dialectic deserves some thought.

In the US, to take the case with which I am most familiar, private property and inheritance, market exchange, commodification and monetization, the organization of economic security and social power, all place a premium upon personalized private property vested in the self (understood as a bounded entity, a non-porous individual), as well as in house, land, money, means of production, and the like, as the elemental socio-spatial forms of political-economic life. The organization of production and consumption forges divisions of labor and of function upon us and constructs professionalised personas (the planner, the professor and the poet as well as the proletarian, all of whom, as Marx and Engels point out in *The Communist Manifesto*, 'have lost their halo' and become in some way or another paid agents of bourgeois power). We live, according to this argument, in a social world that converts all of us into fragments of people with particular attachments, skills, abilities integrated into those powerful and dynamic structures that we call a 'mode of production'. Furthermore, the fierce spatio-temporalities of contemporary daily life – driven by technologies that emphasize speed and rapid reductions in the friction of distance and of turnover times – preclude time to imagine or construct alternatives other than those forced unthinkingly upon us as we rush to perform our respective professional roles in the name of technological progress and endless capital accumulation. The material organization of production, exchange and consumption rests on and reinforces specific notions of rights and obligations and affects our feelings of alienation and of subordination, our conceptions of power and powerlessness. Even seemingly new avenues for self-expression (multi-culturalism being a prime recent example) are captive to the forces of capital accumulation and the market (for example, love of nature is made to equal ecotourism, and ethnicity is reduced to a matter of restaurants or authentic commodities for market).

The net effect is to limit our vision of the possible. Our 'positionality' or 'situatedness' as beings is a social construct in exactly the same way that the mode of production is a social creation. And this 'positionality' defines who or what we are (at least for now). 'Where we see it from' within that process provides much of the grist for our consciousness and our imaginary. From the fund of our situated experience we draw certain conclusions as to possibilities and in that relation lies a limitation: we cannot see much further than the horizon broadly dictated by where we already are.

Even Adam Smith considered that 'the understandings of the greater part of men are necessarily formed by their ordinary employments' and that 'the uniformity of (the labourer's) life naturally corrupts the courage of his mind.' If this is only partially true – as I am sure it is – it highlights how the struggle to think alternatives – to think and act differently – inevitably runs up against the circumstances of, and the consciousness that derives from, a localized daily life and the way the political person gets constructed. Where, then, is the courage of our minds to come from? The embeddedness of persons within larger collectivities (such as those of neighborhood or community) becomes a problem precisely because the norms of behavior and of belonging that define social solidarities operate as constraints which, like our ordinary employments, can just as easily have the effect of limiting the courage of our minds rather than liberating them for more radical styles of action. The fierce social controls imposed by homeowner associations which tolerate very little deviance from social norms which are broadly internalized, accepted and even welcomed by most residents is a case in point. It is hard not to conclude with Paul Knox

(1994, *pace* Foucault) that such associations constituted 'a web of servitude regimes that regulate land use and mediate community affairs in what often amounts to a form of contracted fascism'.

But we can all of us individually desire, think and dream of difference. And we have available to us a wide array of resources for critique, resources from which to generate alternative visions as to what might be possible. Utopian schemas, for example, typically imagine entirely different systems of property rights, living and working arrangements, all manifest as entirely different spatial forms and temporal rhythms. This proposed reorganization (including its social relations, forms of reproductive work, its technologies, its forms of social provision) makes possible a radically-different consciousness (of social relations, gender relations, of the relation to nature, as the case may be) together with the expression of entirely different rights, duties and obligations founded upon collective ways of living. 'Where we learn it from' can be just as important as 'where we see it from'. Communities and neighborhoods are key sites within which explorations occur, both in terms of the learning and construction of new imaginaries of social life as well as their tangible realizations through material and social practices. The tension between conformism and deviation is writ large in the historical geography of community life.

But the deviations which form such a rich seedbed for social change and so often challenge the status quo within the interstices of urban life are not without their internal contradictions either. Voluntary bonding and association to realize some common dreams is one thing but social pressure and the forcing that often occurs as solidarities are formed can sometimes veer towards coercions while charismatic and hierarchical leaderships forge structures of power, influence and control which can become highly centralized within localities. And when such structures become deeply embedded in the city, they have their own fragmenting effect as local leaderships (even when not directly bought off by higher powers), depending crucially upon their positioning for their sense of identity, refuse to merge or submerge their particular interests into the framework of some broader movement. The US is full of community activism of precisely this sort, the effect being to confine a militant grassroots politics within a straitjacket of self-imposed constraints with respect to larger social transformations. Here, too, the fascisms that reign in our heads as well as within our political practices take a toll upon the effectiveness of grassroots movements to radically alter the world.

Looking at matters from this micro-level tells us, however, how hard the practical work will be to get from where we are to somewhere else. To begin with, the chicken-and-egg relation of how to change ourselves through changing our world can at best be set slowly though persistently in motion as a project to alter the forces that construct the political person. This cannot occur as some radical revolutionary break (though traumatic events and social breakdowns – economic crises, uprisings, wars – have often opened a path to radically different conceptions). The perspective of a long revolution is necessary. To construct that revolution, some sort of collectivisation of the impulse and desire for change is necessary. No one can go it alone. And there are plenty of thinkers, armed with resources from, for example, political or utopian traditions who can act as subversive agents, fifth columnists inside of grassroots movements with all their limitations yet with one foot firmly planted en route to some alternative possibilities.

We cannot presume that *anything* personal makes for *good* politics. Nor is it possible to accept the thesis beloved in some radical alternative movements (such as deep ecology and some areas of feminism) that fundamental transformations in personal attitudes and behaviors are sufficient (rather than just necessary) for social change to occur. While social change may begin and end with the personal, therefore, there is much more at stake here than individualized personal growth or manifestations of personal commitment. In reflecting on how local solidarities form, it is of course vital to leave a space for the private and the personal (a space in which doubt, anger, anxiety and despair as well as certitude, altruism, hope and elation may flourish). And in bringing persons together into patterns of social and political solidarities, there are as many traps and pitfalls as open paths to change.

Nevertheless, the construction of local solidarities and the definition of local collectivities and affinities is a crucial means whereby the person becomes more broadly political. The negotiation that always lies at the basis of militant particularism and grassroots activism is, therefore, between political persons seeking to change each other and the world as well as themselves. But what are they seeking to change themselves into and why? It is here that the perspective of what any 'long revolution' occurring through the long history of urbanization needs to be developed.

Species being in the city

The urban sociologist Robert Park (1967: 3) once wrote (in a passage that echoes Marx's observations on the labor process):

> the city and the urban environment represent man's most consistent and, on the whole, his most successful attempt to remake the world he lives in more after his heart's desire. But if the city is the world which man created,

it is the world in which he is henceforth condemned to live. Thus, indirectly, and without any clear sense of the nature of his task, in making the city man has remade himself.

Many species, as Lewontin (1982: 162) points out, adapt to the environments they alter and so initiate a long evolutionary process of dialectical transformations of selves and others. Human beings have proven particularly adept at such a process and the idea that 'Man Makes Himself' (to use Gordon Childe's title of long ago) also has a long and fertile history. We transform ourselves through transforming our world (as Marx insisted). We transform our species capacities and powers through cultural, technological, political and social innovations which have wide-ranging ramifications for the kinds of environments to which we then have to adapt. And it is increasingly the environment defined by urbanization that becomes the central milieu within which this adaptive and transformative process occurs.

While it is plausible to argue for some kind of dialectical co-evolution between human biological characteristics and cultural forms over the long term, the explosion of cultural understandings and practices in the past few hundred years has left no time for biological adaptation. Nevertheless, there are basic possibilities and constraints derived from our species character. Elsewhere I have considered these under the headings of (1) competition and the struggle for existence, (2) diversification and differentiation, (3) collaboration, cooperation and mutual aid, (4) environmental transformations (for example, urbanization), (5) the production and reconfiguration of space, and (6) the transformation of temporalities (see Harvey 2000: 209). If these form a basic repertoire of capacities and powers, then the long-term question is how to mobilize a particular mix of them to shape alternative urban forms with more humane consequences for social life. Cities are, after all, large-scale collaborative enterprises incorporating competitive processes, diversifications (divisions of labor, of function of lifestyles and values), the production of built environments, of spaces and of divergent temporalities.

Can we reasonably aspire to consciously intervene in this process of 'remaking ourselves' through urbanization, even, perhaps, acquire some 'clear sense of the nature of (our) task'? It is at this point that commonalities and universal values enter back into the picture, for without discussion and debate upon them, we are left with nothing other than the cumulative effects of micro-actions, contingencies and chance as central to human evolution. Consideration of 'species being in the city' therefore appears just as important to the argument as any discussion of how the personal might be political. Indeed, the dialectic between particularity and universality here appears as fundamental to human affairs.

The dialectics of the grassroots

So how can this theoretical and somewhat abstract exploration be put to work to understand the limitations and possibilities of grassroots movements and militant particularism in relation to broader urban processes?

To begin with, it is immediately evident that urban social movements internalize effects (political, economic and ideological) from the broader social context (including species being) of which they are a part, and that their character is heavily dependent upon this internalization. But the movements are not merely neutral mediators between, say, the personal on the one hand and broader political-economic and ideological forces (such as those that attach to, say, globalization or some nationalist developmental project) on the other. It is in, and through, this positioning that we have to interpret much of the complex historical geography of such movements in relation to broader currents of change and understand their future potentialities.

From the theoretical perspective I have constructed, it is entirely possible to understand urban social movements as predominantly socio-political reflections, if not overt constructs, of some broader politics or even biological imperatives. This entails mechanisms to procure tangible goods (material or psychic) for grassroots leaders that can be passed on as personal benefits to enough elements in the population to ensure adhesion to some collective politics. There are, in fact, many versions of this. The 'political machine' politics of many cities in the US – a much maligned system of governance that often worked well for immigrants, the poor and even for certain elements of business (making Mayor Daley's Chicago of the 1960s into 'the city that worked') – is a predominantly political version of this. While this is not usually referred to as an 'urban social movement' it forms, as it were, one polar extreme of localized collective action in the urban sphere. It overtly lacks that aspect of autonomy and voluntarism that is more often used to characterize urban social movements but there are strong theoretical grounds in any case for questioning the ideals of voluntarism and autonomy as vital values more generally. The homeowner associations in the US are not orchestrated politically from above but in their acceptance of a dominant mode of a market economy characterized by individualism at one level and class, racial or ethnic interests at another they fall exactly into an ideological line that is just as politically repressive and homogenizing as any political system could construct. And the same could be said for much of the structuring of urban spaces that goes on through the powers of ethnicity, religion or cultural forms. To characterize religious communal movements as autonomous and voluntarist seems to go well beyond what they actually are, in exactly the same way that organized ethnic enclaves (such as the chinatowns that characterize many western cities) can hardly be understood apart from the diasporic activities of ethnically-based business elites that make use of such enclaves for broader purposes.

The central question to which this points is to identify the real relations internalized within *all* urban social movements of whatever stripe (and however autonomous and voluntarist they may claim to be or even appear). Only through such identifications is it possible to understand broader allegiances and potentialities for political action at both local and more general spatial scales. The actual 'how' of the internalization is of more than passing interest, however, since it defines direct as well as subtle ways in which a relative autonomy and relatively voluntarist forms of association can be part and parcel of a process of building political power. I have already alluded to the problem of community forms of organization that become hollow and thereby vulnerable, and it is exactly at this point that the 'how' of the internalization of external influences and powers becomes an indicator of levels, strengths and persistence of local solidarities.

I emphasize the significance of these external relations for urban social movements because the latter flourish most and to greatest effect by drawing nourishment from broader resources (political, economic, ideological, religious, ethnic, cultural). Without such forms of nourishment (often structured by NGOs or other forms of organization such as religion or ethnic kinship structures) they quickly disintegrate or fade away. But putting such resource-structures to the forefront as a long-term condition of survival of such movements also indicates something about their potentially-insurgent qualities. Put simply, if local organizations do arise and find no broader resource-field from which to procure nourishment, then they either have to create such a resource-field through a broad-based insurgent politics or by sheer strength and influence force their co-optation by existing powers (in the way that the Civil Rights movement in the US in the early 1960s forced the federal and then state governments into patterns of support for its actions and agendas).

While all of this may seem to propose a somewhat jaundiced or even negative view of urban social movements and militant particularism as sources of social change or even of urban life, I want to suggest that contextualizing them accurately in this way provides a means to assess their extraordinary strengths and importance as mediating agents in the urban process. To begin with, ensuring the vitality of such movements becomes a crucial element in political participation more generally, and vitality cannot be ensured by repressive and hierarchically-structured forms of governance. If the Workers Party in Porto Alegre seeks more active political and economic participation of marginalized populations, then it must set up structures (like the participatory budget process) to ensure relative autonomy and vitality at the grassroots and itself learn to adapt to what the grassroots are about. Dialectical relations of this sort are vital to the construction of any kind of viable democracy in contemporary urban settings. Even more important is the way in which general problems (of economic development, qualities of environment and of life) get recognized in grassroots settings and become politically sensitized as issues that must necessarily be addressed through a broader politics. The internalization of external forces at the local level frequently entails intense contradictions which demand resolutions at the local level which in turn place pressures upon external powers to change their ways (consider the classic case of capitalist developers being intensely resisted by bourgeois residents). Questions of environmental justice, of discrimination in land and housing markets, of discriminatory police violence, of social integration and education, have arisen in this way and been propagated through broader and broader circles from the felt needs of individuals willing and able to give expression to those needs through collective modes of action in local settings. The way in which the personal becomes political and translates back into broader political realms is, in the final analysis, just as important as the internalization of broader powers in local collective movements.

The essential point is to see urban social movements as mediators and militant particularism as a translation from the personal to a broader terrain of politics. Plainly, democratic procedures and governance in general, as well as in urban settings, already do rely and will continue to rely into the foreseeable future upon the mediating institutions of local action and the formation of local solidarities. Whether or not such mediating institutions will play a positive or negative role in relation to the democratization of urban governance remains, of course, to be seen. But broader political-economic forces ignore this dimension to human action at their peril. Hollowed-out local institutions are even more of a threat than a militant particularism characterized by relative autonomy and a charismatic vitality seeking broader-scale reforms. The dialectics of the grassroots and the powers of militant particularism are vibrant forces in urban life in particular and socio-political life in general.

CHAPTER 11

Cartographic identities: geographical knowledges under globalization

First presented to the conference on Social Sciences at the Millennium sponsored by Hong Kong Baptist University in June 2000; the revised version, published here, was presented to the twenty-ninth International Geographical Congress in Seoul in August 2000.

Preliminary observations

The all-encompassing political-economic process we have come in recent years to call 'globalization' has depended heavily upon the accumulation of certain kinds of geographical knowledge (indeed it did so from its very inception which dates back well before 1492 in the case of western capitalism). The further development of this political-economic system will undoubtedly influence Geography as a distinctive discipline as well as geography as a distinctive way of knowing that permeates social thought and political practices. Reciprocally, geographical understandings may affect future paths of political-economic development (through, for example, the recognition of environmental constraints, the identification of new resources and commercial opportunities or the pursuit of juster forms of uneven geographical development). A critical geography might go so far as to challenge contemporary forms of political-economic power, marked by hyper-development, spiralling social inequalities, and multiple signs of serious environmental degradation.

My interest is to look at this dialectical relationship between political-economic and socio-ecological change on the one hand and geographical knowledges on the other. I begin with three basic observations.

First, though the history of this dialectical relationship is a fascinating area of enquiry (as, for example, in the whole relationship between geographical knowledges, state formation, colonization, military operations, geopolitics and the perpetual seeking-out of commercial and economic advantages), I shall largely ignore any explicit discussion of this historical record here. Nevertheless, I recognize that this past legacy weighs heavily upon contemporary geographical knowledges and that any broad-based attempt to transform the latter must, at some point, confront the particularities of past achievements.

Secondly, I use the plural 'knowledges' because I think it dangerous to presume there is some settled way of understanding or a unified field of knowledge called 'geography' even within the academy. A 'discipline that ranges from palaeo-ecology and desert morphologies to postmodernist and queer geography' obviously has an identity problem. The presumption that there is some yet-to-be discovered 'essentialist' definition of geography's subject matter, its methods, and its 'point of view' has to be challenged, though it is a long time since anyone dared write so confident a book as Hartshorne's *Nature of Geography*.

This strategic position becomes even more important in relation to my third point: there is a significant difference between geographical knowledges held (often instrumentally) in different institutional settings (for example state apparatuses, the World Bank, the Pentagon and the CIA, the Vatican, the media, the public at large, NGO's, the tourism industry, multinational corporations, financial institutions, and so on) and the geography taught and studied within departments that operate under that name. The tension between Geography as a distinctive discipline and geography as a way of assembling, using, and understanding information of a certain sort in a variety of institutional settings is important. Geographical knowledges of the latter sort are widely dispersed throughout society. They deserve to be understood in their own right (for example how the tourism industry or cable television has created and promoted a certain geographical sense in society). Different institutions, furthermore, create a demand for different kinds of geographical knowledge (the tourism industry is not interested in highlighting the geography of social distress). If academic geography does not or cannot meet these various demands, then someone else surely will.

From these preliminary remarks I draw some immediate conclusions:

1. We need general studies in comparative historical and geographical settings to better understand how the dialectical relationship between forms of geographical knowledge and socioeconomic and ecological development occurs.
2. We need careful studies of how geography as a mode of understanding is formulated, used and applied in different institutional settings (for example the military, Greenpeace, the state apparatus, multinational corporations, and so on).
3. We need to better understand the links between geographical discourses which emanate from particular institutions and the way geographical knowledges are created and taught both within and without the specific discipline of Geography.
4. We need to think through the principles that might govern the

'proper' application of 'sound' geographical knowledges in specific settings. Here the discipline of geography has a potential role of considerable importance, as both arbiter and judge of appropriate uses of properly-formulated geographical knowledges.

Cosmopolitanism and its geography

In a recent paper in *Public Culture*, I looked at how claims about global governance, management and regulatory activity are now being mobilized through ideals of 'cosmopolitanism'. Writers like David Held have argued eloquently that such a cosmopolitan perspective is essential to the evolution of democratic institutions of global governance to regulate neoliberalism. But what kind of geographical knowledge is presupposed in such an argument?

Nussbaum, one of the main proponents of the cosmopolitan ideal in the US, complains how 'the United States is unable to look at itself through the lens of the other, and, as consequence, [is] equally ignorant of itself' precisely because the population is so 'appallingly ignorant of the rest of the world'. In order to conduct any adequate global dialogue, she continues,

> we need knowledge not only of the geography and ecology of other nations – something that would already entail much revision of our curricula – but also a great deal about their people, so that in talking with them we may be capable of respecting their traditions and commitments. Cosmopolitan education would supply the background necessary for this deliberation.

Cosmopolitanism without a 'sound' and 'proper' understanding of 'geography and anthropology is, she implies, an empty ideal.

In making this assertion, Nussbaum follows no less a figure than Kant whose founding arguments on a cosmopolitan ethic are frequently appealed to in the general literature. Kant recognized both geographical and anthropological understandings as 'necessary preconditions' for the discovery and application of all other forms of knowledge, including that of a cosmopolitan ethic. Nussbaum (along with almost everyone else who writes on cosmopolitanism) leaves the nature of the necessary geographical knowledge unspecified. But Kant taught his course on Geography no less than forty-nine times (it was the second most important course he taught). A study of Kant's Geography reveals a serious problem. For not only is Kant's account unsystematic and incoherent (in marked contrast to the rigor of his philosophical works), but it is also prejudicial in the extreme. 'Humanity,' he says, 'achieves its greatest perfection with the white race. The yellow Indians have somewhat less talent. The negroes are much inferior and some of the peoples of the Americas are well below them.' The Hottentots are dirty and you can smell them from far away, the Javanese are thieving, conniving and servile, sometimes full of rage and at other times craven with fear, the Samoyeds are timid, lazy and superstitious, Burmese women wear indecent clothing and like to get pregnant by Europeans . . . it goes on and on in this vein.

Geographical knowledge of this sort appears deeply inconsistent with Kant's universal ethics and cosmopolitan principles. It immediately poses the problem: what happens when universal ethical ideals get inserted as principles of global governance in a world in which some people are considered inferior and others are thought indolent, smelly, or just plain untrustworthy? Either the smelly Hottentots, the lazy Samoyeds, the thieving Javanese, and the indecent Burmese women have to reform themselves for consideration under the universal ethical code (thereby flattening out all kinds of geographical differences), or the universal principles operate across different geographical conditions as an intensely discriminatory code masquerading as the universal good.

What appears so dramatically with Kant has, unfortunately, widespread ramifications for contemporary politics. If, as is the case, geographical knowledge in the public domain in, for example, the US is either lacking or of a similar prejudicial quality to that which Kant portrayed, then it becomes all too easy for the US to portray itself as the bearer of universal principles of justice, democracy and goodness while in practice operating in an intensely discriminatory way. The easy way in which various spaces in the global economy can be 'demonized' in public opinion (Cuba, China, Libya, Iran, Iraq, to say nothing of the 'Evil Empire' of the ex-Soviet Union, to use Ronald Reagan's favorite phrase) illustrates all too well how geographical knowledge of a certain sort is mobilized for political purposes while sustaining a belief in the US as the bearer of a global ethic.

So what kind of geographical knowledge is adequate to a cosmopolitan ethic? The question is as deep as it is broad. But there are abundant signs of how significant the relationship might be. A recent poll in the US showed that the more knowledgeable people were about the conditions and circumstances of life in a given country, the less they were likely to support US government military interventions or economic sanctions. Conversely, it then follows that there may be a vested interest for certain kinds of political economic power in leaving the mass of the population in a chronic state of geographical ignorance (or at least feeling no impulsion to cure existing states of such ignorance). Biased or 'empty' geographical knowledges, deliberately constructed and maintained, provide a license to pursue narrow interests in the name of universal goodness and reason.

Cosmopolitanism bereft of geographical specificity remains abstracted and alienated reason, liable, when it comes to earth, to produce all manner of unintended and sometimes explosively evil consequences (which can provoke whole populations to revolt against the universal principles to which they are expected to comply). A hefty dose of geographical enlightenment is therefore a necessary precondition for any kind of reasoned global governance. But what kind of geographical knowledge might be implied here? Geographers tend to be suspicious of cosmopolitan ideals (in part for good reason). But geography uninspired by any cosmopolitan vision either becomes a matter of mere description or a passive tool of existing powers (military, administrative, economic). Liberating the dialectic between cosmopolitanism and geography seems a critical precondition for the achievement of any juster and saner socio-ecological order for the twenty-first century. How can geographical knowledges be reconstituted to meet the needs of democratic global governance inspired by a cosmopolitan ethic of, for example, justice, fairness and reason?

These are big questions, but essential to contemplate not only from the narrow standpoint of Geography as a discipline, but more importantly from the standpoint of the role of geographical knowledges (no matter where produced) in affecting the future trajectory of the global socio-ecological order and its associated patterns of political-economic power. So what kinds of geographical knowledge are presently available to us as we contemplate that question?

Sites for the production of geographical knowledges

Professional geographers, like economists, sociologists and political scientists, do from time to time generate their own data sets and produce novel information to fuel their enquiries. But much of their work rests on the analysis of data, information and perspectives developed elsewhere. There is, curiously, very little formal recognition within Geography of how the geographical knowledges assembled in different institutional settings vary according to distinctive institutional requirements, cultures and norms. If Geography as a discipline aspires to be judge and arbiter of the proper application of sound geographical knowledges, then a first step down that path is to provide principles to evaluate the production of geographical knowledges in different institutional settings. Many geographers attach themselves to external institutions. But this is often viewed as a private or personal matter. Rarely do we sit back and reflect upon the consequences of such attachments for the discipline as a whole. Consider, for example, some of the primary sites for the production of geographical knowledges and how the qualities of such knowledges vary from site to site.

The state apparatus

With its interests in governmentality, administration, taxation, planning, and social control, the state apparatus has steadily been built up from the eighteenth century onwards as a primary site for the collection and analysis of geographical information. The process of state formation was, and still is, dependent upon the creation of certain kinds of geographical understandings (everything from mapping of boundaries to the cultivation of some sense of national identity within those boundaries). For the last two centuries, the state has been perhaps *the* primary site for the production of geographical knowledges necessary for the creation, maintenance and enhancement of its powers. Governmentality rests, however, on a certain set of precepts concerning individuality and objectivity (individuating, counting and locating – hence the importance of mapping – are primary operations in everything from censuses to social security administration). 'Facts' are generated by a variety of means and analyzed accordingly. Furthermore, different departments within the state apparatus develop specialized expertise on, say, agriculture, forestry, transportation, fishing, industry, and the like. Insofar as the state is itself organized hierarchically, it will typically produce geographical knowledges at different spatial scales (local, regional, national). The effect is to fragment the geographical knowledges held within the state apparatus, even while preserving a certain hegemonic attitude (of objectivity and 'facticity') as to how that information is to be collected, analyzed and understood. The state, through planning mechanisms, likewise institutes normative programs for the production of new geographical configurations and in so doing becomes a major site for orchestrating the production of space, the definition of territoriality, the geographical distribution of population, economic activity, social services, wealth and well-being. Through its influence over education, the state can actively produce national and local identities as means to secure its power. When geographers situate themselves within these frameworks of geographical knowledge production they become, sometimes without recognizing it, tacit agents of state power. At the same time, the interests of particular states lead to particular kinds of geographical knowledges (producing identifiable 'national schools' of geography) related, interestingly, to geographical and geopolitical conditions. The 'hidden geography' of geographical knowledges has rarely been addressed except eliptically and occasionally.

Military power

While obviously part of the state apparatus, military power deserves to be categorized separately because it is in this arena that the connection between privileged geographical knowledges and the pursuit of power becomes most obvious. Geographical knowledge is here often held in secret. Access to it is a matter of national security. Getting the maps or geographical information system right is crucial to attaining military superiority while reading them wrong (as in the bombing of the Chinese Embassy in Belgrade in 1999) can produce serious consequences. The connection between geographical knowledges and the military has always been extremely strong (it goes back at least to the Romans, if not before). The conventions and the norms which attach to military requirements affect the nature of the geographical knowledge produced. Engineering perspectives, like the evaluation of terrain conditions affecting vehicular movement, tend to take precedence over evaluating cultural conditions in the population, for example. Only when it is a matter of designing counterinsurgency or civilian control programs do we typically find appeal being made to anthropological and human geographical understandings.

Supranational institutions

These have increasingly become major sources of new geographical knowledges, particularly since 1950. The World Bank, the UN Development Program, the ILO, the WHO, the WTO, UNESCO, FAO, and the like form a huge and rapidly growing domain for the production of a variety of geographical knowledges (often of a specialized sort on topics such as world health, agriculture, labor, and the environment). Traditions of governmentality pioneered within the state apparatus tend to live on in these institutions, giving a certain objectivity and individuality to data forms and frameworks of analysis. The main effect is to produce qualitatively similar information to that compiled within the state apparatus but at a more supranational and global scale. Other supranational institutions, like the European Union and the OECD, take less of a global perspective but nevertheless also operate as key sites for the production of particular geographical knowledges at that geographical scale. A cursory look at, for example, World Bank reports, shows that geographical knowledge structures within the bank have changed significantly over time as different policy directions have taken root (environmental information is now much more prominent while an interest in decentralization and the institutions of civil society as vehicles for promoting economic development have introduced a much greater sensitivity to local cultures and geographical conditions in World Bank reporting). This point can be generalized: geographical knowledges produced within institutional settings can and do change significantly over time.

Non-governmental organizations

In recent decades, NGOs have proliferated, making the production of geographical knowledges throughout civil society at large a much more complicated affair, in part because the objectives of such organizations vary greatly. Organizations like OXFAM or CARE incorporate vast amounts of geographical knowledge, as do human rights groups like Amnesty International, environmental groups like the World Wildlife Fund or Greenpeace, and the vast array of organizations dealing with specific issues (violence, the situation of women and children, education, poverty, health, refugees, and so on). While it may seem inappropriate in some respects, I think we should also include within this arena that vast array of religious organizations (from the Catholic Church to Islamic, Hindu and Protestant groups), community and ethnic organizations (for example diasporas of various sorts) and political parties. These all constitute elements within civil society that contribute to governance and all of which produce geographical knowledges in particular ways (the Catholic Church, for example, not only pioneered territorialized forms of administration in the early Middle Ages, but it has also evolved strong geopolitical strategies for proseletysing and social control ever since). Insofar as such organizations seek to engage with the state or with supranational organizations, they must perforce produce geographical knowledges that are broadly compatible with those held in these more dominant institutions simply for purposes of argument and negotiation.

Corporate and commercial interests

These have their own ways of assembling and analyzing geographical knowledge for their own particular purposes. The vast business of consultancy (sometimes in-house but mainly not) today operates with particular force as corporate and commercial interests seek out expert opinion on marketing possibilities, locational preferences, resource availability (both natural and human), environmental constraints, security of investment, business climate, amenities for personnel, and the like. By the same token, such institutions produce a wide array of geographical knowledge subjected to a certain style of geographical analysis (all the way from real-estate analysis and market-survey information through the grading of governmental bonds to remote sensing of crop yields as a speculative aid in crop futures markets).

The media, entertainment and tourism industries

These industries are a prolific source of geographical knowledges. In this instance, however, we are largely concerned with the projection of images and representations upon a public at large and the predominant effects of those images and representations upon the populations subjected to them. The impact is primarily aesthetic and emotive rather than 'objective'. The selectivity entailed in the choice of images is often problematic. Commercial requirements introduce a bias towards the immediate, the spectacular, the aesthetically acceptable and associative thinking (sexuality, nature and the authenticity of the product, for example). But the variation in images and representations within the media, entertainment and tourism industries is enormous and it forms a highly problematic but influential field within which geographical knowledges get shaped and reshaped in all manner of ways. It is not hard to see the ways in which geographical misinformation gets purveyed in this arena, nor is it hard to see that here, above all, there is a vital role for geographical principles that encourage a broad-based and popular capacity for evaluation and judgement on the nature of the geographical knowledges being constructed and presented.

Education and research institutions

These generate a lot of disciplinary-specific geographical information. Economists, sociologists, anthropologists and political scientists all produce and modify information that has geographical content and often reshape that knowledge to their own disciplinary purposes. Those working on general circulation models of the atmosphere, turbulence in estuaries, biodiversity, environmental history, diffusion of diseases, epidemiology, healthcare delivery, the interpretation of novels, the history of ethnicity or cultural forms, all need to compile geographical knowledges of a specific sort to which geographers can appeal or contribute. Geographical knowledges are found throughout the whole educational and research system. It is quite proper that such knowledges become widely diffused rather than circumscribed within one unified disciplinary frame. This can be perceived within the discipline of Geography either as a threat or as a marvelous opportunity to engage in constructive dialogue about the proper use of sound geographical knowledges in many distinctive spheres of endeavor.

Institutionalized geographical knowledges of the sorts I have mentioned above are particularly important to Geography as an academic discipline. But there are far wider and more general kinds of geographical knowledge embedded in language, local ways of life, the local symbiosis achieved between nature, economy and culture, local mythologies and diverse cultural practices and forms, common-sense prescriptions and dynamic sociolinguistic traditions. Specialized geographical knowledges (everything from the urban knowledge of the taxi driver to the particular knowledge of amateur ornithologists or local antiquarians) abound. Local knowledges, for example, often amount to relatively complete geographical descriptions albeit structured from a certain parochialist perspective. Local and regional identities, conversely, are themselves built (as is the nation state) around the formation and articulation of certain kinds of geographical (often strongly colored by environmentalist sentiments) understandings. Geographers (along with anthropologists) have traditionally paid close attention to these localised 'structures of feeling' and ways of life and in so doing have helped frequently to highlight the conflict between institutionalized knowledges directed towards governmentality and localized knowledges that guide affective loyalties and socio-environmental identities. If I pay scant attention here to these traditional forms of geographical knowledge, this in no way implies lack of respect for them or their importance. They have traditionally provided, and continue to provide, the backbone of argument for an authentically independent discipline of (human) geography. But to insist upon these perspectives and knowledges as the exclusionary basis upon which to exercise geographical judgement is to isolate the discipline from its much broader potentialities.

So what general conclusions can we draw? To begin with, I find it odd that various discussions of the nature of geography pay such scant critical or reflective attention to the ways in which different geographical knowledges generated across such a wide array of institutional bases course through our own disciplinary structure. It is years now since Foucault taught us that knowledge/power/institutions lock together in particular modes of governmentality, yet few have cared to turn that spotlight upon the discipline of Geography itself. They have been unmindful of Foucault's other key observation on the importance of discipline, surveillance and punishment to the functioning of all institutions (from the prison and the factory to the World Bank, the university and even individual disciplines – the double meaning of this last word should alert us to the problem). There are exceptions to this comment. The connection between geographical knowledge and empire has been a strong topic of commentary in recent years, but the relations to state-building, the military apparatus, covert operations, multinational strategies, and even easily targeted institutions like the World Bank or the World Wildlife Fund pass by largely ignored as a topic for critical analysis and commentary, even as interest groups or individuals within the discipline of Geography avidly court connections and sometimes work closely with those institutions.

Furthermore, it is intriguing to examine how conflicts can get articulated as conflicts between geographical understandings. When, for example, Greenpeace attacks the projects of multinational corporations or the

World Bank, it often does so by providing radically different geographical descriptions (emphasizing biotic communities, cultural histories and heritages, distinctive ways of life) compared to the technical specifications laid out in, say, World Bank or company prospectuses. Similarly, when Oxfam disputed State Department policy towards Central America in the 1980s, it did so in part by describing a quite different geographical socio-environmental situation relative to the geopolitical crudities offered by the State Department. Persuading the public politically often proceeds via geographical education.

But particular studies on these topics, useful though they may be, will not do the trick. For what we have to recognize is that Geography as a discipline is situated at the confluence of a vast array of geographical discourses, constructed at quite different institutional sites with often seemingly incomparable (and some would argue incommunicable) rules of operation. Much of the confusion as to what Geography in general might be about rests upon the different allegiances that individual practitioners or groups may have to external institutions, their cultures and their dominant modes of thought (the state apparatus, NGOs, the 'scientific community' or whatever). The inability to find a common language through which to communicate across the innumerable subgroups that typically comprise a geography department (with the 'two cultures' of science and the humanities forming a particularly significant divide) in part has its origins in these multiple allegiances. Hence, I suspect, the reluctance even to contemplate the idea that geography may have a 'nature', an 'essence', or a basic mission as a discipline, and the reduction of most historiography of the discipline in recent years to an account of divergent trends and different schools of thought (with David Livingstone's weak idea of ongoing 'conversations' being the most adventurous unifying theme advanced to date).

We should, I believe, view the confluence of these divergent discourses within the discipline of Geography as an opportunity and an advantage rather than as a source of mystification and confusion. Where else might it be so easy to confront head-on the existence of, say, the 'two cultures' of science and the humanities and in what other setting would it be so easy to pursue not only meaningful conversations but also explore how to translate between and even integrate seemingly incompatible or radically divergent knowledges? I do not argue that such work will be easy (that would be wishful thinking). But it is an interesting zone of endeavor which in its own right is worth struggling for, not in pursuit of some holy grail of a unified field of Geography (with a well defined 'essence') but as a means to explore how unities (general principles and arguments) might be constructed without doing violence to the differences that divide.

Furthermore, the extraordinarily diffuse presence of geographical knowledges across the different disciplines and their dispersal throughout many major institutions provides a ready-made network for the diffusion of '*strong*' geographical ideas, constructed within the discipline. Instead of Geography weakly refracting institutionalized discourses (more or less as a servant of dominant or superior institutions, including other disciplines within the academy to which we all too easily feel inferior) it is surely possible to imagine Geography as a discipline sending strong innovative impulses throughout the academy and across multiple institutional sites based upon the collective work which geographers produce.

This leaves us with a clearer mission. Geography will not survive as a discipline, nor do I think it should survive, unless it develops *strong* geographical ideas expressive of some of the *unities* that we come to identify among the highly-differentiated discourses that converge within our disciplinary frame. Strong ideas will be listened to and command respect elsewhere. And those strong ideas must be born out of experience gained through the specific positionality of our discipline as a convergent point of multiple geographical knowledges. How, then, can we reflect upon the geographical knowledges we hold in order to identify such strong ideas? Furthermore, in the cosmic scheme of things, will such strong ideas be useful and productive in guiding socio-ecological changes in ways that contribute to human emancipation from want, need, suffering and the various forms of alienation and repression that currently surround us? These are the big questions to which we need to find some answers.

The structures of geographical knowledges

Consider, now, the common structural components of geographical knowledges. This may not automatically reveal strong ideas, but it can help identify the unities (if such there are) that underlie highly diverse geographical knowledges and suggest foci around which strong ideas might cluster. Four structural elements stand out.

Cartographic identifications

Map-making and cartography have been central to the history of Geography. Maps have also always been, and continue to be, created and used in an extraordinarily wide range of institutional and disciplinary settings and for a variety of purposes. In the bourgeois era, for example, concern for accuracy of navigation and the definition of territorial rights (both private and collective) meant that mapping and cadastral survey became basic tools for conjoining the geographer's art with the exercise of political and economic power. The exercise of military power and mapping went hand in hand. In the imperialist era, the cartographic basis was laid for the imposition of capitalist forms of territorial rights in areas of the world (Africa, the America, Australasia, and much of Asia) that had previously lacked them. Cartographic definitions of sovereignty (state formation), aided state formation and the exercise of state powers. Cartography laid the legal basis for class-based privileges of land owner-ship and the right to the appropriation of the fruits of both nature and labor within well-defined spaces. It also opened up the possibility for the 'rational' organization of space for capital accumulation, the partition of space for purposes of efficient administration or for the pursuit of improve-ments in the health and welfare of populations (the Enlightenment dream incorporated into rational planning for human welfare).

Cartography is about locating, identifying and bounding phenomena and thereby situating events, processes and things within a coherent spatial frame. It imposes spatial order on phenomena. In its contemporary manifestation, it depends heavily upon a Cartesian logic in which *res extensa* are presumed to be quite separate from the realms of mind and thought and capable of full depiction within some set of coordinates (a grid or graticule). The innovation of thematic, synoptic and iconic maps extended the range of what could be represented cartographically in important respects (synoptic charts in meteorology and climatology becoming basic tools for analysis, for example). Cartographic operations can be found right throughout the academy at the same time as they are fundamental to the work of many institutions (the state, the military, the law, and so on). Information is now often stored digitally and in GIS there exists a powerful tool for automated storage, analysis and instantaneous presentation of data and information in an ordered spatial form.

There is, of course, an extended literature on the limitations of carto-graphic operations and plenty of evaluative materials concerning the uses and abuses of maps, GIS, and the like. Their deployment for propaganda purposes is well known and their function as tools of governance, power and domination has been well portrayed in several settings (particularly that of imperial administration). The history of cartography is now also being written from a broad-based comparative perspective, revealing much about cultural and temporal differences in understandings of human positionality in the world. The evaluation and historiography of carto-graphic forms is well underway by geographers, historians, anthropologists and a wide range of scholars from other disciplines.

Cartography is, plainly, a major structural pillar of all forms of geo-graphical knowledge. Given its fundamental role in Geography as well as in other institutional settings, it provides one thematic point of convergence from which 'strong' ideas about the role of geographical knowledges might derive.

But there is much more to be said about this issue. Locating, positioning, individuating, identifying and bounding are operations that play a key role in the formation of personal and political subjectivities. Who we consider ourselves to be (both individually and collectively) is broadly defined by our position in society and the world. This positioning occurs with or without any formal map of the generally understood sort. There are mental or cognitive maps (perhaps even whole cartographic systems) embedded in our consciousness that defy easy representation on some Cartesian grid or graticule. The mental maps of children, of men and women, of the mentally ill, of adherents to different cultures and religions, of social classes or of whole populations, evidently vary greatly. The intersection of formal mapping procedures with this sense of who we are and how we may locate ourselves is far from innocent. The traces of a new cartographic consciousness are writ large in poetry (for example Shakespeare and the so-called 'metaphysical poets' deploy cartographic imagery to great effect) as well as in literature (even before Daniel Defoe and others made cartographic exploration central to their narrative structures). The effect of reading such literature is to see ourselves in a different positionality, within a different map of the world. The literature on this 'cartographic consciousness' on 'mental' and 'cognitive' maps is now growing by leaps and bounds, suggesting an emergent field of enquiry that links thematics in geography with much of cultural and literary theory (as well as with anthropology and psychology). How urban life is experienced and practiced, for example, has much to do with how we form and reform mental maps of the city.

Plainly the difficulties of communication across these different cartographic modalities is considerable as we imagine placing an expert in techniques of GIS cheek by jowl with a literary critic interested in the cartographic consciousness deployed in Beowulf or Rabelais. Cartography as one central structural support of all forms of geographical knowledge is made up of many intertwining threads. Investigating their intersections provides not only exciting challenges. It also provides some important clues as to how political, personal and psychological subjectivities are sensitive to cartographic endeavours and how changing the map of the world can change not only our modes of thought about that world but also our social behaviors and our sense of well-being (much as the depiction of the earth as a globe from outer space is often credited with affecting the ways in which we think of global problems or even of globalization itself). Cartography, in some or all of these manifestations, provides one central pillar of all forms of geographical knowledge and deserves thorough consideration as such.

The measure of space-time

Maps have traditionally taken the form of two-dimensional spatial representations. They rest, therefore, upon a certain conception of space and an ability to order and locate positions, things and events in that space through precise measurements. The mathematics of map projections (representing a globe upon a flat plane surface) itself has an interesting history. New forms of geometry were first worked out in this context (Gauss devised spherical geometry while conducting a cartographic survey of Hanover, coincidentally providing the first well-grounded estimate of the circumference of the earth).

Ways of representing, understanding and shaping space appear as an element common to all forms of geographical knowledge. Here, too, we encounter a commonality, a unity, within which there exists a whole world of difference. How do the different geographical knowledges that converge upon us conceptualize, understand and represent space?

Many geographers now claim that 'space' is the central, privileged and even defining concept of their discipline. I find this claim rather far-fetched and potentially misleading. Most of the physical sciences (physics and cosmology in particular) and engineering have a long history of dealing with the concept of space (and space-time) and it has likewise been the object of extended reflection in philosophy, literature, anthropology and many of the social sciences. So while the concept of space may be central to the discipline of Geography, it is in part received, like cartography, as vectors of multiple discourses about space, many of which emanate from elsewhere even as they converge within the discipline. To put it this way is not to imply that there is nothing new to be discovered or thought about space (or space-time) within Geography or that there is no indigenous tradition to which we can appeal. Indeed, the recent convergence of multiple discourses about space and space-time within Geography makes this a key point from which 'strong' ideas have emerged. On this point, the discipline plainly does far more than merely refract and reflect what it has derivatively taken from elsewhere.

To the degree that time, space and matter (or process) are fundamental ontological categories in our understanding of the world, Geography internalizes within itself the same problematic as other disciplines. Questions of the absolute, relative and relational conceptions of space (and time) are posed, as is the issue of whether or not time can meaningfully be separated from space. In my own view, 'space-time' or 'spatio-temporality' is the relevant category. This quite properly implies that 'all geography is historical geography' no matter where it is to be found. The importance of this dynamic conception of spatial ordering and spatial form will shortly become apparent. Without it, geographical knowledges tend to become dead and immovable structures of thought and understanding when their most exciting manifestation invariably comes from observing them (or even setting them) in motion.

Spatial structures can, of course, be broken down in terms of nodes, networks, surfaces and flows and the powers of geometric representations can be appealed to as effective means of modelling those structures. The long-standing collaboration between Chorley and Haggett across the boundaries of the social and physical sciences is illustrative of the power of analogical thinking with respect to spatial forms. And there continue to be a host of common problems (both technical and representational) that are of interest. For example, the question of integrating an understanding of processes operating at quite different scales (both temporal and spatial) is a frequent dilemma in research in many areas. The issue of how to understand scale is as important in the modelling of climate change and ecological analysis as it is in understanding the political economy of uneven geographical development. The commonality in this problem is striking and it is surprising to find so little collaborative work on how to handle it.

Armed with the right kitbag of tools, it is possible to set up common descriptive frames and modelling procedures to look at all manner of flows over space, whether it be of commodities, goods, ideas, energy, ecological inputs. The diffusion of cultural forms, diseases, biota, ideas, consumption habits, fashions; the networks of communications, energy transfers, water flows, social relations, academic contacts; the nodes of centralized power, of city systems, innovation and decision-making; the surfaces of temperature, evapotranspiration potential, of population and income potential; all of these elements of spatial structure become integral to our understanding of how phenomena are distributed and how processes work through and across space over time.

But the tendency in this is to construe processes (no matter whether physical, ecological, social or political-economic) as occurring within a fixed spatial frame (absolute space). It is just as important to see the spatial frame itself as malleable and variable (relative and relational), as an actively produced field of spatial ordering that changes sometimes quickly and sometime glacially over time. Space must be understood as dynamic and in motion, an active moment (rather than a passive frame) in the constitution of physical, ecological, social and political-economic life.

Space, like cartography, is as much a mental as a material construct.

This is so not only in the sense that the measurement systems and the mathematical constructs (geometries and calculus) that are used to represent spatiality are products of human thought. The spatial and temporal imaginary, the construction of alternative possible worlds (to use Leibniz's famous formulation) and the senses of space and time that course through consciousness and which present themselves in works of art, poetry, novels, films and multimedia forms – all of these provide a vast array of metaphorical meanings with which it is possible to explore hidden connectivities and analogies. So-called 'mental' or 'imaginary' space and time are rich terrains through which to work in order to understand personal and political subjectivities and their consequences when materialized as human action in space and time.

Attempts to deal with these dynamic aspects of spatiality – generally under the rubric of the 'social construction' or 'production' of space – are now legion. The whole history of capital accumulation which, as Marx long ago observed, has embedded within it an historical tendency towards the annihilation of space through time, points to an evolutionary process in which relevant metrics and measures of both space and time have changed significantly. Speed-up of turnover time and reductions in the friction of distance have meant that spatio-temporality must now be understood in a radically different way from what was operative in, say, classical Greece, Ming Dynasty China or mediaeval Europe. Any search for an alternative to neoliberal globalization must search for a different kind of spatio-temporality.

Here, too, we encounter an arena that demands general reflection both within and without the discipline of Geography. It is an arena of distinctive geographical work within which 'strong' ideas are being generated even if somewhat weakly diffused throughout many other disciplines and across a variety of institutional sites of knowledge production.

Place/region/territory

The 'region' is possibly the most entrenched of all geographical concepts. Within the discipline it has proven the least flexible, mainly because of its central role in those essentialist definitions of the subject which rest exclusively on the study of chorology or regional differentiation. Terms like 'locality', 'territory' and above all 'place' have often been substituted for 'region' in geographical discourses both within and without the discipline. The extensive literatures on 'the local and the global', on 'deterritorialization and reterritorialization', and on the changing significance of 'place' under conditions of hypermobility across space, testify to the vibrancy of the topic and the diversity of conceptual apparatuses with which it is approached.

The central idea is that there is some contiguous space that has the character of an 'entity' of some sort defined by some special qualities. Sometimes the boundaries are clearly demarcated (as with administrative territories) but in other instances they are left ambiguous or even unconsidered (many ideas of 'place' fail to deal at all with the activity of bounding where a place begins and ends). Sometimes the region is defined in terms of homogeneous qualities (for example of land use, soils, geological forms) and sometimes in terms of coherent relations between diverse elements (for example urban functional regions). Sometimes the region is defined in purely materialist terms (physical qualities of terrain, climatological regime, built environments, tangible boundaries) but in others it depends on ideas, loyalties, a sense of belonging, structures of feeling, ways of life, memories and history, imagined community, and the like. In either instance it is important to recognize that regions are 'made' or 'constructed' as much in imagination as in material form and that though entity-like, regions crystallize out as a distinctive form from some mix of material, social and mental processes. The approaches to place/regionality/territory are wondrously diverse no matter where they are found.

The scale problem also enters in, with a hierarchy of labels often deployed that begin with neighborhood, locality and place and proceed to the broader scale of region, territory, nation state, and globe. Region then becomes territorialization at a certain geographical scale. Scaling is not a problem unique to the social side of matters. The bounding of ecosystems, their embeddedness in higher-order systems (hierarchies of systems) and how processes prominent at one scale give way before others at another scale, makes the whole question of 'appropriate' territorial definition as crucial within ecological research as elsewhere.

Whatever the procedure or methodology, once continuous space gets carved up into distinctive regions of whatever sort, the pictures we form of, and the operations we are enabled to conduct upon, geographical information multiply enormously. Comparative studies of geographical differentiation and uneven geographical development become much more feasible.

Furthermore, as human populations frequently organize themselves territorially, so regionality becomes as central to consciousness and identity formation and to political subjectivity as does the cartographic imagination and the sense of space-time. Beyond the obvious cases of nation-state formation and movements for regional autonomy (much more prominent in recent years despite, or perhaps because of, the forces of globalization), the general processes of political articulation resting on everything from community boosterism to 'not in my back yard' politics transforms the world into complex regional differentiations, interregional relations and rivalries.

Geopolitical struggles between territories and regions have therefore been of considerable importance in geographical understandings. The division of the world into distinctive spheres of influence by the main capitalist powers at the end of the nineteenth century, for example, raised serious geopolitical issues. The struggle for control over access to raw materials, labor supplies and markets was a struggle for command over territory. Geographers like Friedrich Ratzel and Sir Halford Mackinder confronted the question of the political ordering of space and its conse-quences head on, but did so from the standpoint of survival, control and domination. They sought to define useful geographical strategies in the context of political, economic and military struggles between the major capitalist powers, or against peoples resisting the incursions of empire or neocolonial domination. This line of work reached its nadir with Karl Haushofer, the German geopolitician, who actively supported and helped shape Nazi expansionist struggles. But geopolitical thinking continues to be fundamental within the contemporary era particularly in the pentagons of military power and amongst those concerned with foreign policy. By force of historical circumstance, all national liberation move-ments must also define themselves geopolitically if they are to succeed, turning the geography of liberation into geopolitical struggles.

But it is not only the interactions between geographical entities that need to be treated in a dynamic way. The processes of region formation are perpetually in flux as social and natural processes reconfigure the earth's surface and its spatially-distributed qualities. New urban regions form rapidly as urban growth accelerates, climate change generates shifts in biotic conditions, water regimes, and the like. Populations shift their perceptions and allegiances, reinvent traditions and declare new regional formations or radically transform the qualitative attributes of the old. Like space-time and the cartographic imagination, the dynamics of the process are by far the most interesting.

Regionality, the dynamics of place and space, the relationship between the local and the global, are all in flux, making the uneven geographical development of the physical, biotic, social, cultural and political-economic conditions of the globe a key pillar to all forms of geographical knowledge.

Environmental qualities and the relation to nature

All societies develop means to evaluate, appreciate, represent and live with-in their surrounding environments (both naturally occurring and humanly constructed, with the distinctions between those two aspects decidedly porous if not increasingly meaningless). Local knowledges concerning the uses of various processes and things, the appreciation of the qualities of local fauna and flora (indigenous resource knowledges), of changing meteorological and climatic conditions, of soil types, of natural hazards, the construction of symbolic meanings and the development of capacities to represent and 'read' the landscape and its signs effectively – these sorts of knowledges have been fundamental to human survival since time immemorial. The nature of such knowledges vary greatly, depending upon technologies, social forms, beliefs and cultural practices all of which instantiate a certain view of the relationship of human life to life and nature in general.

The question of how peoples do and should understand the relationship to environment and nature forms the fourth pillar to all forms of geographical knowledge. But, as with the other structural supports, the issue is not unique to Geography but has a wide-ranging presence across all manner of other institutional sites. Thinking about it within Geography has been strongly influenced by these external institutional needs.

In the bourgeois era, for example, the creation of the world market meant 'the exploration of the earth in all directions' in order to discover 'new useful qualities of things' and the promotion of 'universal exchange of the products of all alien climates and lands' (Marx 1973: 409). The world was consequently understood as a spatially diversified bundle of 'natural' resources waiting to be discovered, exploited and transformed into systems of production of various sorts. Commercial geography reflected this trend. Working in the tradition of natural philosophy but with commercial endeavors omnipresent as a backdrop to their work, geographers such as Alexander von Humboldt set out to construct a systematic description of the earth's surface as a repository of use values, as the dynamic field within which the natural processes that could be harnessed for human action had their being. The accurate description of physical and biotic environments, of climate, soil and water regimes, of resource complexes and possibilities, largely for utilitarian purposes, has remained central to geographical endeavors ever since. This kind of geography was always profoundly materialist but often crassly and a-historically so.

Close observation of geographical variations in ways of life, forms of economy and social reproduction has also been integral to the geographer's practice ever since merchant capitalism came to regard such knowledge as essential to its practices. This tradition degenerated (particularly in the commercial geography of the late nineteenth century) into the mere compilation of 'human resources' open to profitable exploitation through unequal or forced exchange, the imposition of wage labor systems, the redistribution of labor supplies through forced migration (for example indentured labor), and the sophisticated manipulation of indigenous economies and political power structures to extract surpluses. Geographical knowledges were deeply affected by imperial and colonial practices coupled with the exploration of commercial opportunities and markets. The objectification and exploitation of nature under capitalism went hand in hand with the objectification and exploitation of peoples. Many forms of geographical knowledge were complicitous with that politics.

The dialectics of socio-environmental change can, however, take many twists and turns and be understood from a variety of perspectives. The long history of environmental determinism, a doctrine that periodically returns in a variety of guises (as, for example, in the recent work of economists like Jeffrey Sachs and a wide range of other popular authors such as David Landes and Jared Diamond), provides one angle of thought which runs counter to the triumphalist humanism that underlies so-called 'possibilist' doctrines of economic development and change. The resurrection of environmentalist discourses, even within the confines of a major institution like the World Bank (where the issue of 'Is geography destiny?' has been seriously debated in recent years) poses interesting challenges since this is a style of thinking that has long ago been suppressed or abandoned within Geography as a discipline. It would not take much to resurrect the argument and invoke Geography's historical experience with it as evidence in the debate.

The more favored posture within Geography concerns anthropogenic influences in 'changing the face of the earth' (to use a favored title), recognizing the extensive role played by human settlement and action upon everything from the morphology of landscape, habitat transformation to climate change. Instead of seeing humanity as a mere 'object' of evolutionary forces, the trend has been to see ourselves as 'subjects' actively transforming the environments in which we live with all manner of intended and unintended consequences (for ourselves as well as for biotic and physical environments).

We should be prepared to think about this issue in a much more dialectical mode, treating the subject–object distinction as arbitrary, and understanding that in changing the world we change ourselves and that we cannot change ourselves and our society without changing our environmental condition, sometimes in dramatic and radical ways. Social and political projects are always ecological and environmental projects. The fluidity of that idea is constrained, however, by the ways in which relatively permanent features crystallize out to act as barriers to further change. Capitalism, for example, creates a relatively fixed physical and social environment to match its needs at a certain moment in history only to have to face the stressful task of overthrowing those environmental conditions (for example patterns of resource extraction, transport networks and city forms) at a subsequent point in order to create space for further capital accumulation. Conversely, environmental transformations (whether arrived at through human action or occurring by virtue of the dynamic forces always at work within the environment in general) limit socioeconomic transformations (for example nuclear power stations, once constructed, require a certain kind of science and organization in order to be managed over the time-horizon of their existence just as environmental hazards require massive organizational forms if their destructive consequences are to be avoided).

The environmental issue (like that of space and region) gets much more interesting when it is recognized as a dynamic process and when it is treated as a dialectical rather than purely analytic problem. Geographers have already contributed much on an issue that pervades thinking and practices in a wide array of other institutional settings.

Geography among the disciplines

The four structural elements to be found within all forms of geographical knowledge collectively form structural supports for a unified methodological field of activity to be called 'Geography'. A number of points can be made about the positionality of this field among the disciplines.

Work within this field is not confined to the discipline of Geography. A scholar in literary theory studying, say, the works of Wordsworth, might examine his poetry against a cartographic background of the city-country divide, might pay minute attention to the conceptualizations of space and time that symbolize a distinctive way of life, thought and personal subjectivity, might pay close attention to environmental qualities and the portrayal of the relation to nature, and, finally, might examine the way in which the poetry helped to produce the idea of 'the Lakes' as a distinctive region eliding into the creation of a tourist industry (based on Wordsworth's 'Tour Guide' writings) which in turn helped produce a distinctive regionality on the ground.

It is possible to imagine palaeo-ecologists, geomorphologists, sedimentologists, economic geographers, cultural historians and rural sociologists all taking somewhat similar steps in their research design. 'Thinking like a geographer' then entails an understanding in each one of these operations of how the four structural pillars of geographical knowledges can be worked and woven together in specific instances and settings to produce profounder insights into socio-ecological conditions and processes of change. There are some deep commonalities and unities in how seemingly-disparate geographical knowledges are structured and it is surely worthwhile examining more carefully how such structures work.

But what such an examination requires is an approach to 'thinking like a geographer' that is deeply at odds with some of the traditional concerns voiced within the discipline. The problem for Geography as a discipline has been its search for an 'essence' and for an exclusively defined 'nature' which sets it clearly apart from all other disciplines within the social and natural sciences. Taking essentialist definitions of other subject-matters like biology and economics as given, the best that Geography can do is to claim some ' hybrid' status, to hold itself up as some model of higher-order synthesis (a hope that seems futile) or to set itself apart by indulging in 'exceptionalist' claims. The latter can be based on the peculiarities of thinking that derive from deep contemplation of regions and space relations, paying particular attention to the seeming recalcitrance of geographical information in the face of general theory (*ergo* the idea that general laws and universal statements are impossible in geography).

But there is an entirely different mode of thought that avoids essentialist definitions and meanings and which seems far more appropriate to our existing circumstances. Analogical reasoning seeks connections and interrelations, pushes forward metaphors and underlying unities within seemingly disparate phenomena, seeks analogies to illuminate phenomena in one area by examination of another. Above all, it seeks translations between different modes of thought (often emanating from quite different institutions). It is profoundly open and avoids all the turf-wars and exclusions that typify a world dominated by essentialist and purist categories. The moment in the history of geography that was peculiarly fertile in this regard was that led by the collaboration of Chorley and Haggett to produce collective works like 'Models in Geography'. At the heart of that enterprise lay analogical reason opposed to the essentialist definitions earlier sought in, say, Hartshorne's *The Nature of Geography*. What is so impressive about the current situation is the widespread occurrence of analogical reason. Spatial themes, for example, permeate literary and social theory. Of course, there are all sorts of dangers which attach to the wilder use of analogies, metaphors and translations. The organic analogy for the nation state in the work of Ratzel connected to Nazi expansionism understood as a quasi-Darwinian struggle for 'living space' for the nation. Some of the spatial and cartographic metaphors deployed in literary theory today are wildly inappropriate. Part of our scholarly job is to place such transfers of thought and feeling on reasonably solid ground.

But now seems the moment when geographers are superbly placed to be a central guiding-force within the networks of knowledge being created by widespread appeal to analogical reason throughout all spheres of academic activity. But for geographers to take advantage of this position-ality, it is necessary to abandon essentialist attitudes (the negative effects of which are all too plain to see in other spheres of knowledge like multiculturalism, nationalisms, or gender studies). There is, I insist, no 'nature' of geography to be found. The search for such an essence is pro-foundly misplaced if not counterproductive (particularly when individuals or groups believe they have found it). But 'thinking like a geographer' is everywhere. Learning to think 'soundly' and 'properly' as a geographer is a profoundly important attribute in today's world. This is where the unified methodological field of geography is to be found at work. As the example of Kant's cosmopolitanism and its murky tradition all too easily shows, not knowing 'how to think properly like a geographer', how to weave together the four structural pillars of geographical knowledges into a system of geographic wisdom, has long-lasting negative effects upon the collective prospects for emancipatory socio–ecological change.

Political projects

Even the most objectivist and neutral-sounding scientist will acknowledge that the broad context of scientific activity and learning has a great deal to do with human emancipation from want and need, that the improve-ment of human understanding is a necessary condition for the betterment of society (whether it be in material or non-material ways). The claim of objectivity and neutrality is always a circumscribed claim (pertaining to certain limited and carefully defined aspects of the overall learning enterprise).

The supposed neutrality of geographical knowledges has at best proven to be a beguiling fiction and at worst a downright fraud. Geographical knowledges have always internalized strong ideological content. In their scientific (and predominantly positivist) forms, natural and social phe-nomena are represented objectively as things, subject to manipulation, management and exploitation by dominant forces of capital and the state. In their more artistic, humanist and aesthetic incarnations, geographical knowledges project and articulate individual and collective hopes and fears while purporting to depict material conditions and social relations with the historical veracity they deserve. Although it aspires to universal understanding of the diversity of life on earth, Geography has often cultivated parochialist and ethnocentric perspectives on that diversity. It has often been, and still is, captive to special interests and, hence, a formidable, though often covert, weapon in political and social struggle. It has been an active vehicle for the transmission of doctrines of racial, cul-tural, sexual, or national superiority. Cold war rhetoric, fears of 'orientalism' or of some demonic 'other' that threatens the existing order have been pervasive and persuasive in relation to political action. Geographical information can be presented in such a way as to prey upon fears and feed hostility (the abuse of cartography is of particular note in this regard). The 'facts' of geography presented as 'facts of nature' have been used to justify imperialism, neocolonialism, expansionism and geopolitical strategies for dominance.

Many forms of geographical knowledge have been tainted by virtue of their connection to the instrumental ends for which they were designed and the institutional frameworks to which they were beholden. But this is not to say they are useless, irrelevant or too contaminated to be touched (any more than we might dismiss the uses of specific technologies because they were invented for purposes of military domination and destruction). The problem, as much within Geography as without, is to take these varied forms of knowledge, appreciate the circumstances of their origin, evaluate them for what they are, and, if possible, transform them or translate them (with the aid of analogical reason) into different codes where they might perform quite different functions.

Geographical knowledges can be mobilized to humanistic ends. Concerns for the unwise use of natural and human resources, environmental degradations, and inefficient or unjust spatial distributions (of population, industry, transport facilities, ecological complexes, and so on) have led many to consider the question of the 'rational' configurations of geographical distributions and forms. This aspect of geographical practice, which emerged with the early geological, soil and land–use surveys, has increased markedly in the past fifty years as the state has been forced to intervene more actively in human affairs. Even the neoliberal state has continued such practices, though often with different ends in view. Positive knowledge of actual distributions (the collection, coding and presentation of information) and normative theories of location and optimization have proved useful in environmental management and urban and regional planning. These techniques entailed acceptance of a distinctively capitalist definition of rationality, connected to the accumulation of capital and social control. But such a mode of thought also opened up the possibility for planning the efficient utilization of environments and space according to alternative and multiple definitions of rationality.

Geographical knowledges have the largely unrealised *potentiality* to express hopes and aspirations as well as fears, to seek universal understandings based on mutual respect and concern, and to articulate firmer bases for human cooperation in a world marked by strong geographical differences. The construction of geographical knowledges in the spirit of liberty and respect for others, as, for example, in the remarkable work of

Reclus, opens up the possibility for the creation of alternative forms of geographical practice, tied to principles of mutual respect and advantage rather than to the politics of exploitation. Geographical knowledges can become vehicles to express utopian visions and practical plans for the creation of alternative geographies. They can infuse cosmopolitan projects, founded on ideas of justice, tolerance and reason, with geographical understandings that do not automatically negate such worthy universal claims. They can be a vehicle to articulate the legitimate and frequently conflictual aspirations of diverse populations and so become embedded in alternative politics, whether it be through the NGOs or political parties and social movements. They can provide effective means to mobilize knowledge of the world for those emancipatory ends to which all learning and all science has traditionally aspired.

Geographical knowledges occupy a central position in all forms of political action and struggle. They are all the more powerful for being considered so obvious and so banal as to be unworthy of explicit consideration, let alone careful scrutiny. The counter-error to the geographical ignorance of which Nussbaum for one complains, is to insist that we should know everything about everywhere, that we each and every one of us become a walking gazeteer. The impossibility of that leads quickly to the conclusion that there is no solution to the problem other than that which already exists. But a critical geography seeks an alternative path. It seeks out the principles and mechanisms of geographical knowledge production and strives to understand how geographical knowledges are constituted and put to use in political action. It uses this understanding to question how and when different forms of geographical knowledge get deployed in what kinds of political action. It recognizes, in short, the dynamic connections between political powers and geographical knowledges of different sorts. By understanding how the devil so often lies in the geographical details, it offers a means better to counter dominant powers (much as, for example, Greenpeace challenges the World Bank by offering an entirely different geographical interpretation of what, say, the insertion of a large dam in a particular environment really means). But beyond that, a critical geography also recognizes that emancipatory politics depends crucially upon the ability to articulate geographical alternatives in both theory and practice. Geography as we now know it was the bastard child of Enlightenment thought. It either remained hidden or, as with Kant, became the dark side of what the Enlightenment was supposed to be about. It is time to bring it actively into the light of day, legitimize it and recapture its emancipatory possibilities. That is, surely, the strongest of the 'strong ideas' that a critical geography can articulate at this difficult moment in our history.

PART 2

THE CAPITALIST PRODUCTION
OF SPACE

CHAPTER 12

The geography of capitalist accumulation: a reconstruction of the Marxian theory

First published in Antipode, *1975.*

The spatial dimension to Marx's theory of accumulation under the capitalist mode of production has for too long been ignored, This is, in part, Marx's fault since his writings on the matter are fragmentary and often only sketchily developed. But careful scrutiny of his works reveals that Marx recognized that capital accumulation took place in a geographical context and that it in turn created specific kinds of geographical structures. Marx further develops a novel approach to location theory (in which dynamics are at the center of things) and shows that it is possible to connect, theoretically, the general processes of economic growth with an explicit understanding of an emergent structure of spatial relationships. And it further transpires that this locational analysis provides, in albeit a limited form, a crucial link between Marx's theory of accumulation and the Marxian theory of imperialism – a link which many have sought but none have so far found with any certainty, in part, I shall argue, because the mediating factor of Marx's location theory has been overlooked.

In this paper I shall try to demonstrate how the theory of accumulation relates to an understanding of spatial structure and how the particular form of locational analysis which Marx creates provides the missing link between the theory of accumulation and the theory of imperialism.

The theory of accumulation

Marx's theory of growth under capitalism places accumulation of capital at the center of things. Accumulation is the engine which powers growth under the capitalist mode of production. The capitalist system is therefore highly dynamic and inevitably expansionary; it forms a permanently revolutionary force which continuously and constantly reshapes the world we live in. A stationary state of simple reproduction is, for Marx, logically incompatible with the perpetuation of the capitalist mode of production. 'The historical mission of the bourgeoisie,' is expressed in the formula 'accumulation for accumulation's sake, production for production's sake' (1967, vol. 1: 595). Yet this historical mission does not stem from the inherent greed of the capitalist; it arises, rather, out of forces entirely independent of the capitalist's individual will:

> Only as personified capital is the capitalist respectable. As such, he shares with the miser the passion for wealth as wealth. But that which in the miser is mere idiosyncrasy, is, in the capitalist, the effect of the social mechanism, of which he is but one of the wheels. Moreover, the development of capitalist production makes it constantly necessary to keep increasing the amount of capital laid out in a given industrial undertaking, and competition makes the immanent laws of capitalist production to be felt by each individual capitalist, as external coercive laws. It compels him to keep constantly extending his capital, in order to preserve it, but extend it he cannot, except by means of progressive accumulation.
>
> (Marx 1967, vol. 1: 592)

Economic growth under capitalism is, as Marx usually dubs it, a process of internal contradictions which frequently erupt as crises. Harmonious or balanced growth under capitalism is, in Marx's view, purely accidental because of the spontaneous and chaotic nature of commodity production under competitive capitalism (1967, vol. 2: 495). Marx's analyses of this system of commodity production led him to the view that there were innumerable possibilities for crises to occur as well as certain tendencies inherent within capitalism which were bound to produce serious stresses within the accumulation process. We can understand these stresses more easily if we recognize that the progress of accumulation depends upon and presupposes:

1. The existence of a surplus of labor – an industrial reserve army which can feed the expansion of production. Mechanisms must therefore exist to increase the supply of labor power by, for example, stimulating population growth, generating migration streams, drawing 'latent elements' – labor power employed in non-capitalist situations, women and children, and the like – into the workforce, or by creating unemployment by the application of labor-saving innovations.

2. The existence in the marketplace of requisite quantities of, or opportunities to obtain, means of production – machines, raw materials, physical infrastructures, and the like – to permit the expansion of production as capital is reinvested.

3. The existence of a market to absorb the increasing quantities of commodities produced. If uses cannot be found for goods or if an effective demand (need backed by ability to pay) does not exist, then the conditions for capitalist accumulation disappear.

In each of these respects the progress of accumulation may encounter a serious barrier which, once reached, will likely precipitate a crisis of some sort. Since, in well-developed capitalist economies, the supply of labor power, the supply of means of production and of necessary infrastructures, and the structure of demand are all 'produced' under the capitalist mode of production, Marx concludes that capitalism tends actively to produce some of the barriers to its own development. This means that crises are endemic to the capitalist accumulation process.

Crises can be manifest in a variety of ways, however, depending on the conditions of circulation and production at the time. We can see more clearly how this can be so by examining, briefly, how Marx looks at production, distribution, consumption and reinvestment as separate phases (or 'moments') within the totality of the capitalist production process. He argues, for example, that:

> not only is production immediately consumption and consumption immediately production, not only is production a means for consumption and consumption the aim of production . . . but also, each of them . . . creates the other in completing itself and creates itself as the other.
>
> (Marx 1973: 93)

If production and consumption are necessarily dialectically integrated with each other within production as a totality, then it follows that the crises which arise from structural barriers to accumulation can be manifest in each and any of the phases in the circulation and production of value.

Consider, for example, a typical realization crisis which arises because accumulation for accumulation's sake means, inevitably, the 'tendency to produce without regard to the limits of the market' (Marx 1969b: 522). Capitalists constantly tend to expand the mass and total value of commodities on the market at the same time as they try to maximize their profits by keeping wages down which restricts the purchasing power of the masses (Marx 1969b: 492; 1967, vol. 3: 484). There is a contradiction here which periodically produces a realization crisis – a mass of commodities on the market with no purchasers in sight. This overproduction is relative only, of course, and it has nothing to do with absolute human needs – 'it is only concerned with demand backed by ability to pay' (Marx

1969b: 506). Absolute overproduction in relation to all human wants and needs is, in Marx's view, impossible under capitalism.

But such relative overproduction may appear also as underconsumption or as an overproduction of capital (a capital surplus). Marx regards these forms as manifestations of the same basic overaccumulation problem (Marx 1969b: 497–9). The fact that there is a surfeit of capital relative to opportunities to employ that capital means that there has been an over-production of capital (in the form of an overproduction of commodities) at a preceding stage and that capitalists are overinvesting and underconsuming the surplus at the present stage. In all of these cases, overproduction:

> is specifically conditioned by the general law of the production of capital: to produce to the limit set by the productive forces, that is to say, to exploit the maximum amount of labour with a given amount of capital, without any consideration for the actual limits of the market or the needs backed by ability to pay.
>
> (Marx 1969b: 534–5)

This same general law produces, periodically, a:

> plethora of capital [which] arises from the same causes as those which call forth a relative overpopulation, and is, therefore, a phenomenon supple-menting the latter, although they stand at opposite poles – unemployed capital at one pole, and unemployed worker population at the other.
>
> (Marx 1967, vol. 3: 251)

The various manifestations of crisis in the capitalist system – chronic unemployment and underemployment, capital surpluses and lack of investment opportunities, falling rates of profit, lack of effective demand in the market, and so on – can therefore be traced back to the basic tendency to overaccumulate. Since there are no other equilibriating forces at work within the competitive anarchy of the capitalist economic system, crises have an important function – they enforce some kind of order and rationality onto capitalist economic development. This is not to say that crises are themselves orderly or logical – they merely create the conditions which force some kind of arbitrary rationalization of the capitalist pro-duction system. This rationalization extracts a social cost and has its tragic human consequences in the form of bankruptcies, financial collapse, forced devaluation of capital assets and personal savings, inflation, increasing concentration of economic and political power in a few hands, falling real wages, and unemployment. Forced periodic corrections to the course of capital accumulation can all too easily get out of hand, however, and spawn class struggles, revolutionary movements and the chaos which typically provides the breeding ground for fascism. The social reaction to crises can affect the way in which the crisis is resolved so that there is no necessary unique outcome to this forced rationalization process. All that has to happen is that appropriate conditions for renewed accumulation have to be created if the capitalist system is to be sustained.

Periodic crises must in general have the effect of expanding the productive capacity and renewing the conditions of further accumulation. We can conceive of each crisis as shifting the accumulation process onto a new and higher plane. This 'new plane' will likely exhibit certain combined characteristics of the following sorts:

1. The productivity of labor will be much enhanced by the employment of more sophisticated machinery and equipment while older fixed capital equipment will, during the course of the crisis, have become much cheaper through a forced devaluation.
2. The cost of labor will be much reduced because of the widespread unemployment during the crisis and, consequently, a larger surplus can be gained for further accumulation.
3. The surplus capital which lacked opportunities for investment in the crisis will be drawn into new and high profit lines of production.
4. An expanding effective demand for product – at first in the capital goods industry but subsequently in final consumption – will easily clear the market of all goods produced.

It is, perhaps, useful to pick up on the last element and consider how a new plane of effective demand, which can increase the capacity to absorb products, can be constructed. Analysis suggests that it can be constructed out of a complex mix of four overlapping elements:

1. The penetration of capital into new spheres of activity by (1) organizing pre-existing forms of activity along capitalist lines (e.g., that transformation of peasant subsistence agriculture into corporate farming), or by (2) expanding the points of interchange within the system of production and diversifying the division of labor (new specialist businesses emerge to take care of some aspect of production which was once all carried on within the same factory or firm).
2. Creating new social wants and needs, developing entirely new product-lines (automobiles and electronic goods are excellent twentieth-century examples) and organizing consumption so that it becomes 'rational' with respect to the accumulation process (working-class

demands for good housing may, for example, be coopted into a public-housing program which serves to stabilize the economy and expand the demand for construction products of a certain sort).

3. Facilitating and encouraging the expansion of population at a rate consistent with long-run accumulation (this obviously is not a short-run solution but there appears to be a strong justification for Marx's comment (Marx 1969b: 47; and see Marx 1973: 764, 771) that 'an increasing population appears as the basis of accumulation as a continuous process' from the standpoint of expanding the labor supply and the market for products).

4. Expanding geographically into new regions, increasing foreign trade, exporting capital and in general expanding towards the creation of what Marx called 'the world market'.

In each of these respects, or by some combination of them, capitalism can create fresh room for accumulation. The first three items can be viewed really as a matter of *intensification* of social activity, of markets, of people within a particular spatial structure. The last item brings us, of course, to the question of spatial organization and geographical expansion as a necessary product of the accumulation process. In what follows we shall consider this last aspect in isolation from the others. But it is crucial to realize that in practice various trade-offs exist between intensification and spatial extension – a rapid rate of population growth and the easy creation of new social wants and needs within a country may render capital export and an expansion of foreign trade unnecessary for the expansion of accumulation. The more difficult intensification becomes, the more important geographical extension is for sustaining capital accumulation. Bearing this in mind, we will proceed to examine the way in which the theory of accumulation relates to the production of spatial structures.

Transportation relations, spatial integration and the 'annihilation of space by time'

We will start from the proposition that the 'circulation of capital realizes value while living labour creates value' (Marx 1973: 543). Circulation has two aspects; the actual physical movement of commodities from point of production to point of consumption and the actual or implicit costs that attach to the time taken up and to the social mediations (the chain of wholesalers, retailers, banking operations, and the like) which are necessary in order for the produced commodity to find its ultimate user. Marx regards the former as integral to the production process and therefore productive of value (Marx 1967, vol. 2: 150; Marx 1973: 533–4). The latter are regarded as necessary costs of circulation which are not, however, productive of value – they are to be regarded, therefore, as necessary deductions out of surplus, because the capitalist has to pay for them.

The transportation and communications industry which 'sells change in location' (Marx 1967, vol. 2: 52) is directly productive of value because 'economically considered, the spatial condition, the bringing of the product to market, belongs to the production process itself. The product is really finished only when it is on the market' (Marx 1973: 533–4). However, the means of transportation and communication, because they are made up almost entirely of fixed capital, have their own peculiar laws of realization (Marx 1973: 523), laws which stem from the fact that transportation is simultaneously produced and consumed at the moment of its use. Although the transport industry is *potentially* a source of surplus value, there are good reasons for capital not to engage in its production except under certain favorable circumstances. The state is often, therefore, very active in this sphere of production (Marx 1973: 531–3).

The cost of transportation 'is important insofar as the expansion of the market and the exchangeability of the product are connected with it' (Marx 1973: 534). Prices, both of raw materials and finished goods, are sensitive to the costs of transportation and the ability to draw in raw materials over long distances and to dispatch the finished product to a distant market is obviously affected by these costs. The costs of circulation 'can be reduced by improved, cheaper and more rapid transportation' (Marx 1967, vol. 2: 142). One by-product of this is a cheapening of many elements constant capital (raw material inputs) and the extension of the geographical market. Viewed from the standpoint of production as a totality, 'the reduction of the costs of real circulation [in space] belongs to the development of the forces of production by capital' (Marx 1973: 533–4).

Placed in the context of accumulation in general, improvements in transportation and communication are seen to be inevitable and necessary. 'The revolution in the modes of production of industry and agriculture made necessary a revolution . . . in the means of communication and transport' so that they 'became gradually adapted to the modes of production of mechanical industry, by the creation of a system of river steamers, railways, ocean steamers and telegraphs' (Marx 1967, vol. 1: 384). The imperative to accumulate consequently implies the imperative to overcome spatial barriers:

The more production comes to rest on exchange value, hence on exchange, the more important do the physical conditions of exchange – the means of communication and transport – become for the costs of circulation. Capital by its nature drives beyond every spatial barrier. Thus the creation of the physical conditions of exchange . . . becomes an extraordinary necessity for it.

(Marx 1973: 524)

The capitalist mode of production promotes the production of cheap and rapid forms of communication and transportation in order that 'the direct product can be realized in distant markets in mass quantities' at the same time as new 'spheres of realization for labour, driven by capital' can be opened up. The reduction in realization and circulation costs helps to create, therefore, fresh room for capital accumulation. Put the other way around, capital accumulation is bound to be geographically expansionary and to be so by progressive reductions in the costs of communication and transportation.

The opening-up of more distant markets, new sources of raw materials and of new opportunities for the employment of labor under the social relations of capitalism, has the effect, however, of increasing the turnover time of capital unless there are compensating improvements in the speed of circulation. The turnover time of a given capital is equal to the production time plus the circulation time (Marx 1967, vol. 2: 248). The longer the turnover time of a given capital, the smaller is its annual yield of surplus value. More distant markets tie capital up in the circulation process for longer time periods and therefore have the effect of *reducing* the realization of surplus value for a particular capital. By the same token, any reduction in circulation time increases surplus production and enhances the accumulation process. Speeding up 'the velocity of circulation of capital' contributes to the accumulation process. Under these conditions 'even spatial distance reduces itself to time: the important thing is not the market's distance in space, but the speed . . . with which it can be reached' (Marx 1973: 538). There is thus a strong incentive to reduce the circulation time to a minimum for to do so is to minimize 'the wandering period' of commodities (Marx 1967, vol. 2: 249). A dual need, both to reduce the cost and the time involved in movement, thus emanates from the imperative to accumulate:

> While capital must on one side strive to tear down every spatial barrier to intercourse, i.e., to exchange, and conquer the whole earth for its market, it strives on the other side to annihilate this space with time . . . The more developed the capital . . . the more does it strive simultaneously for an even greater extension of the market and for greater annihilation of space by time.

(Marx 1973: 539)

Long-distance trade, because it separates production and realization by a long time interval, may still be characterized by a long turnover period and a lack of continuity in the employment of capital. This kind of trade, and 'overseas commerce in general' thus forms 'one of the material bases, . . . one of the sources of the credit system' (Marx 1967, vol. 2: 251–2). In the *Grundrisse* (1973: 535) Marx develops this argument at greater length:

> It is clear . . . that circulation appears as an essential process of capital. The production process cannot be begun anew before the transformation of the commodity into money. The *constant continuity* of this process, the unobstructed and fluid transition of value from one form into the other, or from one phase of the process into the next, appears as a fundamental condition for production based on capital to a much greater degree than for all earlier forms of production. [But] while the necessity of this continuity is given, its phases are separate in time and space . . . It thus appears as a matter of chance . . . whether or not its essential condition, the continuity of the different processes which constitute its process as a whole is actually brought about. The suspension of this chance element by capital itself is *credit*.

The credit system allows of a geographical extension of the market by establishing continuity where there was none before. The necessity to annihilate space by time can in part be compensated for by an emerging system of credit.

The need to minimize circulation costs as well as turnover times promotes agglomeration of production within a few large urban centers which become, in effect, the workshops of capitalist production (Marx 1967, vol. 1: 352; Marx 1973: 587). The 'annihilation of space by time' is here accomplished by a 'rational' location of activities with respect to each other so as to minimize the costs of movement of intermediate products in particular. 'Along with this concentration of masses of men and capital thus accelerated at certain points, there is the concentration of these masses of capital in the hands of the few' (Marx 1967, vol. 2: 250). The ability to economize on circulation costs depends, however, on the nature of the transportation relations established and here there appears to be a dynamic tendency towards concentration. Improvements in the means of transportation tend:

> in the direction of the already existing market, that is to say, towards the great centres of production and population, towards ports of export, etc. . . . These particularly great traffic facilities and the resultant acceleration 'of the capital turnover . . . give rise to quicker concentration of both the centres of production and the markets.
>
> (Marx 1967, vol. 2: 250)

This tendency towards agglomeration in large urban centers may be diminished or enhanced by special circumstances. On the one hand we find that 'the territorial division of labour . . . confines special branches of production to special districts of a country' (Marx 1967, vol. 1: 353). On the other hand, 'all branches of production which by the nature of their product are dependent mainly on local consumption, such as breweries, are . . . developed to the greatest extent in the principle centers of population' (Marx 1967, vol. 2: 251).

The geographical rationalization of the processes of production is in part dependent upon the changing structure of transport facilities, the raw material and marketing demands of the industry and the inherent tendency towards agglomeration and concentration on the part of capital itself. The latter required a technological innovation to sustain it, however. Hence the importance of the steam engine which 'permitted production to be concentrated in towns' and which 'was of universal application, and, relatively speaking, little affected in its choice of residence by local circumstances' (Marx 1967, vol. 1: 378).

Innovations of this sort, which, relatively speaking, free production from local power sources and which permit the concentration of production in large urban agglomerations accomplish the same purpose as those transport innovations which serve to annihilate space with time. Geographical expansion and geographical concentration are both to be regarded as the product of the same striving to create new opportunities for capital accumulation. In general, it appears that the imperative to accumulate produces concentration of production and of capital at the same time as it creates an expansion of the market for realization. As a consequence, 'flows in space' increase remarkably, while the 'market expands spatially, and the periphery in relation to the centre . . . is circumscribed by a constantly expanding radius' (Marx 1972: 288). Some sort of centre-periphery relation is bound to arise out of the tension between concentration and geographical expansion. We will examine certain aspects of this relation further in the section on foreign trade.

Since the structure of transport facilities does not remain constant, we find 'a shifting and relocation of places of production and of markets as a result of the changes in their relative positions caused by the transformation in transport facilities' (Marx 1967, vol. 2: 250). These transformations alter 'the relative distances of places of production from the larger markets' and consequently bring about 'the deterioration of old and the rise of new centres of production' (Marx 1967, vol. 2: 249).

The emergence of a distinct spatial structure with the rise of capitalism is not a contradiction-free process. In order to overcome spatial barriers and to 'annihilate space with time', spatial structures are created which themselves ultimately act as a barrier to further accumulation. These spatial structures are expressed, of course, in the fixed and immovable form of transport facilities, plant, and other means of production and consumption which cannot be moved without being destroyed. Once the mode of production of capital is brought into being, it 'establishes its residence on the land itself and the seemingly solid presuppositions given by nature themselves [appear] in landed property as merely posited by industry' (Marx 1973: 740). Capital thus comes to represent itself in the form of a physical landscape created in its own image, created as use values to enhance the progressive accumulation of capital on an expanding scale. The geographical landscape which fixed and immobile capital comprises is both a crowning glory of past capital development and a prison which inhibits the further progress of accumulation because the very building of this landscape is antithetical to the 'tearing down of spatial barriers' and ultimately even to the 'annihilation of space by time.'

This contradiction is characteristic of the growing dependency of capitalism on fixed capital of all kinds. With 'fixed capital the value is imprisoned within a specific use value' (Marx 1973: 728) while the degree of fixity increases with durability, other things being equal (Marx 1967, vol. 2: 160). The necessary increase in the use of fixed capital of the immobile sort which the imperative to accumulate implies imposes a further imperative:

> The value of fixed capital is reproduced only insofar as it is used up in the production process. Through disuse it loses its value without its value passing on to the product. Hence the greater the scale on which fixed capital develops . . . the more does the continuity of the production process or the constant flow of reproduction become an externally compelling condition for the mode of production founded on capital.
>
> (Marx 1973: 703)

Capitalist development has to negotiate a knife-edge path between preserving the values of past capital investments in the built environment and destroying these investments in order to open up fresh room for accumulation (for a specific example of this, see Harvey 1975b). As a consequence we can expect to witness a perpetual struggle in which capitalism builds a physical landscape appropriate to its own condition at a particular moment in time, only to have to destroy it, usually in the course of a crisis, at a subsequent point in time. Temporal crises in fixed capital investment, often expressed as 'long-waves' in economic development (see, for example, Kuznets, 1961; Thomas, 1973) are therefore usually expressed as periodic reshapings of the geographic environment to adapt it to the needs of further accumulation.

This contradiction has a further dimension. In part the drive to overcome spatial barriers and to annihilate space with time is designed to counteract what Marx saw as a pervasive tendency under capitalism for the profit rate to fall. The creation of built environments in the service of capitalism means 'a growth of that portion of social wealth which, instead of serving as direct means of production, is invested in means of transportation and communication and in the fixed and circulating capital required for their operation' (Marx 1967, vol. 2: 251). Investment in the means of transportation is bound to increase the organic composition of social capital which tends to generate a fall in the rate of profit at the same time as its effects are supposed to increase the rate of profit. Again, capitalist development has to negotiate a knife-edge between these two contradictory tendencies.

The location theory in Marx is not much more specific than this (although there is much in the analysis of fixed and immovable capital investment which is of interest but which space precludes from considering here). The virtue of these fragmentary analyses lies not in their sophistication. It lies, rather, in the way in which they can be tightly integrated into the fundamental insights into the production of value and the dynamics of accumulation. In this, the Marxian approach is fundamentally different to that typical of bourgeois economic analysis of locational phenomena. The latter typically specifies an optimal configuration under a specific set of conditions and presents a partial static equilibrium analysis. Dynamics are considered at the end of the analysis, usually as an afterthought, and the dynamics never get much beyond comparative statics. Consequently, it is generally acknowledged that bourgeois location theory has failed to develop a satisfactory dynamic representation of itself. The Marxian theory, on the other hand, commences with the dynamics of accumulation and seeks to derive out of this analysis certain necessities with respect to geographical structures. The landscape which capitalism creates is also seen as the locus of contradiction and tension, rather than as an expression of harmonious equilibrium. And crises in fixed capital investments are seen as synonymous in many respects with the dialectical transformation of geographical space. The contrast between the two theoretical stances is important. It suggests that the two theories are really concerned with quite different things. Bourgeois locational analysis is only adequate as an expression of optimal configurations under set conditions. The Marxian theory teaches us how to relate, theoretically, accumulation and the transformation of spatial structures and ultimately, of course, it provides us with the kind of theoretical and material understanding which will allow us to understand the reciprocal relationships between geography and history.

Foreign trade

Marx considers foreign trade from two rather different standpoints: first, as an attribute of the capitalist mode of production, and second, as a historical phenomenon relating an evolving capitalist social formation with precapitalist societies and generating various intermediate social forms (such as colonies, plantation economies, dependent economies, and the like).

Marx invariably abstracts from questions of foreign trade in his analysis of the capitalist mode of production (Marx 1967, vol. 1: 581). He concedes, of course, that 'capitalist production does not exist at all without foreign commerce' but suggests that consideration of the latter merely serves to 'confuse without contributing any new element of the problem [of accumulation], or of its solution' (Marx 1967, vol. 2: 470). He also accepts that foreign trade may counteract the tendency to a falling rate of profit because it cheapens the elements of constant capital as well as necessities and so permits a rising surplus value to be appropriated. But since this raises the rate of accumulation, it merely hastens the fall in the rate or profit in the long run (Marx 1967, vol. 3: 237). The increase in foreign trade, which inevitably arises with the expansion of accumulation, merely 'transfers the contradictions to a wider sphere and gives them greater latitude' (Marx 1967, vol. 2: 408).

Most of Marx's comments on foreign trade relate to it as an historical phenomenon and are therefore peripheral to his main purpose in *Capital*. Foreign trade is treated as a precondition for capitalist accumulation as well as a consequence of the expansion of the market. Since consequences at one stage become preconditions at the next, the development of foreign trade and capitalist social formations are seen as integrally related. 'Special factors' also arise in relation to foreign trade which can confuse, conceal and distort matters. The significance of such factors to actual historical situations is not denied – they are just not regarded as crucial for understanding the inner logic of the capitalist mode of production.

The theoretical and historical analyses intersect at certain points, however. Some of Marx's statements on foreign trade can be interpreted as logical extensions of his theoretical views on how the accumulation process generates transportation relations and locational structures. These views are usually projected into a pre-existing structure of nation states, territories with different natural productive capacities and non-capitalist production systems.

Marx recognizes, for example, that 'the productiveness of labour is fettered by physical conditions' (Marx 1967, vol. 1: 512). In agriculture he expects unequal returns on capital advanced to result from differences in both fertility and relative location (Marx 1967, vol. 3: 650). Natural differences form, therefore a 'physical basis for the social division of labour' (Marx 1967, vol. 1: 514), although they present possibilities only (and not unmodifiable ones at that) because in the last instance the productiveness of labour 'is a gift, not of Nature, but of a history embracing thousands of centuries' (Marx 1967, vol. 1: 512).

Capitalist production and circulation tends to transform these possibilities into an integrated geographical system of production and exchange which serves the purposes of capitalist accumulation. In the process certain countries may establish a monopoly over the production of particular commodities (Marx 1967, vol. 3: 119), while center-periphery relations will be produced on a global scale:

> A new and international division of labour, a division suited to the requirements of the chief centres of modern industry springs up, and converts one part of the globe into a chiefly agricultural field of production, for supplying the other part which remains a chiefly industrial field.
>
>   (Marx 1967, vol. 1: 451)

Capitalists in the advanced countries may also gain a higher rate of profit by selling their goods above their value in competition with 'commodities produced in other countries with inferior production facilities . . . in the same way that a manufacturer exploits a new invention before it has become general' (Marx 1967, vol. 3: 238). Relative productive advantages yield excess profits and if they are perpetuated in the form of a permanent 'technology-gap' it follows (although Marx did not apparently make the point) that technology-rich regions always have the capacity to earn higher profits within a given line of production compared to technology-poor regions.

The international credit system also has a vital role to play in creating the world market and fashioning its structure:

> The entire credit system . . . rests on the necessity of expanding and leaping over the barrier to circulation and the sphere of exchange. This appears more colossally, classically, in the relations between people than in the relations between individuals. Thus, e.g., the English [are] forced to lend to foreign nations in order to have them as customers.
>
>   (Marx 1973: 416; and see Marx 1972: 122)

Capital export – a theme which Lenin (1963: 715–19) elaborates on as crucial to the theory of imperialism as the highest stage of capitalism – can, in Marx's view, provide temporary opportunities for surplus capital. But capital export can take different forms, as we will shortly see, and be engaged in for quite different reasons.

The general drive to overcome all spatial barrier produces a variety of results in relation to non-capitalist forms of production and social organization:

> When an industrial people, producing on the foundation of capital, such as the English, e.g., exchange with the Chinese, and absorb value . . . by drawing the latter within the sphere of circulation of capital, then one sees right away that the Chinese do not therefore need to produce as capitalists.
>
> (Marx 1973: 729)

The interaction of capitalist and non-capitalist modes of production within the sphere of circulation creates strong interdependencies. The circulation of value within the capitalist system becomes dependent on the continued contribution of products and money from non-capitalist societies – 'to this extent the capitalist mode of production is conditional on modes of production lying outside of its own stage of development' (Marx 1967, vol. 2: 110). This is a theme which Luxemburg. (1968) develops at great length in her *The Accumulation of Capital* – she argues, in effect, that the fresh room for accumulation which capitalism must define can exist only in the form of precapitalist societies which provide untapped markets to absorb what is a perpetual tendency for the overproduction of commodities under capitalism. Once these societies are all brought into the capitalist network then, in her view, accumulation must cease.

Marx also argued that the historic tendency of capitalism is to destroy and absorb non-capitalist modes of production at the same time as it uses them to create fresh room for capital accumulation. Initially, the mere penetration of the money form has a disrupting influence – 'where money is not the community, it must dissolve the community' and 'draw new continents into the metabolism of circulation' (Marx 1973: 224–5). In the early stages capital is accumulated out of this 'metabolism of circulation' – indeed, such accumulation is an historical premise for the development of capitalist production. The towns accumulate use values and hence values from the countryside while merchant's capital, as an historically prior form of organization to producer's capital:

> appropriates an overwhelming portion of the surplus product partly as a mediator between commodities which still substantially produce for use

value . . . and partly because under those earlier modes of production the principle owners of the surplus product with whom the merchant dealt, namely, the slave-owner, the feudal lord, and the state (for instance, the oriental despot) represent the consuming wealth and luxury which the merchant seeks to trap . . . Merchant's capital, when it holds a position of dominance, stands everywhere for a system of robbery, so that its development among the trading nations of old and modern times is always directly connected with plundering, piracy, kidnapping, slavery and colonial conquest . . . The development of merchant's capital gives rise everywhere to the tendency towards production of exchange values . . . Commerce, therefore, has a more or less dissolving influence everywhere on the producing organization which it finds at hand and whose different forms are mainly carried on with a view to use value.

(Marx 1967, vol. 3: 331–2)

The resultant forms which emerge from such disruptions depend, however, upon the form of the pre-existing society and the extent of capitalist penetration. One effect, for example, is to create scarcities in the non-capitalist society where there were none before. Necessaries are thereby transformed into luxuries and this:

determines the whole social pattern of backward nations . . . which are associated with a world market based on capitalist production. No matter how large the surplus product, they (the non-capitalist producers) extract from the surplus labour of their slaves in the simple form of cotton or corn, they can adhere to this simple undifferentiated labour because foreign trade enables them to convert these simple products into any kind of use value.

(Marx 1972: 243)

The creation of 'underdevelopment' by means of a capitalist penetration which transforms non-capitalist societies from relatively self-sufficient organizations for the production of use-values to specialized and dependent units producing exchange values, is a theme which has been explored by contemporary writers such as Baran (1957) and Frank (1969). The latter, for example, coins the phrase 'the development of underdevelopment' to call attention to the kinds of processes that Marx had in mind.

These forms of dependency are possible only *after* capitalist production had come to dominate merchant's capital so that the latter now basically serves the purposes of the former. We then find:

the cheapness of the articles produced by machinery, and the improved means of transport and communication furnish the weapons for conquering foreign markets. By ruining handicraft production in other countries, machinery forcibly converts them into fields for the supply of its raw material. In this way, East India was compelled to produce cotton, wool, hemp, jute and indigo for Great Britain.

(Marx 1967, vol. 1: 451)

The manner of such a transformation is of interest and India provides a good example. Originally a field for 'direct exploitation' – the direct appropriation of use values – India was transformed after 1815 into a market for British textile products:

But the more the industrial interest became dependent on the Indian market, the more it felt the necessity of creating fresh productive powers in India, after having ruined her native industry. You cannot continue to inundate a country with your manufactures, unless you enable it to give you some produce in return.

(Marx and Engels 1972: 52)

Capital export in this case served a different purpose from the mere loan of money to finance imports of manufactures. Capital was exported to India to promote commodity production which could, via foreign trade, provide the wherewithal to pay for the goods which were being imported from Britain. Britain had to build up commodity production for exchanges in India if it was to maintain India as an important market.

The same sort of logic, operating under rather different conditions, applies to the development of colonies through settlement. Marx insists here on drawing a distinction:

There are the colonies proper, such as the United States, Australia, etc. Here the mass of the farming colonists, although they bring with them a larger or smaller amount of capital from the motherland, are not *capitalists*, nor do they carry on *capitalist* production. They are more or less peasants who work themselves and whose main object, in the first place, is to produce their own livelihood ... In the second type of colonies – plantations – where commercial speculations figure from the start and production is intended for the world market, the capitalist mode of production exists, although only in a formal sense, since the slavery of Negroes precludes free wage labour, which is the basis of capitalist production. But the business in which slaves are used is conducted by capitalists.

(Marx 1969b: 302–3)

Colonies of the latter sort hold out the prospect for high profits because of the higher rates of exploitation, the lower price of necessaries and, usually, higher natural productivity. Capital may move into such colonies and in the process reduce the excess profit there, but in the process the average rate of profit will rise (Marx 1969b: 436–7). There exists here a positive inducement to the export of capital:

> If capital is sent abroad, this is not done because it absolutely could not be applied at home, but because it can be employed at a higher rate of profit in a foreign country.
>
> (Marx 1967, vol. 3: 256)

With complete mobility, of course, the profit rate will ultimately be equalized although at a higher average rate than before. But colonies of this second sort are still advantageous because they permit the importation of cheap raw materials on the basis of a higher rate of exploitation (which presumes, by the way, certain immobilities to labor power, such as that imposed by slavery).

Colonies of the first sort exist in a very different relation to the capitalist mode of production, however:

> There the capitalist regime everywhere comes into collision with the resistance of the producer, who, as owner of his own conditions of labour, employs that labour to enrich himself, instead of the capitalist. The contradiction of these two diametrically opposed economic systems, manifests itself here practically in a struggle between them. Where the capitalist has at his back the power of the mother-country, he tries to clear out of his way by force, the modes of production and appropriation, based on the independent labour of the producer.
>
> (Marx 1967, vol. 1: 765)

Colonies made up of small independent producers, trading some surplus into the market, are typically characterized by labor shortages and a high wage rate which is not attractive to the capitalist form of exploitation (this is particularly the case where there is an abundance of free land for settlement). Commodity production does not exist in the complete capitalist sense. Colonial forms of this sort may be, therefore, just as resistant to the penetration of the capitalist mode of production as traditional more long-established non–capitalist societies. But since such non–capitalist colonies are created by spin–offs of surplus population and small quantities of capital from the centres of accumulation, and since they also form markets for capitalist production, they are to be viewed as both the result of past accumulation and a precondition for further capital accumulation. The United States prior to the Civil War, for example, provided an important, largely non-capitalist market for the realization of commodities produced under capitalist social relations in Britain.

The final state of capitalist penetration is that which comes with the organization of production along capitalist lines. In 1867 Marx noted how the US was being transformed from an independent, largely non-capitalist, production system into a new centre for capital accumulation. 'Capitalistic production advances there with giant strides, even though the lowering of wages and the dependence of the wage worker are yet far from being down to the European level' (Marx 1967, vol. 1: 773). Marx expected a similar transformation in India:

> When you have once introduced machinery into the locomotion of a country, which possesses iron and coals, you are unable to withhold it from its fabrication. You cannot maintain a net of railways over an immense country without introducing all those industrial processes necessary to meet the immediate and current wants of railway locomotion, and out of which there must grow the application of machinery to those branches of industry not immediately connected with railways. The railway system will therefore become, in India, truly the forerunner of modern industry... (which) will dissolve the hereditary divisions of labour, upon which rests the Indian castes, those decisive impediments to Indian progress and Indian power... The bourgeois period of history has to create the material basis of the new world... Bourgeois industry and commerce create these material conditions of a new world in the same way that geological revolutions have created the surface of the earth.
>
> (Marx and Engels 1972: 85–7)

Such a transformation did not occur in India but it did in the US. The failure to predict correctly in the Indian case has no bearing whatsoever on the validity of the Marxian theory of accumulation under the capitalist mode of production. All the theory says is that capitalism is bound to expand through both an intensification of relationships in the centres of capitalist production and a geographical extension of those relationships in space. The theory does not pretend to predict where, when and exactly how these intensifications and geographical extensions will occur – the latter are a matter for concrete historical analyses. Marx's failure to predict correctly in the case of India was a failure of historical analysis, not of theory.

But it so happens that there are also good *theoretical* reasons for believing that the capitalist *production* system could not and cannot become universal in its scope. For this to be the case would require the equalization of profits, through competition, on a global scale. To begin with, of course, there are all kinds of barriers to be overcome before such an equalization in profit rates could occur. We would have to presume the complete mobility of capital and labour (Marx 1967, vol. 3: 196) and adequate institutional arrangements (free trade, universal money and credit system, 'the abolition of all laws preventing the labourers from transferring from one sphere to production to another and from one locality to another,' and so on). Under capitalism, there are always tendencies pushing in these directions. For example:

> it is only foreign trade, the development of the market to a world market, which causes money to develop into world money and abstract labour into social labour. Abstract wealth, value, money, hence abstract labour, develop in the measure that concrete labour becomes a totality of different modes of labour embracing the world market. Capitalist production rests on the value or the transformation of the labour embodied in the product into social labour. But this is only possible on the basis of foreign trade and the world market. This is at once the pre-condition and the result of capitalist production.
>
> (Marx 1972: 253)

The tendency of capitalism, therefore, is to establish a universal set of values, founded on 'abstract social labour' as defined on a global scale. There is, in like manner, a tendency for capital export to equalize the rate of profit on a global scale. An accumulation process implies a tendency for the penetration of capitalist social relations into all aspects of production and exchange throughout the world.

But different organic compositions of capital between countries, different productivities of labor according to natural differences, the different definition of 'necessities' according to natural and cultural situation, mean that these equalizations will not be accompanied by an equalization in the rate of exploitation between countries (Marx 1967, vol. 3: 150–1). It follows that 'the favoured country recovers more labour in exchange for less labour, although this difference, this excess is pocketed, as in any exchange between capital and labour, by a certain class' (Marx 1967, vol. 3: 238). Marx then notes that:

> Here the law of value undergoes essential modification. The relationship between labour days of different countries may be similar to that existing between skilled, complex labour and unskilled, simple labour within a country. In this case the richer country exploits the poorer one, even where the latter gains by the exchange.
>
> (Marx 1972: 105–6)

These are the kinds of 'special factors' which make of foreign trade a very complex issue, which generate certain peculiarities in the terms of trade between developed and undeveloped societies (Marx 1972: 474–5) and which prevent any direct 'levelling out of values by labour time and even the levelling out of cost prices by a general rate of profit' between different countries (Marx 1972: 201). These kinds of factors are picked up on by Emmanuel (1972) in his analysis of imperialism as 'unequal exchange'.

These complexities do not derive from the failure of capitalist development to overcome the social and cultural barriers to its penetration (although these barriers can be exceedingly resistant). They stem, rather, from the inherent contradictory, and hence imperfect, character of the capitalist mode of production itself. They are to be interpreted, therefore, as global manifestations of the internal contradictions of capitalism. And underlying all of these manifestations is the fact that capitalism ultimately becomes the greatest barrier to its own development. Let us consider how this is manifest on the world stage.

Capitalism can escape its own contradiction only through expanding. Expansion is simultaneously *intensification* (of social wants and needs, of population totals, and the like) and *geographical extension*. Fresh room for accumulation must exist or be created if capitalism is to survive. If the capitalist mode of production dominated in every respect, in every sphere and in all parts of the world, there would be little or no room left for further accumulation (population growth and the creation of new social wants and needs would be the only options). Long before such a situation was reached the accumulation process would slow. Stagnation would set in attended by a whole gamut of economic and social problems. Internal checks within the capitalist mode of production would begin to be felt particularly in the sphere of competition:

> As long as capital is weak, it still itself relies on the crutches of past modes of production, or of those which will pass with its rise. As soon as it feels strong, it throws away the crutches and moves in accordance with its own laws. As soon as it begins to sense itself and becomes conscious of itself as a barrier to development, it seeks refuge in forms which, by restricting free competition, seem to make the rule of capital more perfect, but are at the same time the heralds of its dissolution and of the dissolution of the mode of production resting on it.
>
> (Marx 1973: 651)

Some comments on the theory of imperialism

Marx himself never proposed a theory of imperialism. In his comments on transportation relations, location theory and foreign trade he clearly indicates, however, that he has in mind some sort of general theory of capital accumulation on an expanding and intensifying geographical scale. We have, in the preceding two sections, already sketched in some of the main features of that general theory, to the extent that Marx articulated it.

The theory of imperialism which has emerged post-Marx obviously has something to contribute towards an understanding of that general theory and therefore to an understanding of the ways in which capitalism creates fresh room for accumulation. The trouble is, however, that there is not one theory of imperialism, but a whole host of representations of the matter: Marxist, neo-Marxist, Keynesian, neoclassical and so on.And there are innumerable divergences and differences within each school (Barratt Brown 1974, provides a general overview). I shall confine myself to some general comments.

The problem for the Marxists and neo-Marxists, it is generally argued, is to derive a theory of imperialism out of Marx. And it is generally agreed that no one has yet succeeded in doing so, although many have tried. There is a fairly simple explanation for this state of affairs. Marx constructed a theory of accumulation for a capitalist mode of production in a 'pure' state without reference to any particular historical situation. On this basis, as we have seen, he demonstrates the necessity for intensification and expansion as a concomitant of accumulation. The theory of imperialism, as it is usually conceived of in the literature is, by way of contrast, a theory of history. It is to be used to explain the historical development of capitalist social formations on the world stage. It has to address the way in which conflicting forces and class interests relate to each other in specific historical situations, determine outcomes through their interactions and thereby set the preconditions for the next stage in the evolution of capitalist social formations. Marx never constructed such a historical theory, although there is some evidence that he intended to do so in unwritten books on the state, foreign trade and the world market (Marx and Engels 1955: 112–13).

Marx's theory of the capitalist mode of production plainly cannot be used as the basis for deriving a historically specific theory of imperialism in any direct manner. Yet, as we have seen in the preceding section on foreign trade, Marx's theoretical insights intersect with historical analyses at certain points. And the crucial mediating influence, which most of the writers on imperialism ignore, is the necessary tendency to overcome spatial barriers and to annihilate space with time – tendencies which Marx derives directly from the theory of accumulation. Marx's theories of transportation relations, location and geographical concentration expanding spheres of realization – in short, the general theory of accumulation on an expanding and intensifying geographical scale – in fact comprise Marx's own theory of imperialism (although he did not call it that). Since most writers ignore this general theory embedded in Marx, it would appear that this provides us with the missing link between Marx's theory of accumulation and the various theories of imperialism that have been put forward since.

But even here we cannot make direct derivations. Marx's general theory tells us of the necessity to expand and intensify geographically. But it does not tell us exactly how, when or where. Looking at the intersection of these general arguments with concrete historical analyses, we will usually be able to identify the underlying logic dictated by capital accumulation at work. But the underlying logic does not, and indeed cannot, uniquely determine outcomes. The latter have to be understood in terms of the balance of forces – economic, social, political, ideological, competitive, legal, military, and the like – through which interest groups and classes become conscious of the contradictory underlying logic and seek by their actions to 'fight it out' to some sort of resolution (compare Marx 1970: 21). To specify the relationships between the Marxian theory of accumulation and the theory of imperialism as it is usually construed poses, therefore, a double difficulty. We have to specify how the 'inner logic' of the capitalist mode of production, abstractly conceived, relates to the concrete realities, the phenomenal forms, of the historical process. And we also have to take account of the mediating influence of political, ideological, military and other structures which, although they must be generally organized so as to be coherent with the course of capital accumulation, are not uniquely determined by it.

Most analyses of imperialism usually start in fact from the analysis of actual historical situations. This is particularly true in the work of Third World writers, such as Fanon (1967), Amin (1973) and Frank (1969), whose starting point is the experience of domination and exploitation by the advanced capitalist countries. This experience is then projected into the Marxian framework for understanding exploitation in general. The consequence of this is a variety of representations of the Marxian theory of imperialism. Each representation may be accurate for its own place and time, but each ends up drawing upon just one or two facets of Marx's own theory of capital accumulation for support. By implication, and sometimes quite explicitly, it is suggested that other facets of Marx's theory of accumulation are either irrelevant or wrong.

Luxemburg (1968) is an excellent case in point. She begins her analysis with a concentrated criticism of Marx's reproduction schemes in Volume 2 of *Capital* and, reacting very strongly to the idea implied there that capitalist accumulation can continue in perpetuity, she seeks to show that Marx had failed to demonstrate where the effective demand for commodities was to come from if accumulation was to be sustained. Luxemburg's own solution is that the effective demand has to be found outside of the capitalist system in precapitalist economic formations. Imperialism is to be explained as 'the political expression of the accumulation of capital in its competitive struggle for what still remains open of the non-capitalist environment' (Luxemburg 1968: 446). As evidence, Luxemburg assembles descriptions of the violent penetration of non-capitalist societies, such as China, by capitalists in search of markets as well as descriptions of the various imperialist rivalries amongst the capitalist powers throughout the world.

Luxemburg's argument is, in many respects, both compelling and brilliant. But her analysis amounts to a one-sided development out of Marx. The objection is not that she is wrong – indeed, we have already seen that capitalist development may become contingent upon other modes of production, that the penetration and disruption of non-capitalist societies are implied by the imperative to 'tear down spatial barriers,' and 'that violence, making use of state power, can easily be resorted to. The objection is that Luxemburg sees the consequences of the imperative to accumulate *solely* in these terms. The other means whereby capitalism can create fresh room for accumulation are ignored.

Read as a theoretical treatise on what must happen if all other means for creating fresh room for accumulation are sealed off, Luxemburg's work is a brilliant exposition. Read as a documentation of how the logic of capitalist accumulation underlies the penetration and disruption of non-capitalist societies, the work is compelling. But read as a derivation of the necessity for imperialism out of a correction of Marx's errors in his specification of capitalist reproduction, Luxemburg's work is both erroneous and misconceived. To put the criticism this way is not to say, however, that the processes to which Luxemburg draws attention may not become, at a certain stage in capitalist history, vital to the perpetuation of the capitalist order. Whether or not this turns out to be the case depends however, upon the capacity of the capitalist system to create fresh room for accumulation by other means.

The representation of imperialism in the works of Baran (1957) and Frank (1969) can be considered in a similar way. Clearly implied in Marx's location theory is the emergence of a general structure of center-periphery relations in production and exchange, while the tearing-down of spatial barriers to exchange may create dependency and 'transform necessaries into luxuries' for the economy newly brought into the metabolism of exchange. These kinds of relationships are examined in detail in the work of Baran and Frank and they can relatively easily be integrated into the Marxian frame when the logic of accumulation is projected into an actual historical situation. Baran and Frank are therefore on strong theoretical grounds when they claim that backwardness and underdevelopment can and must be produced and perpetuated by the penetration of capitalist social relations into non-capitalist economies. They may also be on strong factual grounds when they claim that this is the general relationship which exists between the Third World and the metropolitan centres of accumulation. But, as with the work of Luxemburg, the analysis has to be regarded as a single-faceted development out of Marx's theory of accumulation. It would be both erroneous and misconceived to regard this development either as a correction to or a unique derivation out of Marx. Fresh room for accumulation can be created by a variety of stratagems in actual historical situations. Whether or not a different structure of relations to that explored by Baran and Frank is possible depends *not* on the theory but on the possibilities contained in actual historical situations.

Lenin's contribution to the Marxist theory of imperialism is, of course, fundamental. And in some respects it is the most interesting, both with respect to its content and its method. Lenin did not attempt to derive the theory out of Marx. He regarded the phenomena of imperialism as something to be revealed by materialist historical analysis. Specifically, he was concerned to explain the 1914–18 war as an imperialist war 'for the division of the world, for the partition and repartition of colonies and spheres of influence of finance capital, etc.' (Lenin 1963: 673). The method is therefore historical and Lenin uses the term 'imperialism' to describe the general characteristics of the phenomenal form assumed by capitalism during a particular stage of its development, specifically, during the late nineteenth and early twentieth centuries. In this he relies very heavily on the work of a non-Marxist, Hobson (1938). Yet Lenin also seeks to uncover 'the economic essence of imperialism' and to relate the understanding of the phenomenal appearance of imperialism to Marx's theoretical insights into the nature of the capitalist mode of production.

The phenomenal appearance of capitalism in the imperialist stage of its development is summarized in terms of five basic features:

(1) the concentration of production and capital has developed to such a high stage that it has created monopolies which play a decisive role in economic life; (2) the merging of bank capital with industrial capital, and the creation, on the basis of this 'finance capital,' of a financial oligarchy; (3) the export of capital as distinguished from the export of commodities acquires exceptional importance; (4) the formation of international monopolist associations which share the world among themselves, and (5) the territorial division of the whole world among the big capitalist powers is completed.

(Lenin 1963: 737)

The tendency towards concentration and centralization of capital is, in Marx's analysis, integral to the general process of accumulation (Marx 1967, vol. 1, ch. 25). The physical concentration of production to achieve economies of scale in a locational sense is also, in Marx's theory, paralleled by a growing centralization of capital. Lenin also grounds the logic of capital export in Marx's theory. He rebuts the argument that capitalism could ever achieve an equal development in all spheres of production or alleviate the misery of the mass of workers:

> If capitalism did these things it would not be capitalism; for both uneven development and a semi-starvation level of existence of the masses are fundamental and inevitable conditions and constitute the premises of this mode of production. As long as capitalism remains what it is, surplus capital will be utilized not for the purpose of raising the standard of living of the masses in a given country, for this would mean a decline in profits for the capitalists, but for the purpose of increasing profits by exporting capital abroad to the backward countries. In these backward countries profits are usually high, for capital is scarce, the price of land is relatively low, wages are low, raw materials are cheap... The export of capital influences and greatly accelerates the development of capitalism in those countries to which it is exported. While, therefore, the export of capital may tend to a certain extent to arrest development in the capital-exporting countries, it can only do so by expanding and deepening the further development of capitalism throughout the world.
>
> (Lenin 1963: 716–18)

Lenin is here emphasizing certain of the possibilities contained in the Marxian theory of capitalist accumulation when projected into an actual historical situation. Plainly, he is not excluding the development of capitalist production in new centres, although the carving-up of the world into spheres of influence with centres of accumulation and spheres of realization is regarded as a 'managed' rationalization, accomplished by finance capitalism through political manipulations, of the inevitable uneven development of capitalism. But Lenin also argues that imperialism 'can and must be defined differently if we bear in mind not only the basic, purely economic concepts... but also the historical place of this stage of capitalism in relation to capitalism in general, or the relation between imperialism and the two main trends in the working class movement' (1963: 737). Imperialism thus has the effect of 'exporting' some of the tensions created by the class struggle within the centers of accumulation to peripheral areas. The 'superprofits' of imperialist exploitation make it 'possible to bribe the labour leaders and the upper stratum of the labour aristocracy. And that is just what the capitalists of the 'advanced' countries are doing' (1963: 677). This last aspect of imperialism has to be regarded as the joint outcome of the inevitable uneven development of capitalism on a world scale and a corresponding uneven development of the class struggle. Capital becomes mobile in order to escape the consequences of a class struggle waged at a particular place and time or else it repatriates superprofits to buy off its home labour force with material advancement. In either case a geographical expansion of development must occur.

Lenin blends concrete historical analysis, based on the principles of historical materialism, with some fundamental insights from Marx's theory. An evaluation of Lenin's theory must rest, therefore, on an assessment of his historical accuracy and a critical evaluation of the way in which the Marxian theory intersects with the historical materials. On the former score there are grounds for thinking that Lenin's reliance on Hobson and Hilferding led him into some factual errors. In the latter respect, Lenin, like most other writers on imperialism, develops Marx's general theory in a one-sided rather than an all-embracing manner. As a consequence, the connection to the theory of capitalist accumulation is partially obscured from view.

The problem with the Marxist theory of imperialism in general, is that it has become a theory 'unto itself', divorced from Marx's theory of capital accumulation. As a consequence, the argument over what imperialism is, has degenerated into an argument over which of several competing principles should be used to define it. The development of overseas markets? The attainment of cheaper raw materials? The searching out of a more easily exploited and a more docile labor force? Is it primitive accumulation at the expense of non-capitalist societies? Does it involve cheating through exchange? Is it the necessary export of capital to set up new centers of industrial accumulation? Is it the concentration of relative surplus value on a localized basis? Is it the manifestation of monopoly power, expressed through the political organization of a system of nation states? Is it finance capital operating through multinational corporations and government co-optation? Is it simply the international division of labor? Is it a particular combination of any of the above? Under Marx's general theory, all of the above are possible and none are to be excluded. It is, therefore, the task of careful historical analysis to discover which of these manifestations is dominant at a particular stage of development of capitalist social formations. Marx's general theory does not pretend to predict particular forms and manifestations. All it does is to indicate the underlying imperative, contained within the capitalist system, to accumulate capital and to do it, of necessity, on an expanding and intensifying geographical scale.

This is not to say that a theoretical analysis of these various manifestations in relation to capital accumulation is impossible. Indeed, a great deal can be done here. And we can also place one bet. The survival of capitalism is predicated on the continued ability to accumulate, *by whatever means is easiest*. The path of capitalist accumulation will move *to wherever the resistance is weakest*. It is the task of historical and theoretical analyses to identify these points of least resistance, of greatest weakness. Lenin once advised all revolutionary movements to look for the weakest link in capitalism. Ironically, capitalism manages, by trial and error and persistent pressure, to discover the weakest links in the forces opposed to continued accumulation and by exploiting those links to open up fresh pasture for the bourgeoisie to accomplish its historical mission: the accumulation of capital.

Marx's theory of capital accumulation on an expanding geographical scale as a whole

Marx's theory of capital accumulation on an expanding geographical scale is quite complex. We have delved into Marx to try to discover in his writings some of its basic components. But to be appreciated properly these components have to be seen in relation both one to each other and to the various models which Marx devised to understand capitalist production, exchange and realization as a totality. In a rather splendid passage in the *Grundrisse* (1973: 407–10), Marx provides a kind of 'overview sketch' of his general theory:

> The creation by capital of *absolute surplus value* ... is conditional upon an expansion, specifically a constant expansion, of the sphere of circulation ... A precondition of production based on capital is therefore *the production of a constantly widening sphere of circulation*. Hence, just as capital has the tendency on one side to create ever more surplus labour, so it has the complementary tendency to create more points of exchange.

From this, of course, we can derive 'the tendency to create the world market [which] is directly given in the concept of capital itself' and the need, initially at least, 'to subjugate every moment of production itself to exchange and to suspend the production of direct use values not entering into exchange'. Marx then goes on to say that:

> the production of *relative surplus value* ... requires the production of new consumption; requires that the consuming circle within circulation expands as did the productive circle previously. Firstly quantitative expansion of existing consumption; secondly, creation of new needs by propagating

existing ones in a wide circle; thirdly, production of *new* needs and discovery and creation of new use values.

As a result of these expansionary tendencies, capitalism creates:

> a system of general exploitation of the natural and human qualities . . . Hence the great civilizing influence of capital; its production of a stage of society in comparison to which all earlier ones appear as mere *local developments* of humanity and as *nature idolatory*. For the first time, nature becomes purely an object for humankind, purely a matter of utility. . . In accord with this tendency, capital drives beyond national barriers and prejudices as much as beyond nature worship, as well as all traditional, confined, encrusted satisfactions of present needs, and reproductions of old ways of life. It is destructive towards all of this, and constantly revolutionizes it, tearing down all the barriers which hem in the development of the forces of production, the expansion of needs, the all-sided development of production, and the exploitation and exchange of natural and mental forces . . .
>
> But . . . since every such barrier contradicts its character, its production moves in contradictions which are constantly overcome but just as constantly posited. Furthermore, the universality towards which it irresistably strives encounters barriers in its own nature, which will, at a certain stage of its development, allow it to be recognized as being itself the greatest barrier to this tendency, and hence will drive towards its own suspension.

Marx's sketch does not incorporate all of the elements which we have identified in this paper but it does convey a feeling for what he had in mind in constructing a theory of accumulation on an expanding geographical scale. Plainly, the drive to accumulate lies at the center of the theory. This drive is expressed primarily in the production process through the creation of absolute and relative surplus value. But the creation of value is contingent upon the ability to realize it through circulation. Failure to realize value means, quite simply, the negation of the value created potentially in production. Thus, if the sphere of circulation does not expand then accumulation comes to a halt. Capital, Marx never tires of emphasizing, is not a thing or a set of institutions; it is a process of circulation between production and realization. This process, which must expand, must accumulate, constantly reshapes the work process and the social relationships within production as it constantly changes the dimensions and forms of circulation. Marx helps us to understand these processes theoretically. But ultimately we have to bring this theory to bear on existing situations within the structure of capitalist social relations at this point in history. We have to force an intersection between the theoretical abstractions, on the one hand, and the materialist investigations of actual historical configurations on the other. To construct and reconstruct Marx's theory of accumulation on an expanding geographical scale as a totality requires such an intersection. We have indeed to derive the theory of imperialism out of the Marxian theory of accumulation. But to do so we have to move carefully through the intermediate steps. In Marx's own thought it appears that the crucial intermediate steps encompass a theory of location and an analysis of fixed and immobile investment; the necessary creation of a geographical landscape to facilitate accumulation through production and circulation. But the steps from the theory of accumulation to the theory of imperialism, or more generally to a theory of history, are not simple mechanical derivations because down this path we have to accomplish also that transformation from the general to the concrete which comprised the central thrust of Marx's unfinished work. We have to learn, in short, to complete the project which Marx underscores at the beginning of Volume 3 of *Capital*: we have to bring a synthetic understanding of the processes of production and circulation under capitalism to bear on capitalist history and 'thus approach step by step the form which they assume on the surface of society'.

CHAPTER 13

The Marxian theory of the state

First published in Antipode, *1976.*

Introductory remarks

Larry Wolf's paper (1976) raises a variety of questions about the role of the state in relation to capitalist economic development. Some of the questions are practical and concern exactly how and in what ways we can anticipate the intervention of the state in the American economy over the next few years. As in the 1930s, another time of economic troubles, the possibility of centralized national economic planning is being actively considered (together with a more brutal return to 'pure market forces') as a means to rationalize an economic order that has obviously become unbalanced and, perhaps, perilously close – how close we will probably never know – to being totally unhinged. Quite properly, Wolf sees the move towards national economic planning as creating new opportunities as well as new problems for the radical Left. Quite properly too, he argues that the manner in which the move is made will have an effect upon the outcomes. But the issue is perhaps more complex than that. Given the present power structure, I am not as sanguine about even the potential outcomes as he is. I feel I am watching a rerun of a tired movie of the 1930s, with shades of the 1890s, as goals such as 'social justice' and 'conservation' are gradually converted into goals of efficiency and market rationality tinged with not a little socialism for the rich, financial support for shaky corporations and financial institutions, and the like. In each of these two preceding eras, a whiff of national economic policy-making was quickly combined with the drive to rationalize the market system to create the very problems it was designed to get rid of on a higher plane and in more concentrated form in the long run.

Some of the questions which Wolf raises are theoretical, however, and concern the formulation of an appropriate conceptual framework for thinking about state interventionism in general. In the course of these remarks, Wolf takes a few shies at 'dogmatic Marxists' and those who would reduce the state to a 'mere superstructural' form, to a mere manifestation of 'the economic basis'. While these views are not unknown among Marxists, I have the distinct impression that they are frequently figments of bourgeois scholarship, designed to discourage people from trying to understand Marx in all his complexity. Thus we find Marx frequently portrayed as depicting men and women as dominated by rational economic calculation when it was exactly Marx's point that it is the capitalist mode of production which forces such rationality upon us *against* all of the evidence as to what human beings are really all about. We find Marx portrayed as an economic determinist when it was precisely Marx's point that the realm of freedom begins where the realm of necessity ends and that it is only through struggle, political and personal, that we can achieve the command over our social and physical existence which will yield us that freedom. And so it is with Marx's analysis of the state. The essay that follows (which is drawn from a book that seems to take an interminable time to finish) attempts to sort out some of the issues concerning the conception of the state in capitalist society. The essay is rather abstract in nature and for this I apologise, particularly to those who prefer immediate 'down-to-earth' analyses or crushing exposées. But I believe that the practical questions to which Wolf alludes can be understood only against some adequate conceptual and theoretical background. Further, the theory has to be robust enough to help us understand the behaviour of the state under a wide variety of economic, social and political circumstances – in other words, the theory has to help us in Spain, France, Britain, Sweden, Argentina, Chile, Portugal and so on, as well as in the United States. For this reason, it is necessary to resort to a rather abstract mode of analysis and to let concrete investigations take up the matter of how the theory works in actual historical situations. Obviously, the theory remains a mere abstraction until it is put to work. All I can say is that the theoretical statement which follows has been helpful to me in my studies of the urbanization process in Britain and the US and that I have also found it helpful as a means to think about the prospects for state action in the present state of capitalist development. I offer the piece in the hope that others may similarly find it useful and as a partial rebuttal and partial commentary on Wolf's remarks on the Marxist theory of the state in general.

The Marxian theory of state

Marx intended to write a special treatise on the state but never even began the project. His views on the state are scattered throughout his works and, with the help of Engels's more voluminous writings, it is possible to reconstruct, as, for example, Chang (1931) has done, a version of the Marxian theory of the state. Apart from Lenin's (1949 edition) fierce advocacy of what might be called an 'orthodox' Marxist position and Gramsci's perceptive analyses (1971), few Marxists paid attention to the matter until recently, when works by Miliband (1969), Poulantzas (1973; 1975; 1976), Offe (1973), Altvater (1973), O'Connor (1973), Laclau (1975) and others, put the question of the state back into the forefront of Marxist analysis. These contributions have recently been reviewed by Gold, Lo & Wright (1975). This revival of interest in the state has been long overdue. There is scarcely any aspect of production and consumption which is not now deeply affected, directly or indirectly, by state policies. But it would be incorrect to maintain that the state has only recently become a central pivot to the functioning of capitalist society. It has always been there – only its forms and modes of functioning have changed as capitalism has matured. In this essay, I will try to lay a theoretical basis for understanding the role of the state in capitalist societies and show how the state must, of necessity, perform certain basic minimum tasks in support of a capitalist mode of production.

Most of Marx's early writings on the state are specifically directed towards a refutation of Hegel's philosophical idealism by the construction of a materialist interpretation of the state as 'the active, conscious and official expression [of] the present structure of society' (Marx and Engels 1974, vol. 3 (1975): 199). This materialist interpretation of the state broadens somewhat in *The German Ideology* (Marx and Engels 1970: 53–4) to a general conception in which the state is regarded as 'an independent form' which emerges out of 'a contradiction between the interest of the individual and that of the community'. This contradiction is 'always based' in the social structure and in particular 'on the classes, already determined by the division of labour... and of which one dominates all others.' From this it follows 'that all struggles within the state... are merely the illusory forms in which the real struggles of the different classes are fought out among one another.' Engels summarized this view of the state many years later in an oft-quoted passage (which Lenin regarded as fundamental to Marxist orthodoxy):

> The state is therefore by no means a power imposed on society from without; just as little is it 'the reality of the moral idea,' 'the image and the reality of reason,' as Hegel maintains. Rather, it is a product of society at a particular stage of development; it is the admission that this society has involved itself in unsoluble self-contradiction and is cleft into irreconcilable antagonisms which it is powerless to exorcise. But in order that these antagonisms, classes with conflicting economic interests, shall not consume themselves and society in fruitless struggles, a power, apparently standing above society, has become necessary to moderate the conflict and keep it

within the bounds of 'order'; and this power, arising out of society, but placing itself above it and increasingly alienating itself from it, is the state.

(Engels 1941: 155)

The contradiction between particular and community interests give rise, of necessity, to the state. But precisely because the state must assume an 'independent' existence in order to guarantee the communal interest, it becomes the locus of an 'alien power' by means of which individuals and groups can be dominated (Marx and Engels 1970: 54). In the same way that the laborer, through work, creates capital as an instrument for his or her own domination, so human beings create in the form of the state an instrument for their own domination (compare Ollman 1971: 216). These various instruments of domination – in particular the law, the power to tax and the power to coerce – can be transformed by political struggle into instruments for class domination. Engels summarizes Marx's view succinctly:

> As the state arose from the need to keep class antagonisms in check, but also arose in the thick of the fight between the classes, it is normally the state of the most powerful, economically ruling class, which by its means becomes also the politically ruling class, and so acquires new means of holding down and exploiting the oppressed classes. The ancient state was, above all, the state of the slaveowners for holding down the slaves, just as the feudal state was the organ of the nobility for holding down the peasant serfs and bondsmen, and the modern representative state is the instrument for exploiting wage-labour by capital. Exceptional periods, however, occur when the warring classes are so nearly equal in forces that the state power, as apparent mediator, acquires for the moment a certain independence in relation to both.
>
> (Engels 1941: 157)

The use of the state as an instrument of class domination creates a further contradiction: the ruling class has to exercise its power in its own class interest at the same time as it maintains that its actions are for the good of all (Marx and Engels 1970: 106). This contradiction can, in part, be resolved by the employment of two strategies. First, those charged with expressing the ruling will and the institutions through which that will is expressed, must *appear* to be independent and autonomous in their functioning. The officials of the state therefore have to 'present themselves as organs of society standing *above* society... Representatives of a power which estranges them from society, they have to be given prestige by means of special decrees, which invest them with a peculiar sanctity and inviolability.' Consequently, even 'the lowest police officer' has an 'authority' which other members of society do not possess. Vesting state officials with such 'independent authority' poses a further problem. We have to explain how state power can have all the appearances of autonomy vis-à-vis the dominant classes at the same time as it expresses the unity of class power of those classes (cf. Poulantzas 1973: 281). The question of the 'relative autonomy' of the state has consequently been a matter of intense debate among Marxists.

A second strategy for resolving the contradiction builds upon the connection between ideology and the state. Specifically class interests can be transformed into 'the illusory general interest' provided that the ruling class can successfully universalize its ideas as the 'ruling ideas'. That this will likely be the case results from the very process of class domination:

> Each new class which puts itself in the place of one ruling before it, is compelled, merely in order to carry through its aim, to represent its interests as the common interest of all the members of society... it has to give its ideas the form of universality, and represent them as the only rational, universally valid ones. The class making a revolution appears from the very start... not as a class but as the representative of the whole of society.
>
> (Marx and Engels 1970: 65–6)

Marx and Engels in general held that the ruling class:

> rule also as thinkers, as producers of ideas, and regulate the production and distribution of the ideas of their age: thus their ideas are the ruling ideas of the epoch.
>
> (Marx and Engels 1970: 65)

But if these ruling ideas are to gain acceptance as representing the 'common interest' they have to be presented as abstract idealizations, as universal truths for all time. Consequently, these ideas have to be presented as if they have an autonomous existence of their own. Notions of 'justice', 'right', 'freedom' are presented as if they have a meaning independent of any particular class interest. The relationship between the ruling ideas and the ruling class is rendered opaque by a separation and an idealization which, in turn, has the potential to create a further contradiction. Once morality is universalized as 'absolute truth', for example, it is possible for the state, and even the whole mode of production, to be judged immoral (see Marx and Engels 1974, vol. 3 (1975): 108). By the same token, if the state can be represented as an abstract idealization of the common interest, then the state can itself become an abstract incarnation of a 'moral' principle (nationalism, patriotism, fascism, all appeal to this to some degree). The connections between the formation of a dominant ideology, the definition of the 'illusory common interest' in the form of the state and the very specific interests of the ruling class or classes are as subtle as they are complex. Yet, until recently and with the notable exception of Gramsci's quite profound insights, the real relationships have remained as opaque to analysis as they are in daily life. We can reveal the basis of these relationships most easily, however, by analyzing the relationship between the state and the functioning of a capitalist mode of production.

The theory of the state in relation to the theory of the capitalist mode of production

The famous Marxist dictum that 'the executive of the modern State is but a committee for managing the common affairs of the whole bourgeoisie' (Marx and Engels 1952: 44) was in fact meant as a polemical response to the widespread illusory claim that the state expressed the common interests of all. But it is hardly satisfactory as a basis for understanding the real relations between the state and capitalism. We can begin to build such a basic understanding by showing how the state must of necessity fulfill certain basic functions if capitalism is to be reproduced as an ongoing system.

The social relations of exchange and exchange value which lie at the heart of the capitalist mode of production presuppose:

1. the concept of a 'juridical person' or 'individual' (Marx 1973: 243–6), stripped of all ties of personal dependence (such as those characteristic of slavery or the feudal era) and each and all apparently 'free' to 'collide with one another and to engage in exchange within this freedom' (Marx 1973: 163–4);
2. a system of property rights which ensures that individuals can gain command over use values only through ownership or exchange;
3. a common standard of value in exchange (the objectification of which is money) so that only the exchange of equivalents is involved which means that individuals approach each other in the market place essentially as equal as far as the measure of exchange is concerned (Marx 1973: 241). Money is, in short, the great leveller.
4. a condition of reciprocal dependence in exchange (as opposed to personal dependence) which results from the fact that 'each individual's production is dependent on the production . . . and consumption of all others' (Marx 1973: 156 and 242–5). The condition of 'free individuality and equality' are therefore 'socially determined' – they can be achieved 'only within the conditions laid down by

society and with the means provided by society; hence (they are) bound to the reproduction of these conditions and means' (Marx 1973: 156). From this arises the separation of private interests from social necessities, the latter appearing as an 'alien power' (the state) over the individual.

Marx derives a fundamental insight from these propositions:

> Equality and freedom are thus not only respected in exchange based on exchange values but, also, the exchange of exchange values is the productive real basis for all equality and freedom. As pure ideas they are merely the idealized expression of this basis; as developed in juridical, political, social relations, they are merely this basis to a higher power.
>
> (Marx 1973: 245)

The exchange relations embedded in the capitalist mode of production therefore give rise to specific notions concerning 'the individual', 'freedom', 'equality', 'rights', 'justice', and the like. Marx observed that such concepts typically provide the ideological rallying cries of all bourgeois revolutions and he was a consistent critic of those who sought to formulate a revolutionary working-class politics in terms of 'eternal justice' and 'equal rights' since these were concepts reflective of bourgeois social relations of exchange (see, for example, *Critique of the Gotha Programme* (Marx 1938)). Concepts of this sort are more than mere ideological tools, however. They connect to the state by becoming embedded formally in the system of bourgeois law. The capitalist state must, of necessity, support and enforce a system of law which embodies concepts of property, the individual, equality, freedom and right which correspond to the social relations of exchange under capitalism.

The basic paradox which Marx seeks to unravel in *Capital* is how a system of exchange of commodities based in freedom and equality can give rise to a result characterized by 'inequality and unfreedom' (Marx 1973: 249; Marx 1967, vol. 1: ch. 5 and 684). The explanation lies, of course, in the class character of the capitalist relations of production which arose out of a long historical process in which labor power became divorced from control over the means of production which then became the exclusive preserve of the capitalist class. Once created, these relations of production and accumulation must necessarily be fostered, supported and enforced by the use of state power. Private property rights over the commodities being exchanged must be guaranteed so that 'no one seizes hold of another's property by force' and so that 'each divests himself from his property voluntarily' (Marx 1973: 243). Labor power is a commodity which means that it is also a form of private property over which the laborer has exclusive rights of disposal. Money provides the vehicle for accumulation; it permits the individual to carry 'his social power, as well as his bond with society, in his pocket'. (Marx 1973: 157). Capital is nothing more, of course, than money put back into production and circulation to yield more money. If money is to represent real values, the same kind of state regulation of money supply and credit is called for. Also, if the profit rate is to be equalized then both capital and labor must be highly mobile, which means that the state must actively remove barriers to mobility when necessary. In general, the state, and the system of law in particular, has a crucial role to play in sustaining and guaranteeing the stability of these basic relationships. The guarantee of private property rights in means of production and labor power, the enforcement of contracts, the protection of the mechanisms for accumulation, the elimination of barriers to mobility of capital and labor and the stabilization of the money system (via central banking, for example), all fall within the field of action of the state. In all of these respects the capitalist state becomes 'the form of organization which the bourgeois necessarily adopt for internal and external purposes, for the mutual guarantee of their property and interests' (Marx and Engels 1970: 80). The capitalist state cannot be anything other than an instrument of class domination because it is organized to sustain the basic relation between capital and labor. If it were otherwise, then capitalism could not for long be sustained. And because capital is fundamentally antagonistic to labor, Marx regards the bourgeois state as *necessarily* the vehicle by means of which the collective violence of the bourgeois class is visited upon labor. The corollary is, of course, that the bourgeois state must be destroyed if a classless society is to be achieved.

Capitalist production and exchange are inherently 'anarchistic'. Individuals, each in pursuit of his or her private interests, cannot possibly take 'the common interest', even of the capitalist class, into account in their actions. Thus, the capitalist state has also to function as a vehicle through which the class interests of the capitalists are expressed in all fields of production, circulation and exchange. It plays an important role in regulating competition, in regulating the exploitation of labor (through, for example, legislation on minimum wages and maximum hours of employment) and generally in placing a floor under the processes of capitalist exploitation and accumulation. The state must also play an important role in providing 'public goods' and social and physical infrastructures which are necessary prerequisites for capitalist production and exchange but which no individual capitalist would find it possible to provide at a profit. And the state inevitably becomes involved in crisis management and in countering the tendency for the rate of profit to fall. State intervention is necessary in all of these respects because a system based on individual self-interest and competition cannot otherwise express a collective class interest.

We can take this kind of analysis one step further. In the Marxian theory of distribution, the surplus acquired through capitalist production is split into industrial profit, interest to finance capital, and rent to landlords. The homogeneity within the capitalist class breaks down into fractions of capital which are potentially in conflict with each other. Other fragmentations – between merchant capital and industrial capital, for example – can arise out of the divisions of function within the capitalist system. These fragmentations lead to conflicts of interest within the capitalist class as a whole. Factional struggles which from time to time may become highly destructive are therefore to be expected within the capitalist class. The state here plays the role of an arbiter among these conflicting interests. The state need not be neutral in these conflicts because it may be taken over by a fraction of capital under certain circumstances.

We have so far shown that Marx's analysis of the capitalist mode of production can be paralleled at each step by a theoretical derivation of certain minimal state functions: the equality and freedom of exchange must be preserved, property rights must be protected and contracts enforced, mobility preserved, the 'anarchistic' and destructive aspects of capitalist competition must be regulated, and the conflicts of interest between fractions of capital must be arbitrated for the 'common good' of capital as a whole. Strictly speaking, we cannot go much further than this in deriving a theory of the capitalist state. But it is useful to consider two further general points about the state under capitalism, even though we depart from a theoretical derivation.

First, it is easy to see that a particular form of the state – what we may call bourgeois social democracy – is particularly well-equipped to meet the formal requirements of the capitalist mode of production. It embodies a strong ideological and legal defense of equality, mobility and freedom of individuals at the same time as it is highly protective of property rights and the basic relation between capital and labor. A capitalist market exchange economy characteristically thrives on a double-edged freedom which includes freedom of conscience, speech and employment at the same time as it incorporates freedom to exploit, to gain private profit at public expense and to monopolize the means of production. The committment of bourgeois democracy to freedom is in fact a committment to all of these different kinds of freedom simultaneously (compare Polanyi 1968: 74). Under bourgeois democracy too, the separation between private interests and communal needs as represented by the state is typically accomplished by a separation between economic and political power. Private property rights form the basis of economic power but under universal suffrage the privileges of private property are replaced by one-person-one vote which forms the immediate basis of political power. Under these conditions, the relationships between class interests economically conceived, and the state as a political entity are rendered peculiarly opaque which, of course, is advantageous because it is then much easier for the state to maintain the appearance of a neutral arbiter amongst all interests. Under these conditions also, wealth has to employ its power indirectly. Engels argued that:

> It does this in two ways: by plain corruption of officials, of which America is the classic example, and by an alliance between the government and the stock exchange.
>
> (Engels 1941: 1570)

The mechanisms for class domination of the bourgeois democratic state are, as Gramsci (1971) and Miliband (1969) point out, somewhat more pervasive and subtle than this. Also, the fragmentation of the state itself into separate institutions – Miliband (1969: 50) lists, for example, the government, the administrative bureaucracy, the military police, the judicial branch, subcentral government and parliamentary assemblies – make it particularly difficult for any one fraction of capital to gain complete control of all of the instruments of class domination (although the existence of a standing army and police force opens the way to military dictatorship). The formal separation of powers between executive, legislature and judiciary written into the American constitution for example, was specifically designed as a system of checks and balances to prevent the concentration of political power in the hands of any one subgroup. Such a structure ensures that the state can act as an effective arbiter between the various fractional interests within the capitalist class (in this respect, the theory of political pluralism catches one aspect of the truth about bourgeois political structures).

A consideration of the relations between economic and political power lead us to a second point which Gramsci has done much to elucidate. The ruling class has to exercise its hegemony over the state through a political system which it can control only indirectly. In the context of bourgeois democracy, this has certain important consequences. In order to preserve its hegemony in the political sphere, the ruling class may make concessions which are not in its own immediate economic interest. Gramsci argues, however, that 'there is also no doubt that such sacrifices and such compromise cannot touch the essential.' He thus arrives at the following basic conception:

The dominant group is coordinated concretely with the general interests of the subordinate groups, and the life of the State is conceived of as a continuous process of formation and superseding of unstable equilibria (on the juridical plane) between the interests of the fundamental group and those of the subordinate groups – equilibria in which the interests of the dominant group prevail, but only up to a certain point, i.e. stopping short of narrowly corporate economic interests.

(Gramsci 1971: 182)

Bourgeois democracy can survive only with the consent of the majority of the governed while it must at the same time express a distinctive ruling-class interest. This contradiction can be resolved only if the state becomes actively involved in gaining the consent of the subordinate classes. Ideology provides one important channel and state power is consequently used to influence education and to control directly or indirectly, the flow of ideas and information. The relationship between the ideology of the capitalist class and that of administrators and bureaucrats also becomes of great significance (Miliband 1969). More importantly, the state may internalize within itself political mechanisms which reflect the class struggle between capital and labor. Therefore, a key function is to organize and deliver certain benefits and guarantees to labor (minimum living standards and work conditions for example) which may not be, strictly speaking, in the immediate economic interest of the capitalist class. In return, the state receives the general allegiance of the subordinate classes. And, we may note parenthetically, state power can then be used to control the organization of consumption which can be advantageous to the capitalist class in the long run because it stabilizes the market and accumulation. Policies which simultaneously support the dominant ideology and provide material benefits are doubly appropriate of course. We can understand state policies towards working-class homeownership, for example, as simultaneously ideological (the principle of private property rights gains widespread support) and economic (minimum standards of shelter are provided and a new market for capitalist production is opened up).

Under these conditions, the relationships between the state and the class struggle become somewhat ambiguous; it is certainly inappropriate, therefore, to regard the capitalist state as nothing more than a vast capitalist conspiracy for the exploitation of workers. Further, as Gramsci (1971: 182) points out, 'international relations intertwine with these internal relations of nation–states, creating new, unique and historically concrete combinations'. It is in this context, that the role of the state in relation to imperialism, becomes very important. In response to the organized power of labor within its borders, a particular nation–state may seek to export the worst elements of capitalist exploitation through imperialist domination of other countries. Imperialist domination has other functions also – facilitating capital export, preserving markets, maintaining access to an industrial reserve army, and the like. By these means, a nation-state may purchase the allegiance of elements of the working class within its borders at the expense of labor in dependent countries, at the same time as it gains ideological leverage by disseminating the notions of national pride, empire and chauvinism which typically accompany imperialist policies (compare Lenin 1949).

Strictly speaking, these last observations apply to an understanding of the actual history of the state, and of bourgeois social democracy in particular, in the context of capitalist social formations. But theoretical and concrete analyses have to be integrated at some point and the relation between exchange and production under capitalism and the general characteristics of the political system we call bourgeois democracy seems an excellent point to begin upon such an integration. The advantage of a purely theoretical approach to the state under the capitalist mode of production is that it helps us to distinguish, as Gramsci puts it, between what is 'organic' (necessary) and what is 'conjunctural' (accidental) about the particular form assumed by the state in a particular historical situation. And there is clearly a sense in which the capitalist mode of production and bourgeois democracy are organic to each other rather than merely conjuncturally related. In their origins, at least, the relations between the two are not as mysterious as they now seem. The political theory of Locke, for example, which lies at the root of the American constitution and which provides a broad ideological basis for most modern forms of bourgeois social democracy, has a definite economic basis, as MacPherson (1962) has brilliantly demonstrated. We do not have to delve too far into Locke to see the nature of this economic basis: we find, for example, the lineaments of a labor theory of value, a definite principle that only the laborer has the right to dispose of his or her labor power, a defense of property rights accompanied by a moral imperative to use the products of labor for productive purposes and even a recognition that it is money which permits what Locke hypothesised as a 'natural state' of equality to be transformed into a morally justifiable inequality via accumulation. Marx (1969a: 365–7) regarded Locke's political theories very specifically as an ideological and political reflection of the evident needs of a nascent capitalist society. Locke:

> championed the new bourgeoisie in every way, taking the side of the industrialists against the working class and against the paupers, the merchants against the old-fashioned usurers, the financial aristocracy against the governments that were in debt, and he even demonstrated in one of his

books that the bourgeois way of thinking was the normal one for human beings.

(Marx 1972: 592)

Insofar as Locke's political theory provided the ideology for bourgeois democracy and became incorporated in the superstructural forms of the capitalist state, to that degree the bourgeois state champions exactly those same interests. While capitalism can survive under a variety of political institutional arrangements quite well, it appears that bourgeois democracy is a unique product of the economic relations presupposed in this particular mode of production.

The state in capitalist society

We have so far considered the state in abstraction, relating to the capitalist mode of production in particular. Although it is helpful to consider the state in such a manner, it is dangerous to project such understanding into concrete historical analyses uncritically. The danger lies in the tendency to posit the state as some mystical autonomous entity and to ignore the intricacies and subtleties of its involvement with other facets of society. In the *Critique of the Gotha Programme* (1938: 17–18), Marx complains bitterly of the 'riotous misuse' which the program makes of the words 'present-day state'. Marx maintains that such a conception is a mere 'fiction' because the state 'is different in the Prusso-German empire from what it is in Switzerland, it is different in England from what it is in the United States.' He does go on to point out, however, that:

> The different states of the different civilized countries, in spite of their manifold diversity of form, all have this in common, that they are based on modern bourgeois society, only one more or less capitalistically developed. They have, therefore, also certain essential features in common. In this sense it is possible to speak of the 'present-day-state,' in contrast to the future in which its present root, bourgeois society, will have died away.

It is in this last sense that we have so far been considering the state in relation to capitalism. But as we move, as Marx would put it, from the abstract and general to the concrete and particular, so we have to adapt our mode of thinking and analysis. Even theoretically it is important to recognize that:

> the state is not a thing . . . it does not, as such, exist. What 'the state' stands for is a number of particular institutions which, together, constitute its reality, and which interact as parts of what may be called the state system.

(Miliband 1969: 46)

Strictly speaking, Miliband is incorrect in this designation. The state should in fact be viewed, like capital, as a *relation* (Ollman 1971: ch. 30) or as a *process*: in this case a process of exercising power via certain institutional arrangements. It is, for example, the application and enforcement of the law which is of real material significance rather than the structure of law itself. But Miliband is quite correct when he argues that the state is much more than the exercise of power by a government and that it has to include all avenues whereby power can be exercised. In this, the particular structure of institutions is important (though not primary). And it is useful to have some way of categorizing these 'state institutions' if only to draw attention to the diverse channels through which power can be exercised: the judiciary, the executive branch of government, the administration and bureaucracy, the legislature, the military and police, and so on, form various components within this system. And the fragmentations can be taken further: central versus local governments, departmental rivalries and hierarchical structures within the bureaucracy, and the like, all have their part to play. Many of these features may be purely conjunctural, but the net effect of the fragmentation of institutions is probably to make it easier to achieve 'the formation and supersession of unstable equilibria' between fractions of capital and between the dominant and the dominated. It is hardly surprising, therefore, to find contemporary political scientists focusing attention on the processes of exchange within bureaucracies, between bureaucracies and legislatures at the same time as they find it appropriate to analyse collective action and political life in terms of market rationality.

The point to be emphasized here, of course, is that the state as we usually speak of it is an abstract category, which may be appropriate for generalizing about the collectivity of processes whereby power is exercised and for considering that collectively within the totality of a social formation. But the state is not an appropriate category for describing the actual processes whereby power is exercised. To appeal to the category 'the state' as a 'moving force' in the course of concrete historical analysis is, in short, to engage in a mystification.

The conception of the state as a superstructural form which has its basis in a particular mode of production (in this case, capitalism) is perfectly appropriate for purposes of theoretical analysis, but such a conception is singularly inappropriate when naively projected into the study of the history of actual capitalist societies. The bourgeois state did not arise as some automatic reflection of the growth of capitalist social relations. State institutions had to be painfully constructed and at each step along the way power could be, and was, exercised through them to help create the very relations which state institutions were ultimately to reflect. Marx plainly did not regard the state as a passive element in history. The instrumentalities of the state (some of which were feudal in origin) were used to great effect in the early development of capitalism. State power was used to free industrial capital from usurious interest rates (Marx 1972: 468–9), to provide many of the 'necessary pre-requisites' in the form of fixed capital in the built environment – docks, harbors, transport systems, and the like (Marx 1967, vol. 2: 233; Marx 1973: 530–3), to provide mechanisms for concentration of wealth through the mercantile form of imperialism (Marx 1967, vol. 1, ch. 31; and vol. 3, ch. 20). And state power was used indiscriminately and in many instances quite brutally to create the basic relation between capital and labor. Primitive accumulation, the initial divorce of labor from the means of production and from the land was accomplished by force or through the legalized violence of the state via, for example, the enclosure acts in England (Marx 1967, vol. 1, ch. 28). Labor laws and various forms of institutional repression forced the dispossessed labor into the workforce and helped to impose the work discipline necessary for capitalism (Marx 1967, vol. 1: 271). Even whole sectors of production were organized through the exercise of state power in the early stages of capitalist devel-opment (this was the case in nineteenth-century Germany and is epitomized by the Brazilian case in modern times).

Reading Marx, it is very difficult to imagine the birth of capitalism without the exercise of state power and the creation of state institutions which prepared the ground for the emergence of fully-fledged capitalist social relations. Yet we are so lulled by the image of an economic basis and a superstructure which merely reflects in the basis, that we tend to think of the state in a purely passive role in relation to capitalist history. The celebrated statement in *A Contribution to the Critique of Political Economy* (Marx 1970: 21) that 'changes in the economic foundation lead sooner or later to the transformation of the whole immense superstructure' appears particularly misleading if taken at its face value and applied to the state in relation to capitalist history. But even in this passage, Marx quickly counters by pointing out that it is in the 'legal, political, religious, artistic or philosophic' realms that 'men become conscious of conflict and fight it out'. The 'economic basis' and the superstructure come into being simultaneously and not sequentially – there is a dialectical interaction between them. We have been misled, too, into thinking that state interven-tionism is exclusively a phenomenon of late – some would say, decadent – capitalism. 'State capitalism' was in fact very prevalent in the early years of capitalist social formations. Once capitalism matures, of course, and once all the necessary state institutions have been created, the laws written, the interpretations of law established by precedent, then the question of the state appears to fade more into the background simply because bourgeois social relations have become one with it. Indeed, there may be a movement towards the privatization of public functions. But the movement towards *laisser-faire* has always been more ideological than real. It merely amounted to the insistence that certain functions of the market should be allowed to operate freely. It was very easy to demand 'free trade' in nineteenth-century Britain when that country was at the center of capital accumulation and possessed the industrial capacity to dominate the world market. But even at the height of *laisser-faire*, any challenge to the basic capital-labor relation was quickly met with coercion and repression as the British labor movement quickly found out in the years of Chartist agitation. It may well be, of course, that the state has changed its functions with the growth and maturing of capitalism. But the notion that capitalism ever functioned without the close and strong involvement of the state is a myth that deserves to be corrected.

The rise of capitalism was accompanied, and in some respects preceded, by the creation of, and transformation of, state institutions and functions to meet the specific needs of capitalism. The bourgeois state emerged out of a transformation of the feudal state. The forms of the feudal state varied a great deal and because they were, in effect, the raw materials out of which the bourgeois states were fashioned, they have left their mark upon contemporary state forms. There are, of course, some important exceptions. The US, Canada, Australia and New Zealand had no feudal society to overcome (although certain feudal institutions were transplanted) and these states differ quite substantially from Europe (where various forms of feudal state existed) and Latin America (where a curious hybrid form of feudal capitalism was implanted by the Spanish and Portuguese settlement). Within Europe there were substantial differences in feudal structure. The power of the peasant 'estate' in Sweden and the power of agricultural and merchant capital in England after the Dissolution gave to both of these countries a far broader base for political power than was possible in, say, Spain or Prussia. And the process of transformation itself differed markedly from place to place. The violent process of transformation in France effectively eliminated the feudal aristocracy. The slow process of transformation in England after the civil war resulted in the steady integration of aristocracy and landowners first into capitalist agriculture and later, during the nineteenth century, into the industrial power structure. In both cases, the character of the transition has placed an indelible stamp upon the subsequent quality of political life. The political differences between these countries have to be understood against the background of these quite different historical experiences and the cultural and political traditions to which they have given birth. We have also to see the institutions of the state and the relations which are expressed through these institutions as constantly in the process of being reshaped and refashioned. In certain of his historical studies, the *Eighteenth Brumaire of Louis Bonaparte* in particular, Marx provides us with examples of this process at work. We are surely obligated to understand this aspect to the state in the same manner. Yet in the midst of all of the complexities, accidental events, fluid and unstable interactions, which surround political, legal, administrative and bureaucratic life, we cannot afford to lose sight of the essential Marxian insights. Somehow or other, the capitalist state has to perform its basic functions. Should it fail to do so, then it must either be reformed or else capitalism must itself give way to some other method of organizing material production and daily life.

It is perhaps useful to conclude this discussion by posing three unresolved questions – questions which will likely be resolved as much through concrete material investigations of history as through further theoretical analysis.

1. To what degree do the various aspects and instrumentalities of state power yield to the state a relatively autonomous function in relationship to the path of capitalist development, and to what degree can state functionaries act as purely neutral or even self-serving arbiters in class and intra-class conflict? These questions have been in the forefront of much of Poulantzas's recent work.
2. To what degree can the capitalist state vary its forms and structures to give the appearance of quite substantial differentiation amongst the capitalist nations, while fulfilling the basic function of sustaining a capitalist society and ensuring the reproduction of that society? In other words, what variety of institutions is possible given the assumption of a basic underlying purpose to state action.
3. Which structures and functions within the state are 'organic' to the capitalist mode of production and therefore basic to the survival of capitalist social formations and which are, in Gramsci's phrase, purely conjunctural?

These questions are not unrelated to each other and they lie at the heart of any understanding as to how state power can be, and is, used in a society which remains basically capitalist while constantly shifting and changing its institutional forms.

CHAPTER 14

The spatial fix:
Hegel, Von Thünen and Marx

First published in Antipode, *1981.*

I have often wondered why the first volume of Marx's *Capital* ends with a chapter on 'the modern theory of colonization'. The position of such a chapter appears, at first sight, more than a little odd. It opens up the whole question of foreign and colonial trade and settlement in a work which, for the most part, theorizes about capitalism as a closed economic system.[1] Furthermore, it obscures what many would regard as a more 'natural' culmination to Marx's argument in the penultimate chapter. There Marx announces, with a grand rhetorical flourish, the death-knell of capitalist private property and the inevitable 'expropriation of a few usurpers by the mass of the people' (Marx 1967: 762–3). So why not end the volume with this stirring call to arms, so deeply reminiscent of the *Communist Manifesto*? Why append a chapter on what seems a wholly new theme?

I have likewise long been intrigued by Marx's cavalier treatment of Von Thünen. The latter, Marx concedes, asked the 'right question': 'how has the labourer been able to pass from being master of capital – as its creator – to being its slave?' But his answer is, in Marx's opinion, 'simply childish' (Marx 1967: 621). What, then, are the grounds for such an easy dismissal?

In this paper I shall show that Marx's chapter on colonization explains why he thought Von Thünen's solution was so childish. I shall also argue that both Marx and Von Thünen were responding to a challenge thrown down in Hegel's *Philosophy of Right*. Marx's treatment of colonization and Von Thünen's doctrine of the frontier wage constitute their respective answers to a problem Hegel left open: the role of geographical expansion and territorial domination, of colonialism and imperialism, in the stabilization of capitalism. Since this problem is still with us, it seems worthwhile going back to initial formulations of it.

---

[1] Marx occasionally spells out what is otherwise a tacit assumption explicitly. See, for example, *Capital* (1967, vol. 1: 581).

Hegel

Hegel's *Philosophy of Right* is a rich and extraordinary work. In a few trenchant and startling paragraphs in the midst of his exposition, Hegel lays out the lineaments of an economic theory of capitalist imperialism. We first consider how he arrived at such a conception.

The main thrust of the *Philosophy of Right* is to provide an interpretation of law, morality and various aspects of ethical life as 'the objective, institutional expressions of spirit' (Hegel 1967; Avineri 1972: 132). Hegel interprets the *family* as a sphere of ethical life dominated by particular and personal altruism. *Civil society*, on the other hand, is a sphere of 'universal egoism' in which each individual seeks to use others as a means to his or her own ends. This is, above all, the sphere of market competition, the social division of labour and 'universal interdependency' as described in political economy. The evident tension between the family and civil society – between the private and public spheres of social life – can be resolved, in Hegel's view, only through the acquisition of a universalistic consciousness on the part of all and the objective expression of that consciousness through the institutions of *the modern state*. The *rational* state, Hegel claims, can transcend the dualities of private and public life and so restore the broken unity of human existence through synthesis of the roles of 'homme' and 'citoyen' which Rousseau had envisaged as ineluctably split asunder within the complex weave of bourgeois society.[2]

Hegel proceeds, of course, in the grand manner of speculative philosophy. He begins with general abstractions arrived at ideally rather than with any detailed study of how actual social and political institutions work. His conceptual apparatus therefore has no necessary material grounding, while subsequent propositions are rigorously derived out of a dialectical logic ruthlessly applied in the best traditions of philosophical idealism. The intent of the *Philosophy of Right*, however, is to bring the abstractions closer to earth, to provide his logic with a 'political body'. The method of enquiry is, in this regard, exactly opposite to that of Marx who sought, through material historical enquiry, to expose 'the logic of the political body itself' (O'Malley 1970). The politics which Hegel derives sound very conservative because the institutions of 'the rational modern state' which he depicts sound ominously reminiscent of the Prussia of his own time. Yet a strong thread of radical critique also runs through the work and invests it with an intriguing ambiguity.

The passages in the *Philosophy of Right* that are of immediate concern to us, are those in which Hegel depicts the contradictions inherent in

---

[2] On the relationship between the thought of Rousseau and Hegel, see Pelczynski 1962.

bourgeois civil society. Although deeply affected by the writings of the British political economists – particularly Steuart and Adam Smith (Plant 1977) – Hegel rejects the idea that the 'hidden hand' of the market could marvelously harness universal egoism and greed to the benefit of all. Hegel was, after all, scarcely in a position to proclaim the virtues of a free market in a Prussia which clung tenaciously to mercantilist policies administered by a strong centralized state. To keep within the bounds circumscribed by such a politics, Hegel is forced to explain why market coordinations are defective, why they generate contradictions rather than social harmony of the sort proclaimed by Adam Smith. The main difficulty arises, Hegel claims, because labor as the active mediator between 'man and nature' is necessarily the ultimate source of all wealth – the labor theory of value is correct. But private labor is rendered social through a market system founded on universal egoism and greed, while profit necessarily entails the appropriation of the product of someone else's labor. Furthermore, the logic of profit-seeking means a compulsion towards the perpetual transformation of social needs – each seeks to create a new need in the other – and so implies perpetual expansion in both production and consumption. This dynamic produces such rampant contradictions that civil society, left to its own devices and without the interventions of the rational modern state, will surely be brought to the edge of total catastrophe. The interventionism of the state is totally justified.

But let us look a little more closely at the contradictions which build up within civil society under conditions of profit-seeking and free market exchange. Hegel concentrates on the increasing accumulation of wealth at one pole and the increasing mass of the impoverished at the other as the fulcrum of social disruption. Here is how he fashions his argument:

> When civil society is in a state of unimpeded activity, it is engaged in expanding internally in population and industry. The amassing of wealth is intensified by generalizing (a) the linkage of men by their needs, and (b) the methods of preparing and distributing the means to satisfy these needs, because it is from this double process of generalization that the largest profits are derived. That is one side of the picture. The other side is the subdivision and restriction of particular jobs. This results in the dependence and distress of the class tied to work of that sort.
>
> (Hegel 1967: 149–50)

The expansion of production therefore coincides with a decline in the standard of living of the mass of the people below 'a certain subsistence level' and their relative deprivation to the point where they cannot 'feel and enjoy the broader freedoms and especially the intellectual benefits of civil society.' The 'concentration of wealth in a few hands' is associated with 'the creation of a rabble of paupers.' In a telling addition to the original text, Hegel goes on to remark:

> Poverty in itself does not make men into a rabble . . . Against nature man can claim no right, but once society is established, poverty immediately takes the form of a wrong done to one class by another. The important question of how poverty is to be abolished is one of the most disturbing questions which agitate modern society.
>
> (Hegel 1967: 277)

Hegel considers two solutions to this 'disturbing' question. He explores the prospects for preventing the plunge into fateful social disorders by taxing the rich to support the poor, by supporting the poor out of public welfare, or by providing them with new work opportunities. But he concludes that all such solutions would merely exacerbate the problem. For example, the creation of new work would increase the volume of production when 'the evil consists precisely in an excess of production and in the lack of a proportionate number of consumers who are themselves also producers.' For reasons of this sort, it 'becomes apparent that despite an excess of wealth civil society is not rich enough, i.e. its own resources are insufficient to check excessive poverty and the creation of a penurious rabble.'

And so Hegel is forced to consider a second set of solutions. Civil society, he argues, is driven by its 'inner dialectic' to 'push beyond its own limits and seek markets, and so its necessary means of subsistence, in other lands that are either deficient in the goods it has overproduced, or else generally backward in industry.' It must also found colonies and thereby permit a part of its population 'a return to life on the family basis in a new land' at the same time as it also 'supplies itself with a new demand and field for its industry' (Hegel 1967: 150–2).

Imperialism and colonialism are hereby interpreted as necessary resolutions to the internal contradictions that are bound to beset any 'mature' civil society. Hegel is quite explicit that the increasing accumulation of wealth at one pole and the formation of a 'penurious rabble' trapped in the depths of misery and despair at the other, sets the stage for social instability and class war that cannot, according to his analysis, be assuaged by any *internal* transformation in the functioning of civil society. Overproduction and underconsumption, provoked by imbalances in the distribution of income, likewise undermine the internal coherence of industrial enterprise. Civil society is forced to seek an *outer* transformation through geographical expansion because its 'inner dialectic' creates contradictions that admit of no internal resolution.

Having, in a few brief startling paragraphs, sketched the possibility of an 'imperialist' solution to the ever-intensifying contradictions of civil society, Hegel just as suddenly drops the matter. He leaves us in the dark as to whether imperialism and colonialism could stabilize civil society through the elimination of poverty and social distress, in either the short or the long run. Instead, he switches to a detailed analysis of the state as the 'actuality of the ethical Idea'.[3] This seems to imply that he sees the transcendence of civil society by the modern state – an inner transformation – as the only viable solution. Yet he nowhere explains how the problems of poverty and of the increasing polarization in the distribution of wealth and income are to be overcome within the modern state. Are we supposed to believe, then, that those particular problems can readily be dealt with by imperialism? The text is highly ambiguous. This is, as Avineri points out, 'the only time in his system where Hegel raises a problem – and leaves it open' (Avineri 1972: 154; Hirschman 1976: 1–8).

Can civil society be saved from its internal contradictions (and ultimate dissolution) by an *inner* transformation – the achievement of the modern state as the 'actuality of the ethical Idea?' Or does salvation lie in a 'spatial fix' – an *outer* transformation through imperialism, colonialism and geographical expansion? These are the intriguing questions that Hegel leaves open.

Von Thünen

Hegel's *Philosophy of Right* was published in 1821 (with additions in 1833). Von Thünen produced the first draft of *The Isolated State* in 1818–19, revised it extensively in 1824, and published it in 1826.[4] While the concept of 'the isolated state' obviously draws upon the tradition of speculative and philosophical idealism, there is no hint of any direct Hegelian influence in Von Thünen's first published work. He combines an 'ideal construct' – the isolated state – with close empirical observation to produce a fascinating account of the spatial ordering of agricultural production. This account has since been canonized in the folklore of geography, economics and regional science as one of the first systematic attempts to formulate a coherent theory of location and of the social organization of space.[5]

Von Thünen later confessed that as early as 1826 he had abandoned his

---

[3] A brief transition discussion of the corporation takes Hegel from consideration of colonialism into the theory of the state; Hegel 1967: 152–5.

[4] See the introductions in Hall (1966) and Dempsey (1960).

[5] See Isard (1956) and Chisholm (1962). I have also drawn heavily on Barnbrock (1976).

'inherited views, being those of the owning classes', and become possessed of an entirely new vision. He dared not publish these new views, he said, for fear being branded 'a fanatic or even a revolutionary' (Dempsey 1960). But thereafter he concentrated his attention upon the moral and economic principles that determined the *natural*, and therefore the *just wage* of the laborer. His views on this question ultimately saw the light of day in 1850, the year of his death, as Part 2 of *The Isolated State*. Although the object of Von Thünen's enquiry undergoes a profound change, the two parts of *The Isolated State* exhibit certain continuities. The ideal construct of the isolated state is preserved, for example, but is used in Part 2 as a tool to investigate how social stability, continuity and harmony can be maintained in a civil society increasingly threatened by the social disorders stemming from rising class antagonisms and mass poverty.

Although he makes no reference to Hegel, Von Thünen's concerns in Part 2 of *The Isolated State* are almost identical to those expressed in the passage of the *Philosophy of Right* dealing with the internal contradictions of civil society. Like Hegel, Von Thünen rejects the idea that the hidden hand of the market can harness universal egoism to the benefit of all. He indicates an inevitable deterioration in the condition of civil society in the absence of any remedial measures.[6] He also explores both inner and outer tranformations as means to reconcile the contradictions inherent in civil society. And the language and conceptual apparatus is very Hegelian. The parallels are just too close to be accidental. Given Hegel's stature in German intellectual life during the 1820s and 1830s, it is very unlikely that Von Thünen proceeded in ignorance of Hegel's arguments.

Von Thünen's concerns also reflect directly the disturbed social situation in Europe prior to the revolutions of 1848. As early as 1842, he professed himself deeply worried by 'the views and teachings of the communists' who were not satisfied 'to ask for the labourer a *natural* wage, but immediately start with chimerical hopes and unreasonable demands' such as 'the distribution of property and equality of income'. He saw in such views the first signs of 'an incipient struggle' which could ultimately 'bring devastation and barbarism all over Europe'. The exaggerations of the communists would, he feared, inspire the multitude, 'become more popular and take root in the minds of the people especially if these views are proposed and expounded by skillfull if unprincipled writers' (Dempsey 1960: 219). And all of this was written six years before *The Communist Manifesto* burst upon the European scene.

---

[6] 'Self-interest,' he says, 'has found no counterweight in the knowledge of duty and truth.' See Dempsey (1960) pp. 218–20.

Von Thünen did not himself consider that it was 'in the plan of the world spirit' that 'all progress in the development of humanity must be realized only after numerous setbacks and bought by much blood and misery of many generations.' Yet it 'is one of the dismaying results of history that as a rule a mistake is not overcome by truth, nor by justice, nor by reason and right, but by another injustice.' The principle evil in this case arose because meager wages and grinding poverty – the lot of the mass of the people – had no clear moral justification and could therefore provide a fertile ground for social discontent. Doctrines of the subsistence wage or of supply and demand merely replicated reality and provided no solution to the crucial question: 'is the meager wage that the common labourer gets almost everywhere a *natural* one or is it caused by exploitation which the labourer cannot avoid?' The discovery of what constituted a *natural* or *just* wage was imperative because it was only in terms of such a conception that the rights, duties and obligations of the bourgeoisie could be defined. 'In the perception of truth and right and in such control of egoism that the privileged voluntarily give up what they unjustly own lie the means to get humanity peacefully and happily to further development and higher goals.' It was 'the high and sublime task of science' to discover and make known these truths 'not by means of experience or the course of history but by reason itself' (Dempsey 1960: 217–20).

Von Thünen pins down the central contradiction in civil society more precisely than Hegel. The 'source of evil', he argues, lies 'in the divorce of the worker from his product'. This means that two factors of production – capital and labor – which must cooperate to produce anything, exist in an antagonistic relation to each other. 'In this opposing interest, then, lies the reason the proletariat and the owners will perpetually oppose each other as natural enemies and will stay unreconciled as long as the division in their interest is not eliminated.' Low wages, he explains, 'have their origin in the fact that the capitalists and landowners take so large a part of what the laborers produce.' Furthermore, technological changes in no way improve the lot of the laborer: 'in our present social organization the worker will not be affected by this; his condition stays as it was, and the whole increase in income will fall to the entrepreneurs, capitalists, and landlords.' Had the social organization been such, laments Von Thünen, to allow workers but a fifth of the benefits flowing from improved productivity,

> joy and satisfaction would have spread over thousands of families, the disturbances and violence through which the workers forced a higher wage for themselves in the spring of 1848 would not have occurred, and the fine patriarchal bond which in the past existed between the masters and those in their charge would not have been destroyed.
>
> (Dempsey 1960: 327)

How, then, could this patriarchal bond be restored and these opposing interests be reconciled? Could a form of social organization be arrived at which guaranteed a just distribution of the social product between capital and labour, one which also gave the laborer opportunities for education and self-advancement? Answers to these questions depended, in Von Thünen's view, on first answering a single fundamental question: what is the natural, *just share* of labor in the product that labor creates?

Von Thünen's solution is contained in the doctrine of the *frontier wage*. 'On the frontier of the cultivated plain of the Isolated State, where free land is to be had in unlimited quantities, neither the arbitrariness of the capitalists nor the competition of the workers nor the magnitude of the necessary means of subsistence determines the amount of wages, but the product of labour is itself the yardstick for the wages.' If the workers close to the frontier 'are to be kept from setting up a colony' and 'are to be induced to continue to work for their former master for wages, the wages plus the interest they get from lending the capital which would have been necessary for setting up their own little colony must be equal to the product of labour which can be produced by a worker's family in such a colony.' In this way, 'wages and interest forming themselves on the frontier of the Isolated State set the norm for the whole state.' After due and careful consideration of all the relationships involved, Von Thünen concludes that the natural wage throughout the whole Isolated State was fixed by the formula, $\sqrt{ap}$, where a is the essential subsistence needs of the worker and p the product of his labour (Dempsey 1960: Chapters 14–15).

This is the equilibrium wage at which total output, the accumulation of capital, and the wage rate are simultaneously maximized. It is the wage at which both 'workers and capitalists have a mutual interest in increasing production.' It is the wage at which class struggle will dissipate and social harmony be achieved. The 'barrier between the two classes which has existed up until now will be removed,' because the natural wage gives the laborer sufficient access to education and opportunities for self-advancement to enable the more talented, thrifty, and energetic of them to become capitalists. This is the wage at which 'all of these evil conditions which sicken the social situation of Europe disappear.'[7] Blessed with such magical properties, small wonder that Von Thünen regarded $\sqrt{ap}$ as his most signal achievement, to be inscribed upon his tombstone at his death.

Two aspects of Von Thünen's argument interest us. First, his conception of capital and the social conditions that determine its formation.

---

[7] Ibid. pp. 221 and 327. See also Dempsey's introduction, p. 177.

Secondly, his appeal to processes of colonization and spatial expansion as means to justify his argument concerning the equilibrium wage. We take up each in turn.

Von Thünen defines capital as *things*, 'useful tools' produced by human labor and which increase the efficiency of human labor (Dempsey 1960: 245, 251). Capital exists, therefore, without presupposing any class relation between capitalists and laborers. In 'the original condition' everyone works, but there are two kinds of labor: that used to produce tools (capital) and that used to produce subsistence needs. Frugal and more efficient workers can produce a surplus product in the form of tools and, by virtue of the increased efficiency which the tools promote, produce even larger surpluses. The latter can be lent to others who will be willing to give up, in return, a portion of the surplus product which the tools they borrow help to generate. Herein lies the origin of interest. Von Thünen then derives one of his most important theorems: 'the revenue which capital as a whole gives when lent out is determined by the use of the last unit of the capital' (Dempsey 1960: 257). This theorem attracted Alfred Marshall's attention as the true foundation for the theory of the marginal productivity of capital.[8] But Von Thünen takes it in a direction which Marshall found quite unacceptable. In the original condition, workers producing subsistence could 'shift equally well to the production of capital if labor applied to the production of capital received a higher wage than labor applied in available alternative employments.' The transfer of workers would continue 'until equilibrium is reached: that is, until both types of labor are paid the same' (Dempsey 1960: 263).

Free mobility of labor is an essential condition for the realization of equilibrium. Private property and state regulation appear to pose barriers to that mobility. But private property arises, Von Thünen argues, only under conditions of resource (land) scarcity relative to population growth. Wages can fall only to 'a point where it becomes more advantageous to migrate to less fertile country... where there is still free land, and there to till the soil with the help of acquired and imported free capital.' The barrier of private property is checked by free land at the frontier. The nation states posed a more serious difficulty. A truly ethical state – and here Von Thünen appears to draw on and dispute Hegel directly – would not consider itself 'the center of the earth and the other nations but tools

---

[8] Marshall (1949), p. viii. See also Whitaker (1975), pp. 248–9, who doubts whether Marshall really meant it when he wrote 'my own obligations to Thünen are greater than to any other writer excepting only Adam Smith and Ricardo.'

for (its) own benefit,' for to do so would be to remain in a 'strained condition' in relation to 'the world spirit'. It would, therefore, allow the free mobility of both capital and labor, a condition 'so natural that we can consider the spread of humanity through migration over the whole world as being in accordance with the plan for the world' (Dempsey 1960: 267–9).

The existence of a freely accessible and open frontier appears necessary to the achievement of an equilibrium wage. This frontier provides a 'laboratory' (the scientific imagery is not accidental) for 'the determination of the relation between wages and interest'. The opposition between capital and labor is impossible there because all labor reverts to its original condition of producing either tools (capital) or subsistence goods (Dempsey 1960: 251). Von Thünen indeed invests this idea of the frontier with tangible historical meaning. In North America:

> fertile soil is available in unmeasured vastness for little or nothing . . . only the distance from the market place can set limits to the spread of culture. But these limits are pushed ever farther through steamboat traffic on the rivers and the construction of canals and railroads. There the wage $\sqrt{ap}$ is in fact attainable, and has actually been attained, for we find in America a relation between the wage and the rate of interest corresponding to the formula we have developed for fertile soil. As a result of these relations between workers and capitalists we find in North America general well-being which grows with giant steps. No crude division exists there among the various social orders. Even among the lesser classes elementary learning, reading, writing and arithmetic are to be found more generally disseminated than in Europe.
>
> (Dempsey 1960: 328)

This condition – 'a state of paradise' Von Thünen calls it – is threatened by increasing density of population. It is, furthermore, not directly realizable in Europe where no free land, unoccupied by landlords, could be had. How can the equilibrium wage be achieved in the absence of a real frontier? The theoretical artifice of the isolated state comes to the rescue as a means to identify the just wage even under conditions of a closed frontier. Von Thünen can then turn away from the imperialist solution and concentrate instead upon that inner transformation of civil society which will recapture the paradise lost and raise human beings 'to their spiritual estate'. He envisages a societal reconstruction which will transform the violence of class war into the pacific social harmony of cooperation between capital and labor. Profit-sharing is the organizational form, the natural wage (identified by appeal to the theory of the frontier wage) is the goal and the means is 'the higher development of intellectual powers and . . . the subordination of passion to the domination of reason' (Dempsey 1960: 328).

Yet an awkward question remains. Capital, originally conceived of as useful tools produced by a particular kind of labor, has become a class relation. If passion is to give way to reason, then a moral and economic justification must be found for the evident subordination of wage laborers as a class to those who merely own the products of past 'dead labour'. We have to understand how and why workers went 'from the state of freedom to that of need' in the first place and understand why such a transition was necessary to the ultimate recovery of paradise on earth. Otherwise, 'it seems incomprehensible that man could be placed under the mastery of his own product, capital, and become subordinate to it' (Dempsey 1960: 335). This was, of course, the question which Marx thought so pertinent.

Von Thünen's answer, which Marx thought so childish, rests upon a version of what we now know as 'human capital theory'.[9] The original differentiation between capital and labor simply depended upon the frugality and efficiency of some laborers relative to others. In the current circumstance, Von Thünen finds that the labor of 'a free man costs the capitalist nothing but the maintenance and interest on the capital which the rearing of the worker cost.' Although at first sight 'somewhat repugnant', this view allows us to see that the more the workers embody capital in themselves and in their children, the higher their wages become and the more easily they can penetrate the porous boundary, mainly fixed by education, between capitalists and laborers. In this way, the workers themselves can lift themselves to freedom and mastery over suffering. But they are unlikely to do this on their own. Without the discipline imposed by capital, the workers 'instead of using their surplus on the better rearing of their children, would sink into indolence and sloth' (Dempsey 1960: 337).

For this reason, 'the compulsion that the mastery of capital lays upon men to lead them to their higher destiny is necessary, and so need no longer appear as a scourge, but as the instructor of mankind.' The patriarchal bond, so dear to Von Thünen, finds its justification. 'Capital,' he proclaims joyously, 'dominates man, but in a marvellous way' (Dempsey 1960: 336). It impels the worker towards freedom and is therefore a manifestation of divine law, an integral part of the plan of the world spirit, a reflection of the hand of God. The evil that appeared to be the root cause of the misery of most of mankind – the separation of the laborer from the product of labor – becomes the means to promote that greater

---

[9] Dempsey (1960), pp. 143–9. Dempsey in his introduction discusses the idea. A modern statement can be found in Becker (1975).

state of freedom, that state of paradise here on earth, consistent with divine will (Dempsey 1960: 340).

Von Thünen is less ambiguous than Hegel. Social harmony can be achieved directly wherever the frontier is open and the mobility of labor and capital guaranteed. A somewhat romanticized picture of North America is used to illustrate the point. Where the frontier is closed, as in Europe, burgeoning social unrest must be countered by an inner transformation of civil society. Understanding furnished by an economic science in which the frontier is treated ideally and analytically paves the way.[10] The application of marginalist principles leads to a proper appreciation of what constitutes the just wage. The rights and duties which attach to the roles of capitalist and laborer can thereby be defined. The myth of frontier justice can be internalized within the framework of the modern state by *rational* individuals (and the emphasis is very definitely upon rationality of a certain sort). Von Thünen thereby legitimizes and justifies the perpetuation of class relations and the preservation of that 'patriarchal bond' whereby capitalists can fulfill their obligation to the laborer (profit-sharing schemes, education and externally-imposed discipline to form 'human capital').[11] Antagonistic and warring interests stand to be harmonized by such means. God's will, the plan of the world spirit, can be realized here on earth through human agency. Provided, that is, human beings acquire the universalistic consciousness of marginalist economics.

Marx[12]

The relationship between the thought of Hegel and Marx has been the subject of an immense and continuing debate.[13] As far as I can discover, that between Von Thünen and Marx has not been deemed worthy of comment. Yet a comparison shows they all had much in common. They treat human labor as fundamental and see the alienation of labor from its product as the source of evils to be overcome. They focus on class antagonisms and take a common stand against the central thesis of English political economy: the doctrine that the hidden hand of the market automatically harmonizes conflicting interests and harnesses individual selfishness to the benefit of all. They all introduce the idea of inner and outer tranformations as means to restore social stability and

---

[10] The imagery of the frontier is strongly preserved in bourgeois economics in phrases like 'the factor-price frontier', and the like.

[11] Dempsey includes a copy of Von Thünen's will and defends him against those critics who saw the actual profit-sharing scheme as seriously defective even according to Von Thünen's own principles (see pp. 48; 363–7).

fend off capitalist crises. The strong differences between them therefore exist within a common frame of questions and suppositions.

In an Afterword to *Capital*, written in 1873, Marx points out that he came to terms with Hegel nearly thirty years before. The reference is almost certainly to his *Critique of Hegel's Philosophy of Right*, probably written in 1843. The *Critique* is therefore a seminal work, justifiably viewed as the wellspring from which flowed 'the whole program of research and writing which occupied Marx for the remainder of his life' (O'Malley 1970: xiv).

In it, Marx somewhat surprisingly ignores those passages in which Hegel so stunningly depicts the inner contradictions of civil society and their potential resolution through imperialism. He focuses instead on Hegel's theory of the modern state in order to show that Hegel's solution is 'pure mystification' which served only 'to transfigure and to glorify the existing state of things'. Marx does not abandon the dialectic, however. He merely sought, as he later put it, to turn Hegel's dialectic 'right side up' and give it a material base. In this way Marx hoped to capture the 'fluid movement' of history and so arrive at an accurate representation of the 'momentary existence' and 'transient nature' of 'every historically developed social form'. This, to Marx, was the essence of 'critical and revolutionary' modes of dialectical thinking (Marx 1967: vol. 1, 19).

The immense edifice of thought and elaborate conceptual apparatus which Marx evolved in the course of a prolific lifetime of research, writing and political activism, defies simple summary. But much of what he did can be interpreted, superficially at least, as answers to, or transformations of, questions which Hegel posed. Hegel, as Engels observed, was possessed of an 'exceptional historical sense': 'however abstract and idealist the form employed, the development of his ideas runs always parallel to the development of world history . . . he was the first to try to demonstrate that there is development, an intrinsic coherence in history' and 'this epoch-making conception of history was the direct theoretical precondition of the new materialist outlook.' Marx was, Engels asserts, 'the only one who could undertake the work of extracting from the Hegelian logic the kernel containing Hegel's real discoveries'.[14] And so Marx transforms the occult and mysterious qualities of Hegel's 'world spirit' into the mundane materialities of the world market. Social questions are thereby transplanted from the realms of philosophical contemplation to those of political economic practice. The opposition between concrete labor (the actual production of use values) and abstract labor (the social

[14] Marx, K. and Engels, F., *Collected Works* ((1974), vol. 16 (1980): 474). See also O'Malley's 'Introduction' (1970) for a very good overview on this theme.

qualities of labor that render commodities commensurate in exchange) mirrors, in Marx's political economic schema, the opposition between private and public in Hegel's (and Rousseau's) political conception of civil society. And if philosophical consciousness and the acknowledgement of the state as the actuality of the ethical Idea is the solution for Hegel, then for Marx the real potentiality for emancipation lay with the proletariat, that class which could truly claim universal consciousness by virtue of its experience of universal suffering.[15]

Many of these and other key ideas are first broached in the *Critique*. All the more surprising, therefore, is the total lack of any commentary on Hegel's conception of that 'inner dialectic' in civil society which drove it to seek colonial and imperialist solutions. Marx evidently intended to extend his criticism in this direction. But he never did so. Or so it seems.

Yet there is a sense in which the whole of *Capital* can be construed as an effective transformation and materialist representation of part of Hegel's idealist argument. The theme of increasing polarization between the social classes is, after all, writ large in *Capital*. In the 'general law of capitalist accumulation', Marx shows that the necessary consequence of the real processes at work under capitalism is the reproduction of 'the capital-relation on a progressive scale, more capitalists at this pole, more wage-workers at that'. Furthermore these processes also produce a 'relative surplus population', a 'reserve army' of unemployed 'set free' primarily through technological and organizational change. This reserve army helps to drive wage rates down and to control working-class movements and is, therefore, a 'prime lever' for further accumulation. The net effect, as Marx puts it, invoking imagery deeply reminiscent of Hegel, is that the 'accumulation of wealth at one pole is, therefore, at the same time accumulation of misery, agony of toil, slavery, ignorance, brutality, mental degradation, at the opposite pole, i.e. on the side of the class that produces its own product in the form of capital' (Marx 1967: vol. 1, ch. 25).

Marx has plainly recast Hegel's idealist argument in theoretical materialist terms. He also rebuts Malthus and shows that poverty and relative surplus populations arise under capitalism irrespective of the rate of population growth and clarifies some of the issues that had bothered Von Thünen: why, for example, the lot of the laborer deteriorates in spite of the use of machinery.[16] The essential insight is that the increasing

---

[15] These ideas are most effectively spelled out in Marx's *Contributions to a Critique of Hegel's 'Philosophy of Right'*, reprinted in O'Malley (1970).

[16] Marx's polemic against Malthus is fully spelled out in Marx, K., *Theories of Surplus Value* (1972: 13–68). See also Harvey (1977b).

polarization between capital and labor and the progressive relative impoverishment of the working class is to be interpreted materially as the inevitable product of identifiable forces at work within a particular and historically-achieved mode of production known as capitalism.

But Marx's specification of the 'general law' rests on the ability of capitalists to control both the demand for, and supply of, labor power. This 'double action' of capital infringes upon laws of supply and demand which the bourgeoisie was otherwise wont to regard as 'sacred and eternal'. Marx tacitly accepts Von Thünen's criticisms of the operation of such laws in labor markets but presses the argument to a radically different conclusion. Capitalists must control labor supply, create labor surpluses in effect, either through the mobilization of 'latent' labor reserves (women and children, peasants thrown off the land, and the like) or through the creation of technologically induced unemployment. Any threat to that control, Marx notes, is countered 'by forcible means and State interference'. In particular, capitalists must strive to check colonization processes which give laborers open access to free land at some frontier (Marx 1967: 640). Which brings us back to the whole question of how to interpret colonization in relation to capitalism's unstable 'inner dialectic'.

The purpose of the final chapter on colonization is to show how the bourgeoisie contradicted its own myths as to the origin and nature of capital by the policies it advocated in the colonies. In bourgeois accounts, and Von Thünen's was typical, capital (a thing) originated in the fruitful exercise of the producer's own capacity to labor, while labor power arose as a social contract, freely entered into, between those who produced capital through frugality and diligence, and those who chose not to do so. 'This pretty fancy', Marx thunders, is 'torn asunder' in the colonies. As long as the laborer can 'accumulate for himself – and this he can do as long as he remains possessor of his means of production – capitalist accumulation and the capitalist mode of production are impossible.' Capital is not a physical product but a *social relation*. It rests on the 'annihilation of self-earned private property, in other words, the expropriation of the labourer'. Historically, this expropriation was 'written in the annals of mankind in letters of blood and fire', and Marx cites chapter, verse, and the Duchess of Sutherland to prove his point. The same truth, however, is expressed in colonial land policies, such as those of Wakefield, in which the powers of private property and the state were to be used to *exclude* laborers from easy access to free land in order to preserve a pool of wage laborers for capitalist exploitation. Thus was the bourgeoisie forced to acknowledge in the colonies what it sought to conceal at home: that wage labor is based on the forcible separation of the laborer from control over the means of production (Marx 1967: ch. 32).

Marx here confronts Von Thünen's frontier idealism (both as to past origins and present possibilities) with tough-minded historical materialism and opposes the conception of capital as a social relation to the fetishized view of capital as a thing. And the 'childishness' of Von Thünen's proposals now becomes apparent. The abolition of poverty, unemployment and labor surpluses would eliminate the social basis for further accumulation of capital. To pretend that poverty can be abolished without breaking the 'patriarchal bond' between capital and labor is, in Marx's view, vain illusion, a cruel hoax. The insistence that capital can dominate labor in a 'marvelous way' is crass apologetics. The best that profit-sharing schemes of the sort Von Thünen advocated could achieve was an occasional relaxation, and then only for a privileged group of laborers, of the tension in the 'golden chain' which binds capital to labor.[17]

It is not hard to infer the nature of Marx's objections to Von Thünen's apologetic gyrations. But his position with respect to Hegel is more elusive. Certain aspects fall into place readily enough. If, for example, laborers can return to a genuinely unalienated life through migration to some frontier (as Hegel envisaged), then capitalist control over labor supply is undermined. Such a form of expansion may be advantageous to labor but it could provide no solution to capitalism's problems. The new markets and new fields for industry which Hegel saw as vital could be achieved only through the re-creation of capitalist relations of private property and the power to appropriate the labor of others. The fundamental condition which gave rise to the problem in the first place – alienation of labor – is thereby replicated. Marx's chapter on colonization appears to close off the possibility for any permanent spatial fix. It can be seen, then, as a necessary coda to the penultimate chapter in which the expropriation of the expropriators is urged as the only valid solution to the social dilemmas which capitalism poses. Marx seeks a firm closure to the door that Hegel left partially ajar.

But the door will not stay shut. Hegel's 'inner dialectic' undergoes successive representations in Marx's text. And at each point the question of a spatial resolution to capitalism's contradictions can legitimately be posed anew. The chapter on colonization may suffice for the first volume of *Capital* where Marx concentrates solely on questions of production. But what of the third volume, where Marx shows that the requirements of production conflict with those of circulation to produce crises of disequilibrium in accumulation? Polarization then takes the form of 'unemployed capital at one pole, and unemployed worker population at the other'. Can the formation of such crises be contained through geographical

---

[17] Marx does not make these criticisms directly but it is not hard to infer them.

expansion? Marx does not rule out the possibility that foreign trade can counteract the tendency towards a falling rate of profit in the short run (Marx 1967: 237–59). But how long is the short run? And if it extends over many generations, then what does this do to Marx's theory and its associated political practice of seeking for revolutionary transformations in the heart of civil society?

Marx is infuriatingly unsystematic and vague in dealing with such questions. It has therefore proved very difficult to integrate his theory of the long-run dynamics of accumulation and its internal contradictions, specified for a closed system, with themes of imperialism, colonialism, uneven geographical development, unequal exchange, and the like. Marx was not unaware of such issues, but his comments are scattered all over the place. We have to build a framework to synthesize Marx's various comments on the potency of the spatial fix.

To do this requires a firm interpretation of Marx's view of the 'inner dialectics' of capitalism in crisis. This is not an uncontroversial matter since rival interpretations of Marx's theory of crisis abound.[18] I shall work with a highly simplified version in which individual capitalists, locked into class struggle and coerced by intra-capitalist competition, are forced into technological adjustments which destroy the potential for balanced accumulation, and so threaten the reproduction of both the capitalist and working classes. The end-product of such a process is a condition of *overaccumulation* of capital, defined as an excess of capital in relation to the opportunities to employ that capital profitably. This excess of capital can exist as a surplus of commodities, of money, of productive capacity, and also leads to a surplus of labor power (widespread unemployment or underemployment). The only effective resolution to such crises, in the absence of a spatial fix, is the *devaluation* of capital, as money (through inflation), as commodities (through gluts on the market and falling prices), as productive capacity (through idle or under-utilized plant and equipment, physical infrastructures, and the like, culminating in bankruptcy), and the devaluation of labor power (through falling real standards of living of the laborer).

We now have to consider how overaccumulation and devaluation can be remedied by some form of geographical expansion. Marx's comments on such a prospect can be gathered together under three main headings.

External markets and underconsumption

If overaccumulated capital in Britain is lent out as means of payment to Argentina to buy up the excess commodities produced in Britain, then

---

[18] A full version of the theory is spelled out in Harvey (1982).

the relief to overaccumulation is at best short-lived. Pursuit of such a strategy assumes that the crises of capitalism, which are always partially manifest as a lack of effective demand, are entirely attributable to under-consumption. Marx is as firm in his rejection of the geographical version of this as he is of the original (Bleaney 1976). All that happens, he suggests, is that the effects of overaccumulation are spread out over space during the credit-fuelled phase of expansion. The collapse, when it comes, triggers an intricate sequence of events because of the gaps which exist between the imbalance of trade and balance of payments between regions. He describes a typical sequence this way:

> The crisis may first break out in England, the country which advances most of the credit and takes the least, because the balance of payments . . . which must be settled immediately, is *unfavourable*, even though the general balance of trade is *favourable* . . . The crash in England, initiated and accompanied by a gold drain, settles England's balance of payments . . . Now comes the turn of some other country . . .
> The balance of payments is in times of crisis unfavourable to every nation . . . but always to each country in succession, as in volley-firing . . . It then becomes evident that all these nations have simultaneously over-exported (thus over-produced) and over-imported (thus over-traded), that prices were inflated in all of them, and credit stretched too far. And the same break-down takes place in all of them.

The costs of devaluation are then forced back onto the initiating region by:

> First, shipping away precious metals; then selling consigned commodities at low prices; exporting commodities to dispose of them or obtain money advances on them at home; increasing the rate of interest, recalling credit, depreciating securities, disposing of foreign securities, attracting foreign capital for investment in these depreciated securities, and finally bankrupt-cy, which settles a mass of claims.
>
> (Marx 1967: 491–2, 517)

The sequence sounds dismally familiar. No prospect here, evidently, for a spatial fix to capitalism's contradictions.

A more intriguing possibility arises with respect to trade with non-capitalist social formations. Circumstances can indeed arise which make the development of capitalism 'conditional on modes of production lying outside of its own stage of development'. The degree of relief afforded thereby depends on the nature of the non-capitalist society and its capacity to integrate into the capitalist system and absorb the excess capital. But crises can be checked only if the non-capitalist countries 'consume and produce at a rate which suits the countries with capitalist production'.[19] And how can that be ensured short of some form of political and economic domination? And even then the resolution is bound to be temporary. 'You cannot continue to inundate a country with your manufactures,' says Marx, 'unless you enable it to give you some produce in return.' Hence, 'the more the [British] industrial interest became dependent on the Indian market, the more it felt the necessity of creating fresh productive powers in India' (Marx and Engels 1972). Which broaches a whole new set of problems.

The export of capital for production

Surplus capital can be lent abroad to create fresh productive powers in new regions. The higher rates of profit promised provide a 'natural' incentive to such a flow and, if achieved, raise the average rate of profit in the system as a whole. Crises are temporarily resolved. 'Temporarily' because higher profits mean an increase in the mass of capital looking for profitable employment and the tendency towards overaccumulation is exacerbated, but now on an expanding geographical scale (Marx 1967: 237, 256; Marx 1969b: 436–7). The only escape lies in a continuous acceleration in the creation of fresh productive powers. From this we can derive an impulsion within capitalism to create the world market, to intensify the volume of exchange, to produce new needs and new kinds of products, to implant fresh productive powers in new regions and to bring all labor, everywhere, under the domination of capital. We can interpret the actual historical geography of capitalism as the product of such an imperative. But the 'inner dialectic' of capitalism ensures that such a process 'moves in contradictions which are constantly overcome but just as constantly posited' (Marx 1973: 410). Crises are phases of intense rationalization in geographical transformation and expansion. The inner dialectic of civil society is perpetually assuaged and reproduced through constant resort to the spatial fix.

There are, presumably, limits to such a process. How long can continuous expansion be sustained before geographically-localized crises, or 'switching crises' (which reverse or radically change the direction of capital flows) merge into global crises? And what internal dilemmas inhere within such a process?

When a particular civil society creates fresh productive powers elsewhere

---

[19] Marx (1967), vol. 2, p. 110; vol. 3, p. 257. The dilemmas which arise have recently been abundantly illustrated by the problems of western banks who sought profitable outlets for surplus capital in Poland.

to absorb its overaccumulated capital, it thereby establishes a rival center of accumulation which, at some point in the future, must also look to its own spatial fix to resolve its problems. Marx thought he saw the first step down such a path as the British exported capital to India (Marx and Engels 1972: 85–7). But the transition which Marx anticipated there was blocked by a mixture of internal resistance to capitalist penetration and imperialist policies imposed by the British. The latter were, by and large, specifically geared to preventing the rise of India as a competitor. We can immediately spot the following dilemma. If, for whatever reason, fully-fledged 'outer transformations' are blocked, then the capacity of the home country to dispose of further overaccumulated capital is also blocked. The spatial fix is denied and crises ensue in the home country. The unconstrained growth of capitalism in new regions is an absolute necessity for the survival of capitalism. These are the fields where excess overaccumulated capitals can most easily be absorbed in ways which create new market openings and further opportunities for profitable investment. But we then encounter another kind of difficulty. The new productive forces in new regions pose a competitive threat to the initiating country. The overaccumulation of capital in new regions demands a spatial fix, perhaps even at the expense of capital in the old regions. The US thus absorbed far more British excess capital than India ever did but by the same token became the great competitor to Britain in the world market. West Germany and Japan similarly absorbed far more surplus capital from the US than did the whole of the 'Third World' after 1945 and likewise subsequently emerged as the main centers of competition to the economic hegemony of the US within world capitalism.

Devaluation appears likely, no matter what. The initiating country is faced with a 'catch 22.' The unconstrained development of capitalism in new regions sparked by capital exports brings devaluation at home through international competition. Constrained development abroad limits international competition but blocks the dynamism which creates opportunities for profitable capital export: overaccumulated capital cannot escape and is devalued internally. Small wonder, then, that the major imperialist powers have vacillated in their policies between 'open-door' free trade, and autarky within a closed trading empire.[20]

Nevertheless, within these general constraints all kinds of options exist. The geographical spread and intensification of capitalism is a long drawn-out revolution accomplished over many years. While local, regional, and switching crises are normal grist for the working-out of this process,

---

[20] See, for example, Gardner (1971).

the building of a truly global crisis of capitalism depends upon the exhaustion of possibilities for further revolutionary transformation along capitalist lines. And that depends not upon the capacity to propagate new productive forces across the face of the earth, but upon the supply of fresh labor power. Which brings us back to the question of primitive accumulation.

The expansion of the proletariat through primitive accumulation

'An increasing population,' wrote Marx, is a 'necessary condition' if 'accumulation is to be a steady continuous process' (Marx 1973: 608, 764, 771; Marx 1969b: 47). He subsequently modifies this to mean growth of population 'freed' from control over the means of production, that is, growth in the wage labor force, including the industrial reserve army. The faster the growth in these aggregates, the more crises will likely appear as pauses within an overall trajectory of expansion.[21] So where does this expansion in the exploitable population come from? Marx divides the relative surplus population into three layers: latent, floating and stagnant. We confine attention to the first two categories. The mobilization of latent elements entails either primitive accumulation (the separation of peasants, artisans, self-employed and even some capitalists from control over their means of production) or the substitution of family for individual labor (the employment of women and children). A floating supply can be produced by any combination of sagging commodity production and labor-saving technological innovations. Taken in the context of natural population growth (itself not immune to the influence of capitalism's dynamic),[22] these mechanisms must provide the fresh supplies of labor power to feed accumulation for accumulation's sake.

Marx does not subject these processes to detailed scrutiny, nor does he deal systematically with spatial aspects. But the flow of his logic points clearly to certain conclusions. *Within* a particular civil society, viewed as a closed system, accumulation will accelerate until all latent elements are absorbed and the limits of natural population growth reached. Floating populations must then be increasingly relied upon as the source of an industrial reserve army. Society shifts from the trouble and turmoil of

---

[21] This aspect for Marx's argument is dealt with in greater depth in Sweezy (1942), pp. 222–6; and Morishima and Catephores (1978), ch. 5.

[22] Marx, curiously enough, accepts most of the conventional wisdom on the economic influences affecting population growth, with the additional observation that laborers had no option except to 'accumulate' the only source of wealth they had, their labor power. See Marx (1967) vol. 1, p. 643.

primitive accumulation and the destruction of precapitalist family relationships to the trauma of technologically-controlled unemployment. Both processes will likely be the focus of intense class struggles, though of a rather different sort. But the latter is more problematic for capitalism. Uncontrolled and rapid technological change sparks overaccumulation and, ultimately, the unemployment of capital as well as labor. Though they may not be aware of it, there is a systemic advantage to capitalists of exploiting latent rather than floating labor reserves. The more they depend on the latter, the more serious will crises of devaluation likely be.

To the degree that geographical expansion opens up access to latent labor reserves, it can indeed serve to mitigate devaluation crises. This means some form of primitive accumulation in the exterior (through penetration of capitalist property relations, money forms, the imposition of state and legal controls, and so on). The labor surpluses so created form a field of action for overaccumulated capital. The exact form of labor process and of social relations achieved can vary greatly according to the initial conditions and the kinds of class struggle set in motion. Marx recognized some of the variation. Plantation colonies, run by capitalists on the basis of slave or indentured labor, could be formally integrated into capitalism without being based on free wage labor. Modes of exploitation in traditional peasant-based societies could also be converted into formal rather than real subsumption of labor under capital. The conversion of state powers into a form of state capitalism opens up other possibilities. By and large, Marx did not pay that much attention to the incredible diversity of possible transitional forms which arise as latent labor reserves are mobilized through primitive accumulation on indigenous populations in non-capitalist social formations.[23] He rests content with a central point: the key role of labor surpluses in the search for a 'spatial fix' to capitalism's internal contradictions.

Labor surpluses can also be imported from abroad. This, for Marx, was the significance of Ireland to English capitalism. Primitive accumulation in the former place furnished labor surpluses to the latter and so helped undermine the organized power of English workers (Marx and Engels 1955: 228–33, 235–8). That many such parallels exist in the contemporary world scarcely needs belaboring. But we here encounter a rather interesting tension. The importation of labor surpluses must rest, in the absence of slavery, upon the free geographical mobility of the laborer. But if that privilege is conceded to labor surpluses on the exterior, it is hard to deny it to floating reserves generated at home. In the face of unemployment,

[23] The exposition in Marx, *Grundrisse* (1973: 459–511) is very useful here.

floating reserves may emigrate, particularly if free land is available at
some frontier. Marx here agrees with both Hegel and Von Thünen, that
the lot of the laborer stood to be improved by free migration to a frontier.
Indeed, he habitually attributes superior wages and work conditions in
the US to the existence of a relatively open frontier. But he diverges from
Hegel and Von Thünen because he sees such a condition as antagonistic
to the real interests of accumulation.

The significance of that last chapter on colonization now strikes home
with redoubled force. Primitive accumulation at the frontier is just as vital
as primitive accumulation and technologically-induced unemployment at
home. Internal and external conditions of class struggle are inextricably
intertwined. Here is how Marx depicts the relation in the case of colonies
founded through the free migration of laborers:

> the capitalist regime everywhere comes into collision with the resistance of
> the producer, who, as owner of his own conditions of labour, employs that
> labour to enrich himself, instead of the capitalist. The contradiction of
> these two diametrically opposed economic systems, manifests itself here
> practically in the struggle between them. Where the capitalist has at his back
> the power of the mother-country, he tries to clear out of his way by force,
> the modes of production and appropriation based on the independent
> labour of the producer.
>
> (Marx 1967: 765)

The search for a 'spatial fix' sparks new forms of class struggle, epitomized by the innumerable populist and radical movements spawned
amongst settlers in frontier regions. Indeed, in this case it is not hard to
spell out a simple theoretical framework to capture the central dynamic of
the so-called 'Atlantic Economy' of the nineteenth century (Thomas 1973).
The absorption of latent reserves at home leads to the creation of floating
reserves through technological change. Such floating reserves are attracted
to any open frontier. By the same token, reliance upon floating reserves
exacerbates problems of overaccumulation and devaluation at home. So
capital, too, is attracted to some open frontier. Unemployed capital and
labor power – the hallmark of Marx's conception of crisis – are both
attracted to the frontier. But if capital accumulation is to be served, then
the laborers that moved to the frontier in search of an unalienated existence must be recaptured as wage laborers. Primitive accumulation and
new forms of class struggle necessarily reassert themselves at the frontier.
This is what Marx's final chapter on colonization truly signals.

The general point remains. Although rapid expansion in the wage labor
force (through primitive accumulation, the migration of floating labor

reserves, and the like) can moderate the tendencies to crisis formation in the short run, the social relations which propel capitalism's inner dialectic are merely recreated on a wider geographical scale. There is, under such circumstances, no long-run 'spatial fix' to capitalism's internal contradictions.

Reflections and alarums

Let us reflect on these writings from the standpoint of the history of ideas in their economic and political contexts. Von Thünen shifts from explicit consideration of spatial organization in his early work towards thoroughly a-spatial formulations later on. Although he argues that the truly 'Ethical State' should not place barriers to the geographical mobility of capital and labour, he is faced with the realities of Prussian state interventionism and mercantilism expressed in a context of incipient German nationalism. Like Hegel before him, he is therefore forced to look for that inner transformation which will assuage class conflict and social polarization within the confines of a particular civil society. And so the frontier becomes an analytical factor-price frontier, calibrated according to the marginal productivity of different factors of production within a closed economy. The real lesson which Marshall and all subsequent neoclassical economists learned from this, was that economic science could seek and spell out principles of social harmony without appeal to the political economy of the spatial fix. Economics, as Walter Isard was later to complain, thereafter abandoned all serious consideration of space and accepted Marshall's dictum that 'the influence of time' is 'more fundamental than that of space' (Isard 1956: 24). But elimination of the spatial fix from consideration was also crucial to dismantling traditional political economy. Spatial relations became the exclusive preserve of political theory, which severed all direct connection with the day-to-day realities of the circulation of capital and its contradictions, and substituted an organicist theory of the state (caught in a struggle for survival, needing *lebensraum*, and so on) and associated doctrines of manifest destiny, white man's burden, racist superiority, and the like.

In late nineteenth-century bourgeois thought, then, the connection between politics and economics as well as between inner and outer tranformations became lost, curiously enough at the very historical moment when the careers of figures as diverse as Joseph Chamberlain in Britain, Jules Ferry in France, and Theodore Roosevelt in the US, provided living testimony to the underlying unities (Julien et al. 1949). Each, desperately concerned to put out the fires of class struggle, turned to the politics of imperialism as they hit the limits to internal social reform. At the very moment when the relations between inner and outer transformations were in a state of acute tension, therefore, bourgeois ideology masked the meaning of the relationship by insisting upon the separation between economic and political theory. It was left to a furtive underworld of dissident bourgeois writers, such as Hobson and Mark Twain, and to the Marxists, to try and preserve the unity of political economy as a tool to interpret a deeply troubled history.

But the Marxists, paradoxically, could not find much comfort from Marx (certainly not from *Capital*). For Marx, though supremely aware of the underlying unity of political and economic affairs as well as of the global dynamics of capitalism, excluded specific consideration of the spatial fix on the grounds that integrating questions of foreign trade, of geographical expansion, and the like, into the theory, merely complicated matters without necessarily adding anything new. Again and again he seeks, as in the chapter on 'Colonization', to close the door on a possibility which Hegel left open. There is enough side commentary in his work (some of which we have already cited) to indicate that he was not always satisfied with the closure. But in a world in which Palmerston's 'Pax Britannica' reigned secure, and positioned as he was at the center of *laisser-faire* capitalism with all its ideological blandishments, Marx had little incentive to go beyond depicting the spatial fix as anything other than the violent projection of the contradictions of capitalism onto the world stage. His supreme concern, and contribution in *Capital*, was to unravel the nature of capitalism's *inner* dialectic.

The awesome realities of late nineteenth-century inter-imperialist rivalries, the struggle for autarky in closed trading empires, the collapse of the 'Pax Britannica', and the seemingly inevitable drift towards global war, coupled as all this was with rising labor militancy in the advanced capitalist countries, forced Marxists to confront directly the dynamic relations between inner and outer tranformations. Bukharin, Lenin, Luxemburg, and others, turned to an explicit analysis of imperialism (Bukharin 1972; Lenin 1963, vol. 1: 667–768; Luxemburg 1968). Marxists, moving in exactly the opposite direction to mainstream bourgeois theorists, struggled to extract their theory from the aspatial mould in which Marx had cast it, and so sought to preserve a political-economic analysis relevant to their time. In so doing, they created a new imagery within the Marxist tradition, an imagery which dramatically unifies themes of capitalist exploitation and the spatial fix. Centers exploit peripheries, the metropolis exploits its hinterland, the first world subjugates and mercilessly exploits the third, and so on. Class struggle *within* a particular civil society is reconstituted as the struggle of peripheral social formations against some central source of oppression. The country revolts against the city, the periphery against the center, the third world against the first. So powerful is this spatial imagery that it threatens on occasion to engulf and quite replace the interpretation of capitalism's inner dialectic which Marx so carefully wrought. And it certainly undermines any simple version of Marx's hopes for proletarian internationalism founded on a universalistic consciousness born out of the universality of working-class suffering.

The door which Hegel so presciently opened still stands open wide. To pass through it is to accept the tension between inner and outer transformations as the focus of theoretical concern. A thorough understanding of that tension bridges the gap between the Marxian theory of accumulation (spelled out, for the most part, in purely temporal terms) and Lenin's view of the historical geography of capitalist imperialism. A reconstruction of Marx's peripheral writings on the spatial fix indicates, however, that he had a far deeper appreciation of its potentialities and limitations than many give him credit for. By the same token, a close reading of Lenin (which we have not attempted here) suggests that his theory of imperialism is more deeply rooted in Marx's theory of accumulation than is immediately apparent.[24] It is not simply that both agree that the contradictions of capitalism cannot, in the long run, be assuaged by resort to imperialism. Both, it turns out, are in broad agreement as to the processes which link inner and outer tranformations. But Lenin adds one crucial insight which Marx lacks, an insight which can easily be grafted onto Marx's theory with quite alarming implications.

Marx's rough and ready denial of the efficacy of any spatial fix to capitalism's internal contradictions permits him to concentrate attention upon the fundamental processes of crisis formation. The theory of over-accumulation-devaluation reveals the height of insanity, the intense destructive power, implicit in the capitalist mode of production. Beneath its facade of market rationality, and counterposed to its creative powers to revolutionize the productive forces, the bourgeoisie turns out to be 'the most violently destructive ruling class in history'.[25] In the depths of crises, capitalists unleash the violence of primitive accumulation upon

[24] The full derivation of Lenin's theory out of Marx requires many steps which I cannot go through here. First and foremost we must provide a material basis for forces which, in the face of geographical mobility of both capital and labor power produce and sustain class alliances within a territorially bounded civil society. I attempt to cover this step in Harvey (1982), ch. 13. The second step is to show how such class alliances, though fundamentally unstable, can crystallize around relatively rigid configurations of political and military power wielded through the state apparatus. This second step has been the focus of an immense and continuing controversy within the Marxist literature on the theory of the state.
[25] The actual phrase is from Berman (1982), p. 100.

each other, destroy vast quantities of capital, cannibalize and liquidate each other in that 'war of all against all' which Hobbes had long before seen as an inherent characteristic of market capitalism. What Marx nowhere anticipates, but Lenin emphasizes, is the conversion of this process into economic, political and military struggles between nation-states. At times of savage devaluation, the search for a spatial fix is converted into inter-imperialist rivalries over who is to bear the brunt of devaluation. The export of unemployment, inflation, and idle productive capacity, become the stakes in an ugly game. Trade wars, dumping, tariffs and quotas, restrictions on capital flow and foreign exchange, interest-rate wars, immigration policies, colonial conquest, the subjugation and domination of tributary economies, the forced reorganization of the division of labor within economic empires, and finally, the physical destruction and forced devaluation of a rival nation's capital through war, are some of the options at hand.

Twice in the twentieth century, after all, the world has been plunged into global war through inter-imperialist rivalries. Twice, in the space of a generation, the world experienced the massive devaluation of capital through physical destruction, the ultimate consumption of labor power as cannon fodder. It has never proved easy to explain this history on the basis of a theory which appeals to the class relation between capital and labor as the fulcrum upon which capitalist history turns. Marx's reluctant dealings with the question Hegel posed brings us to the brink of such an understanding. The grafting of Lenin's insights onto Marx's representations tells a fuller story.

So who is right? If Von Thünen is to be believed, there is nothing inherent in capitalism to dictate accelerating class polarization, conflict, or inter-imperialist war. The bourgeoisie, armed with a proper consciousness of its duties and obligations, as well as of its rights, stands to lead humanity to ever more civilized pastures, a veritable Eden of the rights of man, of pacific social harmony. And if all this fails to materialize, it must simply be attributed to the frailties of human nature, moral failings of the spirit, a failure to grasp and implement the divine plan for the world. Capitalism, for Marx, is much more problematic than that. It constitutes a permanently revolutionary force, sweeping away all older ways of life, unleashing untold powers to expand the productivity of social labor. But it also contains within itself the seeds of its own negation, seeds which grow and ultimately crack open the very foundations in which they are rooted. Crises are inherent in capitalism. At such moments the irrationality and awful destructive power inherent in the capitalist mode of production become more readily apparent: unemployed capital at one pole and unemployed labor power at the other. Resort to the spatial fix partially masks the irrationality of capitalism, however, because it allows us to attribute devaluation through physical destruction, through global war, to purely political failings. A properly constituted Marxian theory of the relations between inner and outer transformations strips away such illusion. It lays bare the roots of crisis formation in both its national and international aspect, in geographical dimension.

The question of who is right and wrong is of immense and immediate import. If the Marxian theory of the spatial fix is right, then the perpetuation of capitalism in the twentieth century has been purchased at the cost of the death, havoc and destruction wreaked in two world wars. But each war has been waged with ever more sophisticated weapons of destruction. We have witnessed a growth in destructive force that more than matches the growth of productive force which the bourgeoisie must also create as a condition of its survival. Our present plight must surely give us pause. As the crisis tendencies of capitalism once more run amok, inter-imperialist rivalries sharpen, and the threat of autarky within closed trading empires looms. The struggle to export devaluation comes to the fore and belligerence dominates the tone of political discourse at all levels. And with this comes the renewed threat of global war, this time waged with weapons of such immense and insane destructive power, that not even the fittest stand to survive. The message which Marx long ago sought to impress upon us appears more urgent than ever:

> The violent destruction of capital not by relations external to it, but rather as a condition of its self-preservation, is the most striking form in which advice is given it to be gone and to give room to a higher state of social production.
>
> (Marx 1973: 749–50)

CHAPTER 15

The geopolitics of capitalism

First published in Social Relations and Spatial Structures, *1985.*

I wish to consider the geopolitical consequences of living under a capitalist mode of production. I shall construct my argument theoretically, but its historical relevance will, I hope, be self-evident enough to encourage debate and, perhaps, political action on a matter of deep and compelling urgency.

The core features of a capitalist mode of production

The phrase 'mode of production' is controversial, but for the purpose of my argument I can put a relatively simple interpretation upon it. We can, I think, all reasonably agree that the reproduction of daily life depends upon the production of commodities produced through a system of circulation of capital that has profit-seeking as its direct and socially accepted goal. The circulation of capital can be viewed as a continuous process in which money is used to buy commodities (labor power and means of production such as raw materials, machinery, energy inputs, and the like) for the purpose of combining them in production to make a fresh commodity that can be sold for the initial money outlay plus a profit. Schematically, this can be represented as a system of circulation of the following sort:

$$M - C \begin{cases} LP \\ \ldots P \ldots C' - M + \Delta m, \text{ etc.} \\ MP \end{cases}$$

The theory that follows is based on an analysis of such a circulation process. I shall also assume an atomistic form of competitive market society in which many economic agents engage in this form of circulation. Deviations from this assumption, except under conditions later to be specified, in no way affect the logic of my argument. I do not mean to imply, however, that everything that happens under capitalism can be reduced to some direct or even indirect manifestation of the circulation of capital. Some commodities are produced and traded without appeal to profit incentives and there are innumerable transactions between economic agents which exist outside the circulation of capital. But I do insist that the survival of capitalism is predicated on the continuing vitality of this form of circulation. If it breaks down because, for example, profits can no longer be had, then the reproduction of daily life as we now know it would dissolve into chaos. Furthermore, I shall also insist that a constant source of preoccupation under capitalism is the creation of social and physical infrastructures that support the circulation of capital. This, again, does not mean that I interpret all such phenomena as tightly functional for the circulation of capital. But legal, financial, educational, and state administrative systems, together with built environments, transportation, and urban systems, to mention but a few of the key arrangements I have in mind, have to be ranged broadly in support of the circulation of capital if daily life is to be reproduced effectively.

A deep, rigorous analysis of the circulation of capital reveals a number of core features. This was, of course, the analytic task that Marx set himself in *Capital* and I shall follow the line of thinking he constructed. Since I have explored, analyzed, and to some degree extended Marx's results at length elsewhere (Harvey 1982), I feel free to summarize without detailed proofs and justifications. At the risk of gross oversimplification I will reduce the core features of the circulation of capital to the ten points I need to ground my argument.

1. The continuity of the circulation of capital is predicated upon a continuous expansion of the value of commodities produced. This is so because the value of commodities produced at the end of the sequence (C') is greater than the value of commodities absorbed in production (C). It is this increase in value that is captured in the money form of profit (Δm). A 'healthy' capitalist economy is, therefore, one with a positive growth rate. The closer we get to a stationary state (let alone actual decline), the more unhealthy the economy is judged to be. This translates into an ideology of growth ('growth is good') no matter what the environmental, human, or geopolitical consequences.

2. Growth is accomplished through the application of living labor in production. To be sure, individual capitalists may gain profits from buying cheap and selling dear, but in so doing their gain is another's loss. Redistributions of social power through unequal exchange may be important to the rise and subsequent reorganizations of capitalism (for example, the initial concentration of wealth through merchant trading and the subsequent centralization of capital in giant corpora-

tions). But redistribution is no adequate basis for the continuous
circulation of capital. A healthy capitalist economy is one in which all
capitalists earn positive profits. And this requires that real value be
added in production. Living labor (as opposed to the 'dead labor'
embodied and paid for in other commodities) is, then, the exclusive
source of real value added in production.

3. Profit has its origin in the exploitation of living labor in production.
    'Exploitation' can here be stripped of its more emotive connotations.
    It denotes a moral condition in which living labor is treated as a
    reified 'factor' of production and a technical condition in which it is
    possible for labor to create more in production than it gets through
    the exchange of its labor power as a commodity. It does not follow
    that the laborer receives as little as possible. Situations arise in which
    the laborer gets more at the same time as the gap between what labor
    gets and what it creates in production also increases. Put another way,
    a rising material standard of living for the laborer is not necessarily
    incompatible with a rising rate of exploitation.

4. The circulation of capital, it follows, is predicated on a class relation.
    'Class' is also a loaded term. But I can here give it a restricted and
    very simple meaning. The circulation of capital entails the buying
    and selling of labor power as a commodity. The separation between
    buyers and sellers opens up a class relation between them. Those who
    buy rights to labor power in order to gain a profit (capitalists) and
    those who sell rights to labor power in order to live (laborers) exist
    on opposite sides of this buyer–seller divide. The division of class
    roles that this implies is not exhaustive of all possible or even impor-
    tant class relations under capitalism. Nor is the buying and selling of
    labor power exclusively restricted to the domain of capital circulation.
    But without the capital–labor relation expressed through the buying
    and selling of labor power, there could be no exploitation, no profit
    and no circulation of capital. Since all the latter are fundamental to
    commodity production and social reproduction, so the class relation
    between capital and labor is arguably the most fundamental social
    relation within the complex weave of bourgeois society.

5. This class relation implies opposition, antagonism, and struggle. Two
    related issues are at stake. How much do capitalists have to pay to
    procure the rights to labor power and what, exactly, do those rights
    comprise? Struggles over the wage rate and over conditions of
    laboring (the length of the working day, the intensity of work, control
    over the labor process, the perpetuation of skills, and so on) are con-
    sequently endemic to the circulation of capital. There are, of course,
    innumerable other sources of tension, conflict, and struggle, not all

of which can be reduced directly or indirectly to a manifestation of the capital-labor antagonism. But class struggle between capital and labor is so fundamental that it does indeed infect all other aspects of bourgeois life.

6. Of necessity, the capitalist mode of production is technologically dynamic. The impulsion to fashion perpetual revolutions in the social productivity of labor lies, initially, in the twin forces of intercapitalist competition and class struggle. Technological and organizational changes give individual capitalists advantages over their rivals and help secure profit in the market. They provide a weapon (not always used with unmitigated success) to control the intensity of work and to diminish the power of workers in production through the replacement of monopolizable skills. They also enable capitalists to exert leverage over the supply of labor power (and consequently the wage rate) through the creation of technologically-induced unemployment. Changes in one sphere necessitate parallel changes in another and create massive reverberations throughout the whole fabric of bourgeois society (particularly in the military sphere). Technological dynamism then appears self-perpetuating. Small wonder that the ideology of progress and its inevitability becomes deeply rooted in bourgeois life and culture.

7. Technological and organizational change usually requires investment of capital and labor power. This simple truth conceals powerful implications. Some means must be found to produce and reproduce surpluses of capital and labor to fuel the technological dynamism so necessary to the survival of capitalism.

8. The circulation of capital is unstable. It embodies powerful and disruptive contradictions that render it chronically crisis-prone. The theory of crisis formation under capitalism is complex and controversial in its details. But consideration of the preceding seven points reveals a central contradiction. The system has to expand through the application of living labor in production whereas the main path of technological change is to supplant living labor, the real agent of expansion, from production. Growth and technological progress, both necessary features of the circulation of capital, are antagonistic to each other. The underlying antagonism periodically erupts as fully-fledged crises of accumulation, total disruptions of the circulation process of capital.

9. The crisis is typically manifest as a condition in which the surpluses of both capital and labor which capitalism needs to survive can no longer be absorbed. I call this a state of *overaccumulation*. Surplus capital and surplus labor power exist side by side with apparently no

way to bring the two together to accomplish socially-useful tasks. The irrationality that lurks at the heart of a supposedly rational mode of production comes to the surface for all to see. This is the kind of irrationality, with massive unused productive capacity and high unemployment, into which most western economies have sunk these past few years.

10. Surpluses that cannot be absorbed are devalued, sometimes even physically destroyed. Capital can be devalued as money (through inflation or default on debts), as commodities (unsold inventories, sales below cost price, physical wastage), or as productive capacity (idle or underutilized physical plant). The real income of laborers, their standard of living, security, and even life chances (life expectancy, infant mortality, and the like) are seriously diminished, particularly for those thrown into the ranks of the unemployed. The physical and social infrastructures that serve as crucial supports to the circulation of capital and the reproduction of labor power, may also be neglected. Crises of devaluation send deep shockwaves throughout all aspects of capitalist society. They often spawn acute social and political tensions. And out of the associated ferment new political forms and ideologies can spring.

I hold that crises are inevitable under capitalism no matter what measures are taken to mitigate them. The tension between growth and technological progress is just too powerful to be contained within the confines of the circulation of capital. It is, however, open to human ingenuity and political action to alter both the timing, spatial extent, and form of manifestation of the crisis. We shall, in what follows, examine some of these possibilities. It is also open to human ingenuity and political action to convert crises into catalytic, though traumatic, moments in human progress, rather than letting them dissolve into barbarism, testimony to the frailty and futility of all enlightened human aspirations. To seize the moment of crisis as the opportunity for creative revolutionary change, however, requires deep understanding of how crises form and unfold.

Surpluses of capital and labor power: the pivot of capitalist development

The historical geography of capitalism can best be viewed from the standpoint of the triple imperatives of production, mobilization and absorption of surpluses of capital and labor power. Without the prior creation and mobilization of such surpluses, the circulation of capital could not even begin nor expansion be sustained. On the other hand, the continuous production of potential capital surpluses in the form of profit, coupled with revolutions in technology that throw people out of work, just as perpetually pose the problem of how to absorb such surpluses without devaluation. The probability of crisis perpetually brews within this tension between the need to produce and the need to absorb surpluses of capital and labor power.

Original accumulation, according to Marx, rested upon the violent expropriation of the means of production that put capital surpluses in the hands of the few while the many were forced to become wage laborers in order to live. The migration of surplus labor power from the country to the city, the urban concentration of wealth by merchants (looting the world through unfair exchange) and usurers (undermining landed property and converting it into money wealth) together with the extraction of a surplus product from the countryside for the benefit of the city, facilitated both the social and geographical concentration of the surpluses. The important point, however, is to recognise that capital and labor power surpluses can be generated *outside* the circulation of capital and mobilized through diverse processes of primitive accumulation and geographical concentration.

The necessary surpluses can also be produced *within* the circulation process of capital. Profit can be converted into capital. Indeed, a necessary condition for the realization of profit in the present is a conversion of a part of past profit into fresh capital investment. Only in this way can the necessary expansion upon which the survival of capitalism depends be sustained (Harvey 1982: ch. 3). The production of labor power surpluses poses a deeper problem. Unemployment can be created by technological change but the maintenance of a constant pool of surplus laborers by such a mechanism means that the crises sparked by the tension between technological change and growth would be frequent and deep. Primitive accumulation, the mobilization of 'latent' reserves (women and children, workers from non-capitalist sectors) and population growth provide alternative sources of surplus labor power. Within a purely capitalist society, it appears that a positive rate of population growth is the securest long-term foundation for relatively trouble-free accumulation, although in the short run the massive movement of, say, women into the workforce can also suffice (Harvey 1982: ch. 6). But here we encounter a problem because the reproduction of labor power is not under the direct control of the capitalist. The latter may pay a social wage sufficient to reproduce and expand the labor force, improve its qualities even. They may create all kinds of social means to try and influence workers to have or not to have children. But the workers' response cannot be guaranteed. Labor power is not, therefore, a commodity like any other. How the dynamics of accumulation mesh with population growth cannot be predicted in advance and the whole relation of the circulation of capital to the reproduction of labor power remains a thorny, perhaps unresolvable problem.

Capital and labor power surpluses, however produced, have to be absorbed. Under normal conditions, we might expect the capitalist penchant for accumulation to take care of matters, albeit with strong cyclical rhythms and occasionally uncomfortable discontinuities. There are two general circumstances where this is not the case and both merit discussion. First, strong disproportionalities in the ratio of capital to labor power surpluses can leave one or the other devalued. Secondly, during crises, produced surpluses of both capital and labor power cannot be absorbed and both are then devalued.

The processes whereby surpluses of capital and labor power are produced do not guarantee that they can be assembled in time and space in exactly the right proportions to be absorbed into a given process of circulation of capital. To some degree, the technologies embedded within the circulation of capital can adjust to accommodate such differences, though often at the cost of radical restructuring. Free geographical mobility of unevenly-distributed surpluses can also help. Nevertheless, situations arise and even persist, in which surpluses of one sort cannot be absorbed because surpluses of another sort are not present in the requisite quantities and qualities. Either capital or labor power are devalued, but not both. To the degree that the dominant power relation favours capital, so the most likely persistent condition will be that of capital shortage and labor power surpluses with all the attendant social devastation that attaches to the devaluation of labor power.

The condition that interests me most, however, is that in which unemployed surpluses of capital and labor power exist side by side. This is the condition of crises into which capitalism is periodically and inevitably plunged because its technological dynamic undermines its capacity to sustain growth. Both capital and labor power are then devalued. Is there no way to avoid such an unmitigated social, economic and perhaps even political disaster? To pose that question is, in effect, to ask: are there ways to absorb the surpluses productively through opening up new conduits and paths for the circulation of capital? In what follows, I shall argue that spatial and temporal displacements offer ample opportunities to absorb the surpluses with, however, dramatic consequences for the dynamics of accumulation. I shall then go on to show that neither stratagem offers a permanent resolution to the inner contradictions of capitalism but that resort to either (or both) fundamentally alters the way in which crises are expressed.

Temporal displacement through
long-term investments

The circulation of capital has to be completed within a certain time span. This I call the 'socially necessary turnover time', the average time taken to turn over a given quantity of capital at the average rate of profit under normal conditions of production and circulation. Individual capitalists who turn over their capital faster than the social average earn excess profits. Those who fail to make the average suffer relative devaluation of their capitals. Competition then generates pressures to accelerate turnover times through technological and organizational change. Any aggregate acceleration releases surpluses of both capital and labor power. By the same token, abnormal conditions of devaluation are usually signalled by a general slow down (we speak of a 'sluggish' economy, with unsold inventories piling up, and the like).

But some capital necessarily circulates at a much slower pace, such as fixed capital (machinery, physical plant and infrastructures) and within the consumption fund (consumer durables, housing, and so on). The production of science and technology, and the provision of social infrastructures of education, healthcare, social services, judiciary, state administration, law enforcement and military protection, define areas in which the gestation time of projects is typically long and the return of benefits (if any) spread out over many years. Investments of this sort depend upon the prior creation of surpluses of both capital and labor power relative to current consumption needs. We now encounter the happy circumstance that such surpluses are continuously being generated within the circulation process of capital. What better way to absorb them than to shift them into long-term projects in the formation of physical and social infrastructures? Indeed, investment in science and technology and in the habituation of workers (through education or repression) to more intensive work rhythms, as well as in new machinery, transport and communications systems, information systems, new distribution capacities and the like, can all help to promote faster aggregate turnover times. A part of the circulation of capital slows down in order to promote accelerating turnover times for the remainder.

The possibility exists here for a dynamic equilibrium in which surpluses are absorbed in the creation of physical and social infrastructures that facilitate the creation of even more surpluses. Such a 'spiral' form, I believe, largely accounts for those phases of capitalist development where internal growth appears self-sustaining. It will also almost certainly be marked by massive transformations in employment structures because increasing productivity in basic production is accomplished by increasing absorption of surpluses in the production and maintenance of social and physical infrastructures.

But at some point the spiral encounters barriers that cannot be overcome. It is then typically interrupted by a crisis in which the labor power and capital deployed everywhere are subject to devaluation. I now have to show how and why such interruptions are unavoidable. To do this, I must first explain how surpluses are actually shifted from current production and consumption into long-term investments in physical and social infrastructures.

Consider, first, the reallocation of labor power. There are serious frictional problems because redundant shoemakers cannot instantaneously become scientists and it would be a very talented roadmender indeed who could switch easily into teaching as conditions dictate. Labor power is not qualitatively homogeneous, and surpluses of one sort cannot usually be instantaneously absorbed elsewhere. The transformation of employment and occupational structures is inevitably slow and this itself can check the persistence of any spiral form of development.

The reallocation of surplus capital likewise poses problems. The surplus can exist as money, commodities or productive capacity. If it exists as particular use values (shoes and shirts) or productive capacity (lathes and lasts) it cannot be converted directly into a railroad or a new educational service. It must first be converted into money. This presents the first barrier to be overcome because overaccumulation defines a state in which the smooth conversion of capital from one form to another, and into money in particular, has become impossible. Credit can surmount this barrier but it does not do so unambiguously. Why would that credit not be used to create even more surpluses of the same sort? Why would it be attracted into long term investment when every capitalist is deeply conscious of the need to conform to the dictates of socially-necessary turnover time? Here is the second barrier to be overcome. The answer to it is as old as capitalism itself. It lies in the creation of 'fictitious capital': bonds, mortgages, stocks and shares, government debt, and the like (Harvey 1982: chs 9 and 10). What fictitious capital does is to convert a long-drawn-out circulation process (the capital embedded in a railroad, for example) into an annualized rate of return. It does this by facilitating the daily buying and selling of rights and claims to a share in the product of future labour. The rate is sometimes fixed (bonds) or variable, according to what labor actually produces year by year (shares). But it is measured in terms exactly comparable to the rate of profit over socially-necessary turnover time in current production.

Through the use of credit and fictitious capital, surplus capital can flow from one sphere to another. When, for example, the annualized yield on some fictitious capital (railroad shares, government debt, and the like) exceeds the rate of profit in current production, then there is an incentive for capital to switch from present to future uses. The switch is unlikely to be smooth, however, because of the 'lumpiness' of many of the investments involved (railroads, hospitals, and the like) and the different working periods required to make a project operational. Furthermore, the nature of many of the investments – public uses and the difficulty of charging directly for their use – often precludes action on the part of individual capitalists, so that new special organizational forms (joint stock companies, state and quasi-public enterprises) have to be created if railroads, ports, universities, scientific and educational centres, and the like, are to be created. Capital markets must also be well organized if clear market signals as to differentials in the annualized rate of return are to be defined. And, finally, the fictitious qualities of investments tied to the product of labor stretched out long into the future introduces strong elements of risk, uncertainty, human judgement and anticipation.

In this intricate world, where myriad investors make decisions as to how best to deploy their capital within a financial system where commodity futures trade side by side with government and corporate debt, property mortgages, stocks and shares, and so on, dynamic equilibrium between short- and long-term circulation processes could be achieved through the purest fluke. There lurk within this world all manner of traps and pitfalls, so many opportunities for errors of judgement to compound into configurations of savage disequilibrium. Credit plus fictitious capital may be the magic potion to make all capitals instantaneously convertible, but plainly it is a volatile mix, capable of almost instantaneous combustion in the fires of crisis formation. Yet we cannot here define the *necessity* for crises, only so many possibilities that could still, in principle at least, be kept in check through compensating oscillations. The necessity for crisis must be otherwise established.

Consider, then, dynamic equilibrium at its smoothest and simplest. The credit equivalent of surplus use values (commodities and productive capacity) is added to surplus money capital arising out of current production and invested as fictitious capital in long-term projects. Surplus labour power then finds employment. The extra demand for wage goods and means of production matches the surplus use values in current production. Inventories shrink and capacity utilization rises. Prices and profits recover, reinvestment in current production resumes and further surpluses of both capital and labor power are generated to be absorbed once more through fictitious capital formation and further investment in long-term projects. Such a process can plainly continue *ad infinitum* provided there is no limit to the volume of fictitious capital formation.

Fictitious capital, however, is a claim on future labor. If its value is to be realised, then future labor must be deployed in such a way as to ensure a rate of return on the initial investment. What happens, in effect, is that present problems are absorbed through contracting future obligations. To the degree that the problem is absorbed rather than eliminated, dynamic equilibrium means continuous temporal displacement through accelerating fictitious capital formation. The volume of indebtedness increases and future labor is increasingly imprisoned within a framework of contractual obligations (see Figure 15.1). At some point, the debts must be paid. Exactly when depends upon the turnover time of the capital deployed in particular physical and social infrastructures. But accelerating fictitious capital formation – the true heart of the spiral of development – means that more and more living labor in current production has to be given over to working off past obligations.

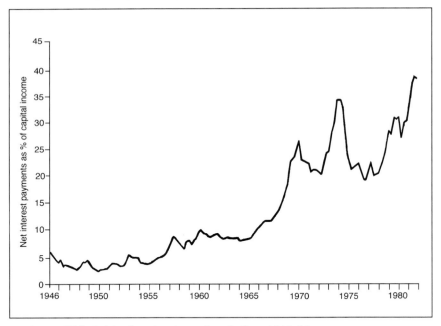

Figure 15.1 Indebtedness in advanced capitalism, 1946–80

Two possibilities for crisis formation then arise. Under the first, over-accumulated capital stored in physical and social infrastructures is realised through an active growth in current production (facilitated in part through improved infrastructures). But then overaccumulated capital flows back from storage to combine with excess capital in current production to create ever-greater pools of surplus capital. The capacity for further fictitious capital formation is blocked either by labor or resource constraints or by the circulation of capital in the existing infrastructures which cannot be disturbed before their life is out without devaluation. A general crisis ensues with surpluses everywhere subject to devaluation. Under the second, the capital stored in physical and social infrastructures is not realized and is devalued. The crisis now appears to be provoked through the lag of productivity (and perhaps through shortages of both capital and labor) in current production relative to the volume of debts contracted. Devaluation then focuses on the debts. These can be devalued socially through monetization (inflation) or individually through default on privately-contracted obligations.

In the long run, crises cannot be avoided. But how long is the long run? By stretching out the tendency to overaccumulate far into the future, crises can perhaps be staved off for many years. But the longer the crises are staved off, the greater the quantity of fictitious capital, the more the overaccumulation problem itself accumulates in pent-up form, and the deeper the ultimate crisis. But 'ultimate' has no strict date. Even in the midst of crisis, debts can be restructured and stretched out to avoid the full impact here and now.

The form of crisis can also change. For example, the absorption of surpluses of capital and labor power in bouts of speculative railroad and urban building, so typical of the nineteenth century, produced periodic crises of overaccumulation of such assets. The timing of the crises was largely dictated by the typical turnover times of such projects. The fictitious capital (railroad stocks and shares, builders' debts) was devalued, debts written off, enterprises bankrupted and labor laid off. Though increasingly intolerable from the standpoint of both capital and labor, this system had the virtue of leaving the use-value of the asset behind while cleaning out overaccumulated capital vigorously and unambiguously. By contrast, the massive absorption of surpluses through state action (highway construction, healthcare, education) so characteristic of the period since 1945, together with state support for private indebtedness, has more recently put the accent on state-backed debt. The construction cycle has all but disappeared and traditional restraints to fictitious capital formation have been removed through state action that effectively underwrote an extended economic boom that lasted a whole generation. The state could monetise the debt away by printing money. But this produced inflation, a form of crisis that builds slowly and spreads devaluation across the whole of society. The trouble, of course, is that any attack upon inflation reveals the chronic debt problem and any attack upon that

reveals that productivity has not kept pace with accelerating debt formation (how could it?). The end-result is the conversion of an inflationary
crisis into a more conventional deflation in which devaluation has to be
administered by the state.

While it is not my intention to clothe this theoretical argument in
historical verisimilitude, I do not believe it would be hard to do so. The
postwar boom, for example, was in part fuelled through accelerated fictitious capital formation and increased indebtedness backed by state power.
The effect has been to create such a pent-up force for devaluation that
it is hard to see how capitalism can work its way out of it. We now see
capital and labor power being devalued in production, investment and
maintenance of social and physical infrastructures neglected and part of
the debt written off (usually through restructuring and stretching out, as
has been done with New York City, Mexico, Brazil, Poland, and others).
And state policy is caught between the Scylla of accelerating inflation
and the Charybdis of savage devaluation. The prospect looms that we may
be paying for the 1945–69 boom through stagnation and depression until
the end of the century. We cannot, however, push such an argument much
further without exploring the spatial dimension.

Theorizing the historical geography
of capitalism

The question now to be resolved is whether the interior dilemmas of capitalism can be resolved through geographical expansion or restructuring.
Is there, in short, a 'spatial fix' to the internal contradictions of capitalism?
The export of surpluses of labor power and capital, after all, appears an
easy enough way to avoid devaluation. All kinds of possibilities exist here
to stave off crises, sustain accumulation, and modify class struggle
through geographical shifts and restructurings. But the end-result, I shall
conclude, is that crises become more global in scope at the same time as
geopolitical conflicts become part and parcel of the processes of crisis
formation and resolution.

The path to this conclusion is strewn with all number of difficulties.
The issue of space and geography is a sadly neglected stepchild in *all*
social theory in part, I suspect, because its incorporation has a numbing
effect upon the central propositions of *any* corpus of social theory.
Microeconomists working with a theory of perfect competition encounter
spatial monopolies, macroeconomists find as many economies as there are
central banks and a peculiar flux of exchange relations between them, and
Marxists looking to class relations find neighbourhoods, communities,
regions and nations. Marx, Marshall, Weber, and Durkheim all have this

in common: they prioritize time and history over space and geography and, where they treat of the latter at all, tend to view them unproblematically as the stable context or site for historical action. Changing space relations and geographical structures are accommodated by ad hoc adjustments, externally imposed redefinitions of regions and territories within and between which the perpetual flow of the social process takes place. The way in which the space-relations and the geographical configurations are produced in the first place passes, for the most part, unremarked, ignored.

There is something extremely unsatisfactory about all this. To begin with, the suspicion lurks and many a theory of imperialism asserts that the survival of capitalism into the twentieth century has been assured only through the transformation of space-relations and the rise of distinctive geographical structures (such as core and periphery, First and Third worlds). The 'innovation waves' that others, impressed by Schumpeter, see as so fundamental to the absorption of surpluses of capital and labor power over time have often had everything to do with the transformation of space – railroads and steamships, the automotive industry, aerospace and telecommunications. The multinational corporation, with its capacity to move capital and technology rapidly from place to place, to tap different resources, labor markets, consumer markets and profit opportunities, while organizing its own territorial division of labor, derives much of its power from its capacity to command space and use geographical differentials in a way that the family firm could not. In any case, the implications of the dramatic transformations wrought in the geography of production, consumption and exchange throughout the history of capitalism are in themselves, surely, worthy of study.

Direct confrontation of that task might help heal divisive and wounding schisms within the Marxian tradition. Marx himself boldly sketched a theory of capitalist history powered by the exploitation of one class by another. Lenin, on the other hand, spawned a different tradition in which the exploitation of people in one place by those in another (the periphery by the centre, the Third World by the First) takes centre stage. The two rhetorics of exploitation coexist uneasily and the relation between them remains obscure. The theoretical foundation of Marxism-Leninism is thereby rendered ambiguous, sparking savage debates over the right to national self-determination, the national question, the prospects for socialism in one country, the universalism of class struggle, and the like.

To be sure, I caricature to some degree the thought of both Marx and Lenin to highlight a fundamental problem. Marx frequently admits of the significance of space and place in his writings. The opposition of town and country, the significance of the territorial division of labor, the concentration of productive forces in urban agglomerations, geographical differentials in the value of labor power and even in the operation of the law of value, the importance of reducing spatial barriers through innovations in transport and communications, are all present in his works (Harvey 1975, 1982). And historically he has to admit that the transition to capitalism (and the prospects for socialism) differs from place to place even within western Europe (to say nothing of Russia and Asia). The politics of the Irish question also forced him to confront regional and cultural divergence as fundamental to waging class struggle. But none of this is really integrated into theoretical formulations that are powerful with respect to time but weak with respect to space. Geographical variation is excluded as an 'unnecessary complication'. His political vision and his theory are, I conclude, undermined by his failure to build a systematic and distinctively geographical and spatial dimension into his thought.

At first sight this is the gap that Lenin appears to fill. Overtly he roots his arguments deep in Marx, but his study of the origins of capitalism in Russia and the fact of interimperialist rivalries culminating in the First World War led him directly to introduce geographical and spatial dimensions into his arguments. But the modifications, it turns out, are ad hoc adjustments that say no more than that capitalism undergoes its own specific course of development depending upon conditions in this or that territory and that the fundamental dynamic of capitalism forces the major capitalist powers into geopolitical struggles and confrontations. To convert the Marxian insights into a geopolitical framework, Lenin introduced the concept of the state which, to this day, remains *the* fundamental concept whereby territoriality is expressed. But in so doing, Lenin largely begged the question as to how or why the circulation of capital and the deployment of labor power should be national rather than global in their orientation and why the interests of either capitalists or laborers should or even could be expressed as national interests. Lenin gave geographical expression to the dynamics of capitalism at the expense of reopening the historical question of the relation between civil society and the state.

I do not accept the idea that space-relations and geographical structure can be reduced to a theory of the state, or that a prior theorisation of the rise of the capitalist state is necessary to reconstruct the historical geography of capitalism. Our task is, rather, to construct a general theory of space-relations and geographical development under capitalism that can, among other things, explain the significance and evolution of state functions (local, regional, national, and supranational), uneven geographical development, interregional inequalities, imperialism, the progress and forms of urbanisation and the like. Only in this way can we understand how territorial configurations and class alliances are shaped and reshaped, how territories lose or gain in economic, political, and military power, what the external limits are to internal state autonomy (including the transition to socialism), or how state power, once constituted, can itself become a barrier to the unencumbered accumulation of capital, or a strategic centre from which class struggle or interimperialist struggles can be waged.

The historical geography of capitalism has to be the object of our theorizing, historico-geographical materialism the method of enquiry. Bravely said, but hard to do. To begin with, we encounter an incredible variety of physical and biotic environments across the surface of the earth, many of which have been substantially modified by centuries of human action. The diversity of that action has produced a variegated geographical landscape in which cultural and socio-structural differentiations have taken deep root. Such particularistic geographical differentiation may be encompassed but by no means totally crushed under the homogenizing heel of the circulation of capital. Viewed abstractly, space also possesses more complex and particularistic properties than time. It is possible to reverse field and move in many different directions in space whereas time simply passes and is irreversible. The metric for space is also less easily standardized. Time or cost of movement over space do not necessarily match each other and both yield different metrics to simple physical distance. Compared with this, the chronometer and the calendar are wondrously simple. Geographical space is always the realm of the concrete and the particular. Is it possible to construct a theory of the concrete and the particular in the context of the universal and abstract determinations of Marx's theory of capitalist accumulation? This is the fundamental question to be resolved.

The production of spatial organization

Marx was not necessarily wrong to prioritize time over space. The aim and objective of those engaged in the circulation of capital must be, after all, to command surplus labor *time* and convert it into profit within the *socially-necessary turnover time*. From the standpoint of the circulation of capital therefore, space appears in the first instance as a mere inconvenience, a barrier to be overcome. Capitalism, Marx concludes with remarkable insight, is necessarily characterized by a perpetual striving to overcome all spatial barriers and 'annihilate space with time' (Marx 1973: 539). But it transpires that these objectives can be achieved only through the production of fixed and immobile spatial configurations (transport systems, and so on). In the second instance, therefore, we encounter the contradiction: spatial organization is necessary to overcome space. The task of spatial theory in the context of capitalism is to construct dynamic representations of how that contradiction is expressed through historical-geographical transformations.

The starting-point for such a theory must lie at the interface between transport and communications possibilities on the one hand and locational decisions on the other. Marx, for example, argued strongly for the idea that the capacity to overcome spatial barriers and annihilate space with time through investment and innovation in transport and communications systems belongs to the productive forces of capitalism. I note in passing that G. A. Cohen lists space but not the capacity to overcome space in his otherwise definitive list of productive forces (Marx 1973: 533–4; Cohen 1978). The impulsion to revolutionise productive forces is as strong in this field as in any other. The history of capitalism has therefore been marked by dramatic reductions in the cost or time of movement together with improvements in continuity of flow. Space relations are thereby continuously subject to transformation. Other forms of technological change can achieve the same objective but by a different route. There are abundant contemporary examples of changes that liberate production from dependence upon localized labor skills, raw materials, intermediate products, energy sources and the like. By increasing the range of possible substitutions within a given production process, capitalists can increasingly free themselves from particular geographical constraints.

But since technologically-defined spatial constraints of some sort always exist, the question remains: what happens within their confines? Obviously, capital and labor power must be brought together at a particular point in space for production to proceed. The factory is such an assembly point while the industrial form of urbanization can be seen as a specific capitalist response to the need to minimize the cost and time of movement under conditions of inter-industry linkage, a social division of labor, and the need for access to both labor supplies and final consumer markets. Individual capitalists, by virtue of their particular locational decisions, shape the geography of production into distinctive spatial configurations.

The upshot of all such processes is a tendency towards what I will call a *structured coherence* to production and consumption within a given space. This structured coherence, as Aydalot notes, embraces the forms and technologies of production (patterns of resource use interindustry linkages, forms of organization, size of firm), the technologies, quantities, and qualities of consumption (the standard and style of living of both labor and the bourgeoisie) patterns of labor demand and supply (hierarchies of labor skills and social reproduction processes to ensure the supply of same) and of physical and social infrastructures (on which more anon) (Aydalot 1976). The territory within which this structured coherence prevails is loosely defined as that space within which capital can circulate without the limits of profit within socially-necessary turnover time being exceeded by the cost and time of movement. An alternative definition would be that space within which a relatively coherent labor market prevails (the space within which labor power can be substituted on a daily basis – the commuter range defined by cost and time of daily labor movement – is a very important spatial disaggregation principle under capitalism). The territorial coherence becomes even more marked when formally represented by the state. Policies regulating the labor process, labor organizing, standards of living of labor (welfare policies, and so on), appropriate regulation and remuneration of capital and the like, apply across the whole territory. The coherence is reinforced informally, though no less powerfully, through the persistence or creation of national, regional or local cultures and consciousness (including traditions of class struggle) that give deeper psychic meaning to territorial perspectives.

There are processes at work, therefore, that define *regional spaces* within which production and consumption, supply and demand (for commodities and labor power), production and realization, class struggle and accumulation, culture and lifestyle, hang together as some kind of structured coherence within a totality of productive forces and social relations.

But there are processes at work that undermine this coherence. They are contained in the core features of capitalism identified at the start of this essay. First, accumulation and expansion, together with the need to produce and absorb labor power and capital surpluses, build pressures within a region that spill outwards (for example, capital export) or pull inwards (for example, immigration). Secondly, revolutions in technology that liberate both production and consumption from spatial constraints, together with improved capacity to overcome spatial barriers and annihilate space with time, render the boundaries of a region highly porous and unstable. Territorial specialization and interregional linkages grow with increased facility of spatial integration. Thirdly, class struggle within a territory may force capitalists or laborers to look elsewhere for conditions more conducive to their respective survival. Fourth, revolutions in capitalist forms of organization (the rise of finance capital, multinational corporations, branch plant manufacturing and so on) permit greater command over progressively larger spaces by associated capitalists.

Such forces tend to undermine any structured coherence within a territory. They may emphasize the international rather than the locally-integrated division of labor and make interregional interdependence more significant than a regionally defined coherence. They may render state territorial boundaries inappropriate and force their modification. They may even undermine the power of the local or nation-state through the production of a fiscal crisis that demands its remedy in a state-backed-attack upon standards of living of labor, upon traditional hierarchies within the labor force, upon the power of local capitalists *vis-à-vis* the multinationals. Regional consciousness and culture may likewise be under-mined, transformed into pale shadows of their former selves.

The persistence of any kind of structured regional coherence, in the face of such powerful forces, appears surprising. It is due in part to the peculiar infrastructural requirements to improve the spatial mobilities of capital and labor power. Since improvements of this sort are correctly judged as eminent threats to regional coherence, we evidently have a paradox on our hands that deserves further explication.

Consider, first, the mobility of capital. This, I have shown elsewhere, must first be disaggregated into the mobility of capital of different sorts (Harvey 1982: ch. 12). The cost and time of money movement in these days of sophisticated credit systems and telecommunications is phenomenally low. Here more than anywhere else we can see the state of perfection achieved under capitalism in annihilating space with time. The cost and time taken to move commodities has also fallen over the past century and a half to the point where transport costs no longer play an important role in the location decisions of all but a handful of industries. The geo-graphical mobility of production capacity, on the other hand, faces tougher constraints. The more an industry depends upon fixed and immobile capital equipment of relatively long life, the less easily it can move with-out devaluation. These differential capacities for geographical mobility of capital in different states within the overall circulation process of capital introduces all kinds of tensions *within* that circulation process in space.

I will lay these aside for the moment in order to get to the fundamental point. Each form of the geographical mobility of capital requires fixed and secure spatial infrastructures if it is to function effectively. The incredible power to move money around the world, so characteristic of the contemporary era, demands not only a well-organized telecommunications system but, as a minimum, secure backing of the credit system by state, financial, and legal institutions. The territoriality of money and the signif-icance of state power to guarantee the quality of money within its territory come into their own. The capacity to move commodities likewise depends upon the construction of a sophisticated, efficient, and stable transport system backed by a whole set of social and physical infrastructures (from legal services to warehouses) to facilitate and secure exchange. Production, for its part, uses not only the fixed and immobile capital directly employed by it, but also depends upon a whole matrix of physical and social services (from sewers to scientists) that must be available *in situ*. Producers, it then follows, can improve their capacity to move to the degree that agents other than themselves (primarily the state) become responsible for greater and greater portions of the fixed and immobile infrastructural costs. The enhanced mobility of production capital over the past two decades has derived exactly from such stratagems.

Consider, now, the geographical mobility of labor power. All kinds of crosscurrents of complexity prevail here which nevertheless bring us to a similar basic result. From the standpoint of the capitalist development process as a whole, the free geographical mobility of labor power and its easy adaptation to the shifting circulation of capital in space appears a necessary condition. On the other hand, individual capitalists plainly prefer a stable, reliable workforce and captive labor supplies (with adequate labor power surpluses to ensure capitalist control over both the labor process and wage rates). To this end, they may actively support basic social reproduction processes (education, religion, healthcare, social services, even welfare) geared to the production and preservation of labor power of a certain quantity and quality within a given territory. They may support state actions that constrain the free mobility of labor power. The laborers, for their part, face a similar dilemma. If they cannot escape from the wages system altogether, they will presumably move in order to improve their real wages, work conditions, and the like. The irony here is that the capitalist development process relies on exactly such behavior to coordinate the demand and supply of labor power in space. But on the other hand, workers can also improve their lot if they stay in place, collectively organize, and fight for a better life. To this end, they may build their own social and physical infrastructures (or co-opt those promoted by the bourgeoisie), struggle to control the state apparatus and thereby enhance their power to improve their lives. And to the extent they succeed, they too may support measures that constrain the free geographical mobility of labor power (immigration in particular). The tension between free geographical mobility and organized reproduction processes within a confined territory exists for both capitalists and laborers alike. And how that tension is resolved for either depends crucially on the state of class struggle between them. Capital flight (and the consequent undermining of territorial coherence and state power) is as typical a response to working-class victories within a territory as is individual worker mobility to escape the more vicious forms of capitalist exploitation. Again, I will lay these tensions aside for the moment in order to get to my immediate point: none of this can occur oblivious of the immobile social and physical infrastructures necessary to ensure the reproduction of labor power of a certain quantity and quality.

We can now draw a fundamental conclusion. The ability of both capital and labor power to move at short order and low cost from place to place depends upon the creation of fixed, secure, and largely immobile social and physical infrastructures. The ability to overcome space is predicated on the production of space. But the required infrastructures absorb capital and labor power in their production and maintenance. We here approach the heart of the paradox. A portion of the total capital and labor power has to be immobilized in space, frozen in place, in order to facilitate greater liberty of movement for the remainder. But the argument now can be brought full circle because the viability of the capital and labor committed to the production and maintenance of such infrastructures can be assured only if the remaining capital circulates down spatial paths and over a time-span consistent with the geographical pattern and duration of such commitments. If this condition is not met – for example, if insufficient traffic is generated to make the railroad profitable, or expansion of production does not follow on massive investment in education – then the capital and labor committed are subject to devaluation. Geographical shifts in the circulation of capital and the deployment of labor power can have an equally devastating, though geographically specific impact upon physical and social infrastructures as the temporal disruptions described earlier in this essay.

Let me now summarize the argument. The structured regional coherence towards which the circulation of capital and the interchange of labor power tends under technologically-determined spatial constraints itself tends to be undermined by powerful forces of accumulation and overaccumulation, technological change and class struggle. The power to undermine depends, however, upon the geographical mobilities of both capital and labor power and these depend, in turn, upon the creation of fixed and immobile infrastructures whose relative permanence in the landscape of capitalism reinforces the structured regional coherence being undermined. But then the viability of the infrastructures is in turn put at risk through the very action of the geographical mobilities they facilitate.

The result can only be a chronic instability to regional and spatial configurations, a tension within the geography of accumulation between fixity and motion, between the rising power to overcome space and the immobile spatial structures required for such a purpose. This instability, I wish to stress, is something that no amount of state interventionism can cure (indeed, it has the habit of generating all manner of unintended consequences out of seemingly rational state policies). Capitalist development must negotiate a knife-edge between preserving the values of past commitments made at a particular place and time, or devaluing them to open up fresh room for accumulation. Capitalism perpetually strives, therefore, to create a social and physical landscape in its own image and requisite to its own needs at a particular point in time, only just as certainly to undermine, disrupt and even destroy that landscape at a later point in time. The inner contradictions of capitalism are expressed through the restless formation and re-formation of geographical landscapes. This is the tune to which the historical geography of capitalism must dance without cease.

The formation of regional class alliances and the instability thereof

All economic agents (individuals, organizations, institutions) make decisions on the circulation of their capital or the deployment of their labor power in a context marked by a deep tension between cutting and running to wherever the rate of remuneration is highest, or staying put, sticking with past commitments and recouping values already embodied. How this tension between fixity and motion is worked out is fundamental to our theory. This is the conceptual bridge that allows us, if properly constructed, to integrate Marx's history with Lenin's geography of capitalist dynamics.

What I will try to show is that regional class alliances, loosely bounded within a territory and usually (though not exclusively or uniquely) organized through the state, are a necessary and inevitable response to the need to defend values already embodied and a structured regional coherence already achieved. The alliance can also actively promote conditions favourable to further accumulation within its region. But I shall also show that such alliances are bound to be unstable. They cannot contain the fundamental forces making for crises while they internalize potentially explosive class and factional divisions. Their boundaries are also highly porous and subject to modification.

Different factions of capital and labour have different stakes within a territory depending upon the nature of the assets they control and the privileges they command. Some are more easily drawn into a regional class alliance than are others. Land and property owners, developers and builders, those who hold the mortgage debt, and state functionaries have most to gain. Those sectors of production which cannot easily move (by virtue of the fixed capital they employ or other spatial constraints) will tend to support an alliance and be tempted or forced to buy local labor peace and skills through compromises over wages and work conditions. Factions of labor that have through struggle or out of scarcity managed to create islands of privilege within a sea of exploitation will also just as surely rally to the cause of the alliance to preserve their gains. If a local compromise between capital and labor is helpful to both accumulation and the standard of living of labor (which it can be for a time), then most factions of the bourgeoisie and the working classes may support it. Nor, I want to stress, is the alliance purely defensive in posture. Experience shows that an efficiently organized regional economy (that structured coherence to which we have already referred) replete with adequate social and physical infrastructures, can be beneficial to most. Community and regional boosterism becomes very much part of the game as all elements within the alliance seek to capture and contain the benefits to be had from channelling flows of capital and labor power through the territory under their control. The struggle for community, regional or national solidarity as the ideology behind the alliance, may support, reconstitute or in some cases (as I believe can be shown for the United States) actively create local and regional cultures and traditions. The conclusion is inescapable: if regional structures and class alliances did not already exist, then the processes at work under capitalism would necessarily create them.

I advance this proposition independently of any appeal to the concept of the state. I do so because I want to stress that the drive towards state formation and dissolution under capitalism has to be understood in the context of forces making for the formation and dissolution of regional class alliances. The state is different, however, from other agents in a variety of respects. First, territory and the integrity of territory is the objective of its personnel to a degree uncharacteristic of other agents. Secondly, by virtue of its authority, it can give firmer shape and cohesion to regional class alliances through the institutions of law, governance, political participation and negotiation, repression and military might. Thirdly, it can impose relatively firm boundaries on otherwise porous and unstable geographical edges. Finally, by virtue of its powers to tax and to control fiscal and monetary policy, it can actively promote and sustain that structured regional coherence to production and consumption to which capitalism in any case tends and undertake infrastructural investments that individual capitalists could not tackle. It can also become a central agent for the promotion of nationalist ideology. For all these reasons, the state becomes the key to the expression of the tendency to form regional class alliances and adds its own specific rationale to this fundamental underlying process.

The upshot is a regional class alliance that typically builds upon the apparatus of state power, engages in community boosterism and strives for community or national solidarity as the means to promote and defend an amalgam of various class and factional interests within a territory. Spatial competition between localities, cities, regions and nations takes on new meaning as each regional alliance seeks to capture and contain benefits in competition with others. Global processes of class struggle appear to dissolve before our eyes into a variety of interterritorial conflicts. Lenin is vindicated.

But the stability of any regional class alliance is undermined by exactly the processes that Marx described so well. Accumulation and overaccumulation, class struggle and technological change disrupt and transform regional alliances in much the same way that they affect all fixed spatial configurations. Even the most solid partners in a regional alliance may be tempted to move even at the best of times, and, at the worst of times, individual behaviour becomes very unpredictable. Competition forces all economic agents to be on the alert for the main chance to make a geographical move that gives them an advantage over their rivals. The instability in part results simply because individuals do not have the luxury of knowing exactly what their rivals will do. Similar problems arise in the realms of class struggle. While capital and labor may move into an alliance on some issues (barriers to cheap imports, for example) and compromise on others (collective bargaining procedures, for example) the antagonism between them can never totally disappear. And when class struggle sharpens, the alliance becomes more and more fragile. Factions of capital may be tempted to flee the region altogether or strike back at the power of labor by threats to move or threats to open the floodgates to cheap imports or low-wage immigrant labor. Such threats can antagonise other factions of capital who cannot so easily escape local commitments. Financiers, producers, merchants, landlords and the like do not necessarily see eye to eye. And labor, which once adopted conciliatory policies to consolidate its position within the class alliance, may be tempted to resuscitate more revolutionary demands. The conditions for breakdown and disintegration of the regional alliance are ever present. The dynamic of capitalism in the end tends to break apart the very alliances it initially promotes. The stresses become particularly fierce under conditions of crisis. It then seems that the only way to keep the alliance intact is to seek an external resolution to the region's problems.

The search for a 'spatial fix'

We now return to the initial question, suitably modified to take account of the general geographical conditions under which accumulation occurs. In the face of an 'inner dialectic' that tends towards disequilibrium, can a regional alliance maintain its cohesion and stave off overaccumulation and devaluation through geographical expension and restructuring? Can the surpluses of capital and labor power be disposed of and remunerated by entering into external relations with other regions?

An expansion of foreign trade does little or nothing to resolve the problem. Surplus commodities are traded away and their equivalent in value shortly received back in the form of other commodities. This does nothing to relieve a condition of general surplus. If, however, the trade is credit-financed (or the country concerned is prepared to run a negative trade balance ad infinitum) then matters look rather differently. A region can lend surplus money capital to another and thereby finance the purchase of its own surplus commodities, thus ensuring full employment of both its productive capacity and labor power. This combination of temporal and spatial displacement can work well, often for extended periods of time, until the debts fall due. The only way they can be paid is by expanding commodity imports that can only exacerbate the overaccumulation problem at home. Either that, or the debts cannot be paid and the money lent is lost.

Surplus labor power can be sent abroad to found colonies. There are two problems with this solution. First, if labor can freely move to an unalienated existence on some frontier, the capitalist control over internal labor supply breaks down and an important condition for the perpetuation of capitalism is undermined. Secondly, the export of excess labor power does nothing for the surplus capital left behind unless the latter is absorbed through a rising demand from the colonies. But then the colony must pay for the goods it buys through commodity production. And that means more surplus commodities and capital in the long run.

The export of capital unaccompanied by labor power or the reverse flow of labor power without capital can have a half-hearted and temporary palliative effect on the tendency towards overaccumulation. The benefit arises because the rapid expansion of labour supply forms, as we saw earlier in this essay, a more secure basis for relatively trouble-free accumulation than would be the case under conditions of slow population growth. Processes of primitive accumulation outside the region are here mobilized as means to manage and control the supply of labor power in relation to the available capital within the region. The drive to create labor surpluses within the region by processes internal to the circulation of capital is thereby diminished. The effect is 'halfhearted' from the standpoint of most laborers, though for the privileged groups within a regional alliance the effects may be positive because relatively full employment can be maintained at home. The latter may support controlled guest-worker programs and external neocolonialism as a matter of immediate self-interest. But in the long run, the higher rates of exploitation and expansion produce more and more capital. So important though the processes of primitive accumulation on the exterior are, they provide no permanent solution to the problem even if there were no limit to the population available or to the resistance encountered.

If, however, the excess capital and labor power are both put to creating new productive capacity in new areas, then the surpluses stand to be absorbed for much longer periods. Investment in basic infrastructures, as we have seen, is long-term while the continuous expansion of a whole new regional capitalist economy creates a continuous and rising demand for the surpluses of capital and labor power produced at home. The only problem with this solution is that the new regional economy tends to achieve its own internal structured coherence, to fashion its own regional class alliance to promote and protect its interests, and is itself bound to become expansionary, technologically dynamic, beset by class struggle and inherently unstable. It, too, begins to produce surpluses of capital and labor power that become increasingly difficult to absorb. It, too, in the long run is forced to look to its own 'spatial fix'. In so doing, it inevitably finds itself in competition with the home country on the world market and, if it wins, it can force devaluation onto the home economy through international competition. To take an obvious example, in the nineteenth century, massive quantities of surplus capital and labor power from Britain were syphoned off to the US, but in the end it was the US that defeated Britain on the world market.

To avoid such an eventuality, the home country may impose dependent forms of development on the new region. The subservient economy then produces only what the home country wants and in the quantities it needs. The free development of a new regional capitalism is kept in check and whatever regional class alliance arises is kept firmly under the control of the home country. But the dependent territory cannot then expand fast enough to absorb the surpluses being generated at home. The export of capital quickly subsides into a mere trading relation that can do nothing to relieve the underlying problems of overaccumulation. Thus India, under British domination from the start, mounted no competitive challenge to British industry but by the same token it was far less significant as a field for the absorption of surpluses than, for example, the US. The same principle was at work after the Second World War. Surpluses from the US found a far more accommodating home in western Europe and Japan than they did in the Third World, but it was the former that mounted the major competitive challenge to the US in world markets.

There is, evidently, a 'catch 22' of the following sort: if the new region is to absorb the surpluses, then it must be allowed to develop freely into a fully-fledged capitalist economy that is bound in the end to produce its own surpluses and so enter into international competition with the home base. If the new region develops in a constrained and dependent way, then the rate of expansion is not fast enough to absorb the burgeoning surpluses in the home economy. Devaluation occurs: unless, of course, new growth regions can be opened up. The effect, however, as both Marx and Lenin long ago observed, is to spread the contradictions of capitalism over ever-wider spheres and give them even greater latitude of operation.

But, *nota bene*, capitalism can open up considerable breathing-space for its own survival through pursuit of the 'spatial fix', particularly when combined with temporal displacements of the sort described earlier. It is rather as if, having sought to annihilate space with time, capitalism buys time for itself out of the space it conquers. So although we can continue to assert that crises cannot, in the long run, be avoided, we have to countenance the possibility that the long run might be very long. Yet the long run must also be punctuated by what I have elsewhere called intense 'switching crises', cataclysmic moments that reshape the whole geography of capital accumulation, break down rigid spatial structures and regional class alliances, even undermine the power of state formations and reconstitute them all in a new geographic configuration that can better accommodate the powerful expansionary, conflictual and techno-logical dynamic of a restless, shifting, capital flow (Harvey 1978; 1982: ch. 13). But the question perpetually hovers: what happens when, for whatever reason, the 'spatial fix' is stymied and the debts incurred by temporal displacement fall due?

Marx's exclusion of any spatial fix permits him to concentrate attention on the fundamental processes of crisis formation. The theory of overac-cumulation-devaluation reveals the intense destructive power that lurks beneath capitalism's façade of technological progress and market ratio-nality. In the course of a crisis, vast quantities of capital are devalued and destroyed, the laborers and their labor power suffer a like fate, and capi-talists cannibalise and liquidate each other in that 'war of all against all' that is the ultimate hallmark of a capitalist mode of production.

What Marx nowhere anticipates but Lenin emphasizes, is the conversion of this process into economic, political and military struggles between nation states. We have now constructed a more general proposition. In the face of the inexorable processes of crisis formation, the search for a spatial fix converts the threat of devaluation into a struggle between unstable regional alliances over who is to bear the brunt of the crisis. Faced with the prospect of splintering into a thousand warring pieces, a regional alliance may consolidate itself and turn its destructive tendencies towards the outside. The export of unemployment, inflation and idle productive capacity become the stakes in an ugly game. Trade wars, dumping, tariffs and quotas, restrictions on capital flow and foreign exchange, interest-rate wars, immigration policies, colonial conquest, the subjugation and domination of tributary economies, the forced reorganization of the territorial division of labor within economic (even corporate) empires and, finally, the physical destruction and forced devaluation achieved through military confrontation and war, can all be caught up as part and parcel of the processes of crisis formation and resolution. The search for a 'spatial fix' takes a viciously competitive and perhaps even violent turn.

The geopolitics of capitalism

The year 1980 ushered in a difficult and dangerous decade in the historical geography of capitalism. By 1983, unemployment had soared above 10 per cent in most industrial nations (with the notable exception of Japan) and unutilized productive capacity and unsold inventories had risen to unprecedented levels. The inflationary surge of the late 1960s and the stagflation of the 1970s in retrospect appear as but preludes to a classic crisis of devaluation of both capital and labor power under conditions of rampant overaccumulation.

The interregional and international division of labor is now in the course of rationalization and reconstruction through a mixture of strong processes of technological change and geographical mobility of capital. Previous patterns of regional coherence are thrown into disarray and traditional regional class alliances either disintegrate or forcibly consolidate in an effort to project devaluation onto the exterior. The new geographical differentials in productivity that have opened up generate in turn dramatic transformations in global and regional trade patterns and money flows, creating conditions of chronic national and international monetary instability. The geographical uncertainties force time horizons to shorten, thus exacerbating an already serious debt problem (private and public, local, national and international) accumulated out of many years of rapid and, in retrospect, excessive fictitious capital formation. The capacity to absorb surpluses of capital and labor power through temporal and geographical displacement, at least under the general conditions laid down in the immediate postwar period, appears to have run out. The only solution consistent with those conditions is for the deficit-ridden to bail out the bankrupt and the bankrupt to pay off the deficits.

Short of such an unreal expectation, we can expect only a rising tide of devaluation that disrupts regional class alliances and sours relations between them. Protectionist responses of all sorts (not just at the national level and by no means confined to tariffs and other conventional devices) abound. Aggressive moves are made to export devaluation to other regions. The US steel industry in alliance with trades unions, to take a recent example, forces restrictions on cheaper imports from Europe and Japan which in turn restrict imports from Brazil and South Korea. But then US steel yields to temptation and decides to import cheaper British slab steel, provoking charges from its competitors and union alike that it is undermining national interest and exporting jobs for narrow commercial gain.

By myriad processes of this sort, regional and international shifts in economic and political power occur, shifts which the policies of particular governments appear powerless to prevent by normal means. Indeed, both national and international policies lose whatever coherence they may once have had. Plans to improve the competitiveness of industry within a regional alliance entail accelerating technological changes that remove living labor from production at home while exporting unemployment abroad. Policies designed to export devaluation to the Third World not only spark riots in Sao Paulo and Santiago, but put at risk the vast debt owed by those countries. That debt, which in a classic case of combined geographical and temporal displacement rose from just over $20 billion to almost $200 billion for the three largest borrowers (Brazil, Mexico and Argentina) between 1972 and 1983, now appears largely uncollectable. The top ten US banks that hold nearly $40 billion of it face financial ruin in the event of a default. Has the Federal Reserve any choice under such circumstances except to loosen the money supply in the US, bail out the banks, and reignite domestic and international forces of inflation? And in any case, the only way the debt can be paid is by expanding imports from the Third World, which means, at a time of general devaluation, importing unemployment into the US. If the deficit-ridden bail out the bankrupt, how can the bankrupt pay off the deficits without deepening the problems of the deficit-ridden?

Geopolitical realignments and conflicts appear inevitable under such conditions. Even NATO, the geopolitical centrepiece of postwar capitalism, is threatened by internal economic rivalries and disaffection. The Pentagon may seek to enhance NATO's solidarity but the Federal Reserve just as surely undermines it by monetary policies judged appropriate to control inflation but which also force unacceptable levels of devaluation upon western Europe. Policies directed overtly at the Soviet bloc by the US affect adversely those countries, such as West Germany, which sought outlets for surplus capital in east-west trade (squabbles over credit to the Soviet Union, the gas pipeline, and the Polish debt are recent cases in point). Some governments in western Europe seek another round of the spatial fix by the export of unchecked capitalist growth to the Third

World (as proposed in the Brandt Report). They envisage a geopolitical realignment of western Europe with more dynamic regions in the Third World and, in so doing, run up against the US which still, for example, so interprets the Monroe Doctrine as to mean the right to neocolonial domination (in the name of anti-Sovietism and anti-communism) throughout much of Latin America. The Japanese, perceived correctly as the main competitive threat to the commercial hegemony of the US and western Europe, have constructed a special kind of capitalist economy that is highly dynamic and expansionary but also inflexible downwards, with little capacity to absorb the devaluation of either capital or labor power. The Japanese look fearfully to consolidate their hold on Third World markets while levelling off their inroads into the industrialized world. Meanwhile, the US finally forgets Pearl Harbour and in a bid to reduce its own defence burden and budget deficit urges rearmament on Japan where militarism is in any case quietly resurgent.

The seemingly solid presuppositions upon which the postwar boom was built have just melted into air. Gone is the strong and stable dollar as the pivot of the international monetary system (Bretton Woods). Gone, too, are the open spaces for surplus capital through the reconstruction of war-devastated economies (the Marshall Plan) and the commitment to expand world trade by elimination of barriers to commodity exchange (GATT) and capital flow. Heightened international and interregional competition and accelerating technological change undermine the expansionary dynamic and put the whole global economy into a tailspin. The 'disintegration of the west' as Mary Kaldor so graphically depicts it, proceeds apace (Kaldor 1978). Can the disintegration be arrested, depression, revolution, war (or some combination of the three) be averted?

Under such conditions, we can but look back nervously at the economic and diplomatic history of the 1930s, that tortured prelude to a global intercapitalist war that did more to transform the historical geography of the world than any other sequence of events in history. Can that happen again? And, if so, how and why? Pure analogies may never satisfy, but they can provoke serious analysis and reflection. First, we should note how rapidly geopolitical and economic alignments shifted in the face of economic chaos. While there were many telltale signs of fragility in the 1920s (hyperinflation in Germany, grumbling unemployment in Britain, speculative bombast in the US) the main geopolitical cleavage in the world was certainly that between the Soviet Union and the capitalist powers. But by 1933, the capitalist world had split asunder into so many hostile camps, the British sheltering behind Commonwealth Preference, the Japanese within a 'co-prosperity sphere' forcibly appropriated, and the Germans about to embark upon a policy of *Lebensraum* through political, economic and ultimately military domination. Only the US vainly sought (in its own self-interest) to sustain an 'open-door' policy in a world where alliances of regional class alliances (with strong working-class support for the most part) progressively sealed themselves off, politically and militarily, within closed trading empires. But if the strong expansionary dynamic culminating in overaccumulation prevails as inexorably as we have here depicted it, then the regional class alliance faces a dismal choice of depression and (perhaps) revolution at home or military confrontation abroad (the ultimate form of 'the spatial fix').

Second, in spite of all the high drama of 'New Deal' politics in the US or *autobahn* construction in Fascist Germany, there is very little evidence that such inner transformations of civil society in any way actually resolved the internal contradictions of capitalism. Unemployment was rising strongly in the US on the eve of its entry into the war and there had been little or no revival of world trade or of reinvestment over and above that directly created by government expenditures by 1939. Then, as now, the need for fiscal responsibility stymied the best-laid plans to absorb the surpluses of capital and labor power. It was in fact the Second World War that brought full employment and reinvestment, but it did so under conditions where vast amounts of capital stood to be physically destroyed and many idle workers consumed as cannon fodder. And it was precisely the geographical unevenness of that destruction that opened up new spaces in the postwar period for the absorption of surplus US capital under the aegis of that benevolent 'spatial fix' known as the Marshall Plan.

Third, the inner transformations wrought in the 1930s paled into insignificance when compared with the dramatic institutional and geopolitical reconstruction wrought out of the ashes of the Second World War. At the insistence of the US (by then the hegemonic world power), the 'open door' prevailed, buttressed by an array of supranational institutions (such as the World Bank and the International Monetary Fund) under the de facto control of the US and an international monetary agreement that effectively made the US the world's banker. The dissolution of closed trading empires (the British were forced to dismantle Commonwealth preference in return for Lend-lease during the war) and decolonisation spawned numerous independent but economically powerless new states throughout the Third World (in much the same way that new states had been carved out of Europe after the First World War). Everything was ordered to prevent the emergence of rival power blocs within the capitalist world and to facilitate the internationalisation of (mainly US) capital under conditions of fairly restricted geographical mobility of labor power. Co-optation and repression, at home and abroad, to keep the free world free for the circulation of capital, became the dominant political theme. To this end, new geopolitical alliances were forged and new foundations laid for the cohesion of regional class alliances within an internationalist framework. And, of course, the Soviet threat and anti-communism became the central ideological tool to ensure the solidarity of potentially competitive regional class alliances. To the degree that this ideology needed a material base, the geopolitical confrontation of the Soviet Union and the communist bloc became central to the survival of capitalism irrespective of Soviet policies or action.

This is the relatively stable geopolitical framework within which the postwar boom took place. And it is also the framework threatened by the very success of that dynamic. Overaccumulation and devaluation, as we began by noting, are everywhere in evidence and the internal cohesion of the capitalist world as a whole, as well as of regional class alliances within it, threaten to dissolve into a chaos of competing and warring forces. Is there, *can* there be, some way to prevent such disintegration and all its untold associated horrors?

Requiem for a conclusion

The theoretical argument I have here set out is, I hold, as fundamental to the elucidation of our present plight as it is to the interpretation of the historical geography of capitalism. If I am correct, and I hasten to add that I hope I am grossly in error and that history or others will quickly prove me so to be, then the perpetuation of capitalism in the twentieth century has been purchased at the cost of the death, havoc and destruction wreaked in two world wars. But each war has been waged with ever more sophisticated weapons of destruction. The bourgeois era has certainly witnessed a growth in destructive force that more than matches the growth of productive force so essential to the survival of capitalism. That the latter should also require the use of that destructive force appears insane. Yet the ideologists of capitalism shed no tears but, like Schumpeter, sing paeans of praise to what they term the 'creative destruction' through which capitalists so dramatically transform the world. But our present plight must surely give us pause. As temporal and geographical solutions to the inner dialectic of overaccumulation run out, the crisis tendencies of capitalism once more run amok, interimperialist rivalries sharpen and the threat of autarky within closed trading empires looms. The struggle to export devaluation within a disintegrating world order comes to the fore and belligerence dominates the tone of political discourse. With this comes the renewed threat of global war, this time waged with weapons of such immense and insane destructive power that not even the fittest stand to survive. The message which Marx long ago jotted down in that notebook that became the *Grundrisse* impresses upon us more urgently than ever:

> The violent destruction of capital not by relations external to it, but rather as a condition of its self-preservation, is the most striking form in which advice is given it to be gone and to give room to a higher state of social production.
>
> (Marx 1973: 749–50)

Capitalism did not invent war any more than it invented writing, knowledge, science or art. Not all wars, even in the contemporary era, can be truly regarded as capitalist wars. And war will not necessarily disappear from the human scene with the demise of capitalism. But what our theory strongly urges is that we see the replacement of the capitalist mode of production, that expansionary and technologically dynamic process of circulation that we began by examining, as a necessary condition for human survival. And that is a task beyond the prerogative of any single class or community. It is, I submit, a task that should be the immediate focus of every atom of our collective attention.

CHAPTER 16

From managerialism to entrepreneurialism: the transformation in urban governance in late capitalism

First published in Geografiska Annaler, *1989.*

A centerpiece of my academic concerns these last two decades has been to unravel the role of urbanization in social change, in particular under conditions of capitalist social relations and accumulation (Harvey 1973; 1982; 1985a; 1985b; 1989a). This project has necessitated deeper enquiry into the manner in which capitalism produces a distinctive historical geography. When the physical and social landscape of urbanization is shaped according to distinctively capitalist criteria, constraints are put on the future paths of capitalist development. This implies that though urban processes under capitalism are shaped by the logic of capital circulation and accumulation, they in turn shape the conditions and circumstances of capital accumulation at later points in time and space. Put another way, capitalists, like everyone else, may struggle to make their own historical geography but, also like everyone else, they do not do so under historical and geographical circumstances of their own individual choosing, even when they have played an important and even determinant collective role in shaping those circumstances. This two-way relation of reciprocity and domination (in which capitalists, like workers, find themselves dominated and constrained by their own creations) can best be captured theoretically in dialectical terms. It is from such a standpoint that I seek more powerful insights into that process of city-making that is both product and condition of ongoing social processes of transformation in the most recent phase of capitalist development.

Enquiry into the role of urbanization in social dynamics is, of course, nothing new. From time to time, the issue flourishes as a focus of major debates, though more often than not with regard to particular historical-geographical circumstances in which, for some reason or other, the role of urbanization and of cities appears particularly salient. The part that city formation played in the rise of civilization has long been discussed, as has the role of the city in classical Greece and Rome. The significance of cities to the transition from feudalism to capitalism is an arena of continuing controversy, having sparked a remarkable and revealing literature over the years. A vast array of evidence can now likewise be brought to bear on the significance of urbanization to nineteenth-century industrial, cultural and political development as well as to the subsequent spread of capitalist social relations to lesser developed countries (which now support some of the most dramatically growing cities in the world).

All too frequently, however, the study of urbanization becomes separated from that of social change and economic development, as if it can somehow be regarded either as a sideshow or as a passive side-product to more important and fundamental social changes. The successive revolutions in technology, space relations, social relations, consumer habits, lifestyles, and the like that have so characterized capitalist history can, it is sometimes suggested, be understood without any deep enquiry into the roots and nature of urban processes. True, this judgement is by and large made tacitly, by virtue of sins of omission rather than commission. But the antiurban bias in studies of macroeconomic and macrosocial change is rather too persistent for comfort. It is for this reason that it seems worthwhile to enquire what role the urban process might be playing in the quite radical restructuring going on in geographical distributions of human activity and in the political-economic dynamics of uneven geographical development in most recent times.

The shift to entrepreneurialism in urban governance

A colloquium held at Orleans in 1985 brought together academics, businessmen, and policymakers from eight large cities in seven advanced capitalist countries (Bouinot 1987). The charge was to explore the lines of action open to urban governments in the face of the widespread erosion of the economic and fiscal base of many large cities in the advanced capitalist world. The colloquium indicated a strong consensus: that urban governments had to be much more innovative and entrepreneurial, willing to explore all kinds of avenues through which to alleviate their distressed condition and thereby secure a better future for their populations. The only realm of disagreement concerned how this best could be done. Should urban governments play some kind of supportive or even direct role in the creation of new enterprises and if so of what sort? Should they struggle to preserve or even take over threatened employment sources and, if so, which ones? Or should they simply confine themselves to the provision of those infrastructures, sites, tax baits, and cultural and social attractions that would shore up the old and lure in new forms of economic activity?

I quote this case because it is symptomatic of a reorientation in attitudes to urban governance that has taken place these past two decades in the advanced capitalist countries. Put simply, the 'managerial' approach so typical of the 1960s has steadily given way to initiatory and 'entrepreneurial' forms of action in the 1970s and 1980s. In recent years in particular, there seems to be a general consensus emerging throughout the advanced capitalist world that positive benefits are to be had by cities taking an entrepreneurial stance to economic development. What is remarkable, is that this consensus seems to hold across national boundaries and even across political parties and ideologies.

Both Boddy (1984) and Cochrane (1987) agree, for example, that since the early 1970s local authorities in Britain 'have become increasingly involved in economic development activity directly related to production and investment,' while Rees and Lambert (1985: 179) show how 'the growth of local government initiatives in the economic field was positively encouraged by successive central administrations during the 1970s' in order to complement central government attempts to improve the efficiency, competitive powers and profitability of British industry. David Blunkett, leader of the Labour Council in Sheffield for several years, has recently put the seal of approval on a certain kind of urban entrepreneurialism:

> From the early 1970s, as full employment moved from the top of government priorities, local councils began to take up the challenge. There was support for small firms; closer links between the public and private sectors; promotion of local areas to attract new business. They were adapting the traditional economic role of British local government which offered inducements in the forms of grants, free loans, and publicly subsidised infrastructure, and no request for reciprocal involvement with the community, in order to attract industrial and commercial concerns which were looking for suitable sites for investment and trading . . . Local government today, as in the past, can offer its own brand of entrepreneurship and enterprise in facing the enormous economic and social change which technology and industrial restructuring bring.
>
> (Blunkett and Jackson 1987: 108–42)

In the United States, where civic boosterism and entrepreneuralism had long been a major feature of urban systems (see Elkin 1987), the reduction in the flow of federal redistributions and local tax revenues after 1972 (the year in which President Nixon declared the urban crisis to be over, signalling that the federal government no longer had the fiscal resources to contribute to their solution) led to a revival of boosterism to the point where Robert Goodman (1979) was prepared to characterize both state and local governments as 'the last entrepreneurs'. An extensive literature now exists dealing with how the new urban entrepreneurialism has moved center-stage in urban policy formulation and urban growth strategies in the US (see Judd and Ready 1986; Peterson1981; Leitner 1989).

The shift towards entrepreneurialism has by no means been complete. Many local governments in Britain did not respond to the new pressures and possibilities, at least until relatively recently, while cities like New Orleans in the US continue to remain wards of the federal government and rely fundamentally on redistributions for survival. And the history of its outcomes, though yet to be properly recorded, is obviously checkered, pockmarked with as many failures as successes and not a little controversy as to what constitutes 'success' anyway (a question to which I shall later return). Yet beneath all this diversity, the shift from urban managerialism to some kind of entrepreneurialism remains a persistent and recurrent theme in the period since the early 1970s. Both the reasons for and the implications of such a shift are deserving of some scrutiny.

There is general agreement, of course, that the shift has something to do with the difficulties that have beset capitalist economies since the recession of 1973. Deindustrialization, widespread and seemingly 'structural' unemployment, fiscal austerity at both the national and local levels, all coupled with a rising tide of neoconservatism and much stronger appeal (though often more in theory than in practice) to market rationality and privatization, provide a backdrop to understanding why so many urban governments, often of quite different political persuasions and armed with very different legal and political powers, have all taken a broadly similar direction. The greater emphasis on local action to combat these ills also seems to have something to do with the declining powers of the nation-state to control multinational money flows, so that investment increasingly takes the form of a negotiation between international finance capital and local powers doing the best they can to maximize the attractiveness of the local site as a lure for capitalist development. By the same token, the rise of urban entrepreneurialism may have had an important role to play in a general transition in the dynamics of capitalism from a Fordist-Keynesian regime of capital accumulation to a regime of 'flexible accumulation' (see Gertler 1988; Harvey 1989b; Sayer 1989; Schoenberger 1988; Scott 1988; Swyngedouw 1986, for some elaboration and critical reflection on this controversial concept). The transformation of urban governance these last two decades has had, I shall argue, substantial macroeconomic roots and implications. And, if Jane Jacobs (1984) is only half right, that the city is the relevant unit for understanding how the wealth of nations is created, then the shift from urban managerialism to urban entrepreneurialism could have far-reaching implications for future growth prospects.

If, for example, urban entrepreneurialism (in the broadest sense) is embedded in a framework of zero-sum inter-urban competition for resources, jobs, and capital, then even the most resolute and avantgarde municipal socialists will find themselves, in the end, playing the capitalist game and performing as agents of discipline for the very processes they are trying to resist. It is exactly this problem that has dogged the Labour councils in Britain (see the excellent account by Rees and Lambert 1985). They had on the one hand to develop projects which could 'produce outputs which are directly related to working people's needs, in ways which build on the skills of labour rather than de-skilling them' (Murray 1983), while on the other hand recognizing that much of that effort would go for nought if the urban region did not secure relative competitive advantages. Given the right circumstances, however, urban entrepreneurialism and even inter-urban competition may open the way to a non zero-sum pattern of development. This kind of activity has certainly played a key role in capitalist development in the past. And it is an open question as to whether or not it could lead towards progressive and socialist transitions in the future.

Conceptual issues

There are conceptual difficulties to such an enquiry that deserve an initial airing. To begin with, the reification of cities when combined with a language that sees the urban process as an active, rather than passive, aspect of political-economic development poses acute dangers. It makes it seem as if 'cities' can be active agents when they are mere things. Urbanization should, rather, be regarded as a spatially grounded social process in which a wide range of different actors with quite different objectives and agendas interact through a particular configuration of interlocking spatial practices. In a classbound society such as capitalism, these spatial practices acquire a definite class content, which is not to say that all spatial practices can be so interpreted. Indeed, as many researchers have shown, spatial practices can and do acquire gender, racial and bureaucratic-administrative contents (to list just a subset of important possibilities). But under capitalism, it is the broad range of class practices connected to the circulation of capital, the reproduction of labor power and class relations, and the need to control labor power, that remains hegemonic.

The difficulty is to find a way of proceeding that can deal specifically with the relation between *process* and *object* without itself falling victim to unnecessary reification. The spatially grounded set of social processes that I call urbanization produce innumerable artefacts: a built form, produced spaces and resource systems of particular qualities organized into a distinctive spatial configuration. Subsequent social action must take account of these artefacts, since so many social processes (such as commuting) become physically channelled by them. Urbanization also throws up certain institutional arrangements, legal forms, political and administrative systems, hierarchies of power, and the like. These, too, give a 'city' objectified qualities that may dominate daily practices and confine subsequent courses of action. And, finally, the consciousness of urban inhabitants is affected by the environment of experience out of which perceptions, symbolic readings, and aspirations arise. In all of these respects, there is a perpetual tension between form and process, between object and subject, between activity and thing. It is as foolish to deny the role and power of objectifications, the capacity of things we create to return to us as so many forms of domination, as it is to attribute to such things the capacity for social action.

Given the dynamism to which capitalism is prone, we find that these 'things' are always in the course of transformation, that activities are constantly escaping the bounds of fixed forms, that the objectified qualities of the urban are chronically unstable. So universal is this capitalist condition, that the conception of the urban and of 'the city' is likewise rendered unstable, not because of any conceptual definitional failing, but precisely because the concept has itself to reflect changing relations between form and process, between activity and thing, between subjects and objects. When we speak, therefore, of a transition from urban managerialism towards urban entrepreneurialism these last two decades, we have to take cognizance of the reflexive effects of such a shift, through the impacts on urban institutions as well as urban built environments.

The domain of spatial practices has, unfortunately, changed in recent years, making any firm definition of the urban as a distinctive spatial domain even more problematic. On the one hand, we witness the greater fragmentation of the urban social space into neighborhoods, communities, and a multitude of street corner societies, while on the other telecommuting and rapid transport make nonsense of some concept of the city as a tightly-walled physical unit or even a coherently organized administrative domain. The 'megalopolis' of the 1960s has suffered even further fragmentation and dispersal, particularly in the US, as urban deconcentration gathers pace to produce a 'spread city' form. Yet the spatial grounding persists in *some* form with specific meanings and effects. The production of new ecological patternings and structures within a spread city form has significance for how production, exchange, and consumption is organized, how social relationships are established, how power (financial and political) is exercised, and how the spatial integration of social action is achieved. I hasten to add that presentation of the urban problematic in such ecological terms in no way presumes ecological explanations. It simply insists that ecological patternings are important for social organization and action. The shift towards entrepreneurialism in urban governance has to be examined, then, at a variety of spatial scales: local neighborhood and community, central city and suburb, metropolitan region, region, nation state, and the like.

It is likewise important to specify who is being entrepreneurial and about what. I want here to insist that urban 'governance' means much more than urban 'government'. It is unfortunate that much of the literature (particularly in Britain) concentrates so much on the latter when the real power to reorganize urban life so often lies elsewhere, or at least within a broader coalition of forces within which urban government and administration have only a facilitative and coordinating role to play. The power to organize space derives from a whole complex of forces mobilized by diverse social agents. It is a conflictual process, the more so in the ecological spaces of highly variegated social density. Within a metropolitan region as a whole, we have to look to the formation of coalition politics, to class alliance formation as the basis for any kind of urban entrepreneurialism at all. Civic boosterism has, of course, often been the prerogative of the local chamber of commerce, some cabal of local financiers, industrialists and merchants, or some 'round table' of business leaders and real-estate and property developers. The latter frequently coalesce to form the guiding power in 'growth machine' politics (Molotch 1976). Educational and religious institutions, different arms of government (varying from the military to research or administrative establishments), local labor organizations (the building and construction trades in particular) as well as political parties, social movements, and the local state apparatuses (which are multiple and often quite heterogeneous), can also play the game of local boosterism though often with quite different goals.

Coalition and alliance formation is so delicate and difficult a task that the way is open here for a person of vision, tenacity, and skill (such as a charismatic mayor, a clever city administrator, or a wealthy business leader) to put a particular stamp upon the nature and direction of urban entrepreneurialism, perhaps to shape it, even, to particular political ends. Whereas it was a public figure like Mayor Schaefer who played the central role in Baltimore, in cities like Halifax or Gateshead in Britain it has been private entrepreneurs who have taken the lead. In other instances, it has been a more intricate mix of personalities and institutions that have put a particular project together.

I raise these problems not because they are insurmountable or intractable – they are resolved daily within the practices of capitalist urbanisation – but because we have to attend to their manner of practical resolution with a requisite care and seriousness. I shall, however, venture three broad assertions which I know to be true for a city like Baltimore (the case study which underpins much of the argument I offer here) and which may be more generally applicable.

First, the new entrepreneurialism has, as its centerpiece, the notion of a 'public–private partnership' in which a traditional local boosterism is integrated with the use of local governmental powers to try and attract external sources of funding, new direct investments, or new employment sources. The Orleans colloquium (Bouinot 1987) was full of references to the importance of this public–private partnership and it was, after all, precisely the aim of local government reforms in Britain in the 1970s to facilitate their formation (or in the end to bypass local resistance by setting up the urban development corporations). In the US, the tradition of federally backed and locally implemented public–private partnership faded during the 1960s as urban governments struggled to regain social control of restive populations through redistributions of real income (better housing, education, healthcare and the like all targeted towards the poor) in the wake of urban unrest. The role of the local state as facilitator for the strategic interests of capitalist development (as opposed to stabilizer of capitalist society) declined. The same dismissiveness towards capitalist development has been noted in Britain:

> The early 1970s was a period of resistance to change: motorway protest groups, community action against slum clearance, opponents of town centre redevelopment. Strategic and entrepreneurial interests were sacrificed to local community pressures. Conceivably, however, we are moving into a different period in which the entrepreneurial role becomes dominant.
>
> (Davies 1980: 23; quoted in Ball 1983: 270–1)

In Baltimore the transition point can be dated exactly. A referendum narrowly passed in 1978, after a vigorous and contentious political campaign, sanctioned the use of city land for the private development that became the highly spectacular and successful Harborplace. Thereafter, the policy of public–private partnership had a popular mandate as well as an effective subterranean presence in almost everything that urban governance was about (see Berkowitz 1984; Levine 1987; Lyall 1982; Stoker 1986).

Secondly, the activity of that public-private partnership is entrepreneurial precisely because it is speculative in execution and design and therefore dogged by all the difficulties and dangers which attach to speculative as opposed to rationally planned and coordinated development. In many instances, this has meant that the public sector assumes the risk and the private sector takes the benefits, though there are enough examples where this is not the case (think, for example, of the private risk taken in Gateshead's Metrocentre development) to make any absolute generalization dangerous. But I suspect it is this feature of risk-absorption by the local (rather than the national or federal) public sector which distinguishes the present phase of urban entrepreneurialism from earlier phases of civic boosterism in which private capital seemed generally much less risk-averse.

Thirdly, the entrepreneurialism focuses much more closely on the political economy of place rather than of territory. By the latter, I mean the kinds of economic projects (housing, education, and so on) that are designed primarily to improve conditions of living or working within a particular jurisdiction. The construction of place (a new civic center, an industrial park) or the enhancement of conditions within a place (intervention, for example, in local labor markets by retraining schemes or downward pressure on local wages), on the other hand, can have impacts either smaller or greater than the specific territory within which such projects happen to be located. The upgrading of the image of cities like Baltimore, Liverpool, Glasgow or Halifax, through the construction of cultural, retail, entertainment and office centers can cast a seemingly beneficial shadow over the whole metropolitan region. Such projects can acquire meaning at the metropolitan scale of public-private action and allow for the formation of coalitions which leap over the kinds of city-suburb rivalries that dogged metropolitan regions in the managerial phase. On the other hand, a rather similar development in New York City – Southstreet Seaport – constructs a new place that has only local impacts, falling far short of any metropolitan-wide influence, and generating a coalition of forces that is basically local property developers and financiers.

The construction of such places may, of course, be viewed as a means to procure benefits for populations within a particular jurisdiction, and indeed this is a primary claim made in the public discourse developed to support them. But for the most part, their form is such as to make all benefits indirect and potentially either wider or smaller in scope than the jurisdiction within which they lie. Place-specific projects of this sort also have the habit of becoming such a focus of public and political attention that they divert concern and even resources from the broader problems that may beset the region or territory as a whole.

The new urban entrepreneurialism typically rests, then, on a public-private partnership focusing on investment and economic development with the speculative construction of place rather than amelioration of conditions within a particular territory as its immediate (though by no means exclusive) political and economic goal.

Alternative strategies for urban governance

There are, I have argued elsewhere (Harvey 1989a: ch. 1), four basic options for urban entrepreneurialism. Each warrants some separate consideration, even though it is the combination of them that provides the clue to the recent rapid shifts in the uneven development of urban systems in the advanced capitalist world.

First, competition within the international division of labor means the creation of exploitation of particular advantages for the production of goods and services. Some advantages derive from the resource base (the oil that allowed Texas to bloom in the 1970s) or location (for example favored access to the vigor of Pacific Rim trading in the case of Californian cities). But others are created through public and private investments in the kinds of physical and social infrastructures that strengthen the economic base of the metropolitan region as an exporter of goods and services. Direct interventions to stimulate the application of new technologies, the creation of new products, or the provision of venture capital to new enterprises (which may even be cooperatively owned and managed) may also be significant, while local costs may be reduced by subsidies (tax breaks, cheap credit, procurement of sites). Hardly any large scale development now occurs without local government (or the broader coalition of forces constituting local governance) offering a substantial package of aids and assistance as inducements. International competitiveness also depends upon the qualities, quantities, and costs of local labor supply. Local costs can most easily be controlled when local replaces national collective bargaining and when local governments and other large institutions, like hospitals and universities, lead the way with reductions in real wages and benefits (a series of struggles over wage rates and benefits in the public and institutional sector in Baltimore in the 1970s was typical). Labor power of the right quality, even though expensive, can be a powerful magnet for new economic development so that investment in highly trained and skilled workforces suited to new labor processes and their managerial requirements can be well rewarded. There is, finally, the problem of agglomeration economies in metropolitan regions. The production of goods and services is often dependent not on single decisions of economic units (such as the large multinationals to bring a branch plant to town, often with very limited local spillover effects), but upon the way in which economies can be generated by bringing together diverse activities within a restricted space of interaction so as to facilitate highly efficient and interactive production systems (see Scott 1988). From this standpoint, large metropolitan regions like New York, Los Angeles, London, and Chicago possess some distinctive advantages that congestion costs have by no means yet offset. But, as the case of Bologna (see Gundle 1986) and the surge of new industrial development in Emilia Romagna illustrates, careful attention to the industrial and marketing mix backed by strong local state action (communist-led in this instance), can promote powerful growth of new industrial districts and configurations, based on agglomeration economies and efficient organization.

Under the second option, the urban region can also seek to improve its competitive position with respect to the spatial division of consumption. There is more to this than trying to bring money into an urban region through tourism and retirement attractions. The consumerist style of urbanization after 1950 promoted an ever-broader basis for participation in mass consumption. While recession, unemployment, and the high cost of credit have rolled back that possibility for important layers in the population, there is still a lot of consumer power around (much of it credit-fuelled). Competition for that becomes more frenetic while consumers who do have the money have the opportunity to be much more discriminating. Investments to attract the consumer dollar have paradoxically grown apace as a response to generalized recession. They increasingly focus on the quality of life. Gentrification, cultural innovation, and physical upgrading of the urban environment (including the turn to postmodernist styles of architecture and urban design), consumer attractions (sports stadia, convention and shopping centers, marinas, exotic eating places) and entertainment (the organization of urban spectacles on a temporary or permanent basis), have all become much more prominent facets of strategies for urban regeneration. Above all, the city has to appear as an innovative, exciting, creative, and safe place to live or to visit, to play and consume in. Baltimore, with its dismal reputation as 'the armpit of the east coast' in the early 1970s has, for example, expanded its employment in the tourist trade from under one to over fifteen thousand in less than two decades of massive urban redevelopment. More recently, thirteen ailing industrial cities in Britain (including Leeds, Bradford, Manchester, Liverpool, Newcastle and Stoke-on-Trent) put together a joint promotional effort to capture more of Britain's tourist trade. Here is how *The Guardian* (9 May 1987) reports this quite successful venture:

Apart from generating income and creating jobs in areas of seemingly terminal unemployment, tourism also has a significant spin-off effect in its broader enhancement of the environment. Facelifts and facilities designed to attract more tourists also improve the quality of life for those who live there, even enticing new industries. Although the specific assets of the individual cities are obviously varied, each is able to offer a host of structural reminders of just what made them great in the first place. They share, in other words, a marketable ingredient called industrial and/or maritime heritage.

Festivals and cultural events likewise become the focus of investment activities. 'The arts create a climate of optimism – the "can do" culture essential to developing the enterprise culture,' says the introduction to a recent Arts Council of Great Britain report, adding that cultural activities and the arts can help break the downward spiral of economic stagnation in inner cities and help people 'believe in themselves and their community' (see Bianchini 1991). Spectacle and display become symbols of the dynamic community, as much in communist-controlled Rome and Bologna as in Baltimore, Glasgow and Liverpool. This way, an urban region can hope to cohere and survive as a locus of community solidarity while exploring the option of exploiting conspicuous consumption in a sea of spreading recession.

Thirdly, entrepreneurialism has also been strongly colored by a fierce struggle over the acquisition of key control and command functions in high finance, government, or information gathering and processing (including the media). Functions of this sort need particular and often expensive infrastructural provision. Efficiency and centrality within a worldwide communications net is vital in sectors where personal interactions of key decision-makers is required. This means heavy investments in transport and communications (airports and teleports, for example) and the provision of adequate office space equipped with the necessary internal and external linkages to minimize transactions times and costs. Assembling the wide range of supportive services, particularly those that can gather and process information rapidly or allow quick consultation with 'experts', calls for other kinds of investments, while the specific skills required by such activities put a premium on metropolitan regions with certain kinds of education provision (business and law schools, high-tech production sectors, media skills, and the like). Inter-urban competition in this realm is very expensive and peculiarly tough because this is an area where agglomeration economies remain supreme and the monopoly power of established centres, like New York, Chicago, London, and Los Angeles, is particularly hard to break. But since command functions have been a strong growth sector these last two decades (employment in finance and insurance has doubled in Britain in less than a decade), so pursuit of them has more and more appealed as the golden path to urban survival. The effect, of course, is to make it appear as if the city of the future is going to be a city of pure command and control functions, an informational city, a post-industrial city in which the export of services (financial, informational, knowledge-producing) becomes the economic basis for urban survival.

Fourthly, competitive edge with respect to redistributions of surpluses through central (or in the US, state) governments is still of tremendous importance since it is somewhat of a myth that central governments do not redistribute to the degree they used to do. The channels have shifted so that in both Britain (take the case of Bristol) and in the US (take the case of Long Beach-San Diego), it is military and defence contracts that provide the sustenance for urban prosperity, in part because of the sheer amount of money involved but also because of the type of employment and the spin-offs it may have into so-called 'high-tech' industries (Markusen 1986). And even though every effort may have been made to cut the flow of central government support to many urban regions, there are many sectors of the economy (health and education, for example) and even whole metropolitan economies (see Smith and Keller's 1983 study of New Orleans) where such a cut-off was simply impossible. Urban ruling class alliances have had plenty of opportunity, therefore, to exploit redistributive mechanisms as a means to urban survival.

These four strategies are not mutually exclusive and the uneven fortunes of metropolitan regions have depended upon the nature of the coalitions that have formed, the mix and timing of entrepreneurial strategies, the particular resources (natural, human, locational) with which the metro-politan region can work, and the strength of the competition. But uneven growth has also resulted from the synergism that leads one kind of strategy to be facilitative for another. For example, the growth of the Los Angeles-San Diego-Long Beach-Orange County megalopolis appears to have been fuelled by interaction effects between strong governmental redistributions to the defence industries and rapid accrual of command and control functions that have further stimulated consumption-oriented activities to the point where there has been a considerable revival of certain types of manufacturing. On the other hand, there is little evidence that the strong growth of consumption-oriented activity in Baltimore has done very much at all for the growth of other functions save, perhaps, the relatively mild proliferation of banking and financial services. But there is also evidence that the network of cities and urban regions in, say, the Sunbelt or Southern England has generated a stronger collective syn-ergism than would be the case for their respective northern counterparts.

Noyelle and Stanback (1984) also suggest that position and function within the urban hierarchy have had an important role to play in the patterning of urban fortunes and misfortunes. Transmission effects between cities and within the urban hierarchy must also be factored in to account for the pattern of urban fortunes and misfortunes during the transition from managerialism to entrepreneurialism in urban governance.

Urban entrepreneurialism implies, however, some level of inter-urban competition. We here approach a force that puts clear limitations upon the power of specific projects to transform the lot of particular cities. Indeed, to the degree that inter-urban competition becomes more potent, it will almost certainly operate as an 'external coercive power' over individual cities to bring them closer into line with the discipline and logic of capitalist development. It may even force repetitive and serial reproduction of certain patterns of development (such as the serial reproduction of 'world trade centers' or of new cultural and entertainment centers, of waterfront development, of postmodern shopping malls, and the like). The evidence for serial reproduction of similar forms of urban redevelopment is quite strong and the reasons behind it are worthy of note.

With the diminution in transport costs and the consequent reduction in spatial barriers to movement of goods, people, money and information, the significance of the qualities of place has been enhanced and the vigor of inter-urban competition for capitalist development (investment, jobs, tourism, and so on) has strengthened considerably. Consider the matter, first of all, from the standpoint of highly mobile multinational capital. With the reduction of spatial barriers, distance from the market or from raw materials has become less relevant to locational decisions. The monopolistic elements in spatial competition, so essential to the workings of Löschian theory, disappear. Heavy, low-value items (like beer and mineral water), which used to be locally produced are now traded over such long distances that concepts such as the 'range of a good' make little sense. On the other hand, the ability of capital to exercise greater choice over location, highlights the importance of the particular production conditions prevailing at a particular place. Small differences in labor supply (quantities and qualities), in infrastructures and resources, in government regulation and taxation, assume much greater significance than was the case when high transport costs created 'natural' monopolies for local production in local markets. By the same token, multinational capital now has the power to organize its responses to highly localized variations in market taste through small batch and specialized production designed to satisfy local market niches. In a world of heightened competition – such as that which has prevailed since the postwar boom came crashing to a halt in 1973 – coercive pressures force multinational capital to be much more discriminating and sensitive to small variations between places with respect to both production and consumption possibilities.

Consider matters, in the second instance, from the standpoint of the places that stand to improve or lose their economic vitality if they do not offer enterprises the requisite conditions to come to or remain in town. The reduction of spatial barriers has, in fact, made competition between localities, states, and urban regions for development capital even more acute. Urban governance has thus become much more oriented to the provision of a 'good business climate' and to the construction of all sorts of lures to bring capital into town. Increased entrepreneurialism has been a partial result of this process, of course. But we here see that increasing entrepreneurialism in a different light precisely because the search to procure investment capital confines innovation to a very narrow path built around a favorable package for capitalist development and all that entails. The task of urban governance is, in short, to lure highly mobile and flexible production, financial, and consumption flows into its space. The speculative qualities of urban investments simply derive from the inability to predict exactly which package will succeed and which will not, in a world of considerable economic instability and volatility.

It is easy to envisage, therefore, all manner of upward and downward spirals of urban growth and decline under conditions where urban entrepreneurialism and inter-urban competition are strong. The innovative and competitive responses of many urban ruling-class alliances have engendered more rather than less uncertainty and in the end made the urban system more rather than less vulnerable to the uncertainties of rapid change.

The macroeconomic implications of inter-urban competition

The macroeconomic as well as local implications of urban entrepreneurialism and stronger inter-urban competition deserve some scrutiny. It is particularly useful to put these phenomena into relation with some of the more general shifts and trends that have been observed in the way capitalist economies have been working since the first major postwar recession of 1973 sparked a variety of seemingly profound adjustments in the paths of capitalist development.

To begin with, the fact of inter-urban competition and urban entrepreneurialism has opened up the urban spaces of the advanced capitalist countries to all kinds of new patterns of development, even when the net effect has been the serial reproduction of science parks, gentrification, world trading centers, cultural and entertainment centers, large-scale interior shopping malls with postmodern accoutrements, and the like. The emphasis on the production of a good local business climate has emphasized the importance of the locality as a site of regulation of infrastructural provision, labor relations, environmental controls, and even tax policy *vis-à-vis* international capital (see Swyngedouw 1989). The absorption of risk by the public sector and in particular the stress on public-sector involvement in infrastructural provision, has meant that the cost of locational change has diminished from the standpoint of multinational capital, making the latter more, rather than less, geographically mobile. If anything, the new urban entrepreneurialism adds to, rather than detracts from, the geographical flexibility with which multinational firms can approach their locational strategies. To the degree that the locality becomes the site of regulation of labor relations, so it also contributes to increased flexibility in managerial strategies in geographically segmented labor markets. Local, rather than national, collective bargaining has long been a feature of labor relations in the US, but the trend towards local agreements is marked in many advanced capitalist countries over the past two decades.

There is, in short, nothing about urban entrepreneurialism which is antithetical to the thesis of some macroeconomic shift in the form and style of capitalist development since the early 1970s. Indeed, a strong case can be made (see Harvey 1989a: ch. 8) that the shift in urban politics and the turn to entrepreneurialism has had an important facilitative role in a transition from locationally rather rigid Fordist production systems backed by Keynesian state welfarism to a much more geographically open and market-based form of flexible accumulation. A further case can be made (see Harvey 1989a and 1989b) that the trend away from urban-based modernism in design, cultural forms and lifestyle towards postmodernism is also connected to the rise of urban entrepreneurialism. In what follows I shall illustrate how and why such connections might arise.

Consider, first, the general distributive consequences of urban entrepreneurialism. Much of the vaunted 'public-private partnership' in the US, for example, amounts to a subsidy for affluent consumers, corporations, and powerful command functions to stay in town at the expense of local collective consumption for the working class and poor. The general increase in problems of impoverishment and disempowerment, including the production of a distinctive 'underclass' (to use the language of Wilson 1987) has been documented beyond dispute for many of the large cities in the US. Levine, for example, provides abundant details for Baltimore in a setting where major claims are made for the benefits to be had from public-private partnership. Boddy (1984) likewise reports that what he calls 'mainstream' (as opposed to socialist) approaches to local development in Britain have been 'property-led, business and market-oriented and competitive, with economic development rather than employment the primary focus, and with an emphasis on small firms'. Since the main aim has been 'to stimulate or attract in private enterprise by creating the preconditions for profitable investment', local government 'has in effect ended up underpinning private enterprise, and taking on part of the burden of production costs'. Since capital tends to be more, rather than less, mobile these days, it follows that local subsidies to capital will likely increase while local provision for the underprivileged will diminish, producing greater polarization in the social distribution of real income.

The kinds of jobs created in many instances likewise militate against any progressive shift in income distributions since the emphasis upon small businesses and subcontracting can even spill over into direct encouragement of the 'informal sector' as a basis for urban survival. The rise of informal production activities in many cities, particularly in the US (Sassen-Koob 1988), has been a marked feature in the past two decades and is increasingly seen as either a necessary evil or as a dynamic growth sector capable of reimporting some level of manufacturing activity back into otherwise declining urban centers. By the same token, the kinds of service activities and managerial functions which get consolidated in urban regions tend to be either low-paying jobs (often held exclusively by women) or very high-paying positions at the top end of the managerial spectrum. Urban entrepreneurialism consequently contributes to increasing disparities in wealth and income as well as to that increase in urban impoverishment which has been noted even in those cities (like New York) that have exhibited strong growth. It has, of course, been exactly this result that Labour councils in Britain (as well as some of the more progressive urban administrations in the US) have been struggling to resist. But it is by no means clear that even the most progressive urban government can resist such an outcome when embedded in the logic of capitalist spatial development in which competition seems to operate not as a beneficial hidden hand, but as an external coercive law forcing the lowest common denominator of social responsibility and welfare provision within a competitively organized urban system.

Many of the innovations and investments designed to make particular cities more attractive as cultural and consumer centres have quickly been imitated elsewhere, thus rendering any competitive advantage within a system of cities ephemeral. How many successful convention centres, sports stadia, Disney Worlds, harbor places and spectacular shopping malls can there be? Success is often short-lived or rendered moot by parallel or alternative innovations arising elsewhere. Local coalitions have no option, given the coercive laws of competition, except to keep ahead of the game, thus engendering leap-frogging innovations in lifestyles, cultural forms, products and service mixes, even institutional and political forms if they are to survive. The result is a stimulating if often destructive maelstrom of urban-based cultural, political, production and consumption innovations. It is at this point that we can identify an albeit subterranean but nonetheless vital connection between the rise of urban entrepreneurialism and the postmodern penchant for design of urban fragments rather than comprehensive urban planning, for ephemerality and eclecticism of fashion and style rather than the search for enduring values, for quotation and fiction rather than invention and function, and, finally, for medium over message and image over substance.

In the US, where urban entrepreneurialism has been particularly vigorous, the result has been instability within the urban system. Houston, Dallas and Denver, boom towns in the 1970s, suddenly dissolved after 1980 into morasses of excess capital investment bringing a host of financial institutions to the brink of, if not in actual bankruptcy. Silicon Valley, once the high-tech wonder of new products and new employment, suddenly lost its luster but New York, on the edge of bankruptcy in 1975, rebounded in the 1980s with the immense vitality of its financial services and command functions, only to find its future threatened once more with the wave of lay-offs and mergers which rationalized the financial services sector in the wake of the stock market crash of October 1987. San Francisco, the darling of Pacific Rim trading, suddenly finds itself with excess office space in the early 1980s, only to recover almost immediately. New Orleans, already struggling as a ward of federal government redistributions, sponsors a disastrous World Fair that drives it deeper into the mire, while Vancouver, already booming, hosts a remarkably successful World Exposition. The shifts in urban fortunes and misfortunes since the early 1970s have been truly remarkable and the strengthening of urban entrepreneurialism and inter-urban competition has had a lot to do with it.

But there has been another rather more subtle effect that deserves consideration. Urban entrepreneurialism encourages the development of those kinds of activities and endeavors that have the strongest *localized* capacity to enhance property values, the tax base, the local circulation of revenues, and (most often as a hoped-for consequence of the preceding list) employment growth. Since increasing geographical mobility and rapidly changing technologies have rendered many forms of production of goods highly suspect, so the production of those kinds of services that are (1) highly localized and (2) characterized by rapid if not instantaneous turnover time appear as the most stable basis for urban entrepreneurial endeavor. The emphasis upon tourism, the production and consumption of spectacles, the promotion of ephemeral events within a given locale, bear all the signs of being favored remedies for ailing urban economies. Urban investments of this sort may yield quick though ephemeral fixes to urban problems. But they are often highly speculative. Gearing up to bid for the Olympic Games is an expensive exercise, for example, which may or may not pay off. Many cities in the US (Buffalo, for example) have invested in vast stadium facilities in the hope of land- ing a major league baseball team and Baltimore is similarly planning a new stadium to try and recapture a football team that went to a superior stadium in Indianapolis some years ago (this is the contemporary US version of that ancient cargo cult practice in Papua, New Guinea, of building an airstrip in the hope of luring a jet liner to earth). Speculative projects of this sort are part and parcel of a more general macroeconomic problem. Put simply, credit-financed shopping malls, sports stadia, and other facets of conspicuous high consumption are high-risk projects that can easily fall on bad times and thus exacerbate, as the 'overmalling of America' only too dramatically illustrates (Green 1988), the problems of overaccumulation and overinvestment to which capitalism as a whole is so easily prone. The instability that pervades the US financial system (forcing something of the order of $100 billion in public moneys to stabilize the savings and loan industry) is partly due to bad loans in energy, agriculture, and urban real-estate development. Many of the 'festival market places' that looked like an 'Aladdin's lamp for cities fallen on hard times', just a decade ago, ran a recent report in the *Baltimore Sun* (20 August 1987), have now themselves fallen on hard times. Projects in Richmond, Virginia Flint, Michigan and Toledo, Ohio, managed by Rouse's Enterprise Development Co., are losing millions of dollars, and even the South Street Seaport in New York and Riverwalk in New Orleans have encountered severe financial difficulties. Ruinous inter- urban competition on all such dimensions bids fair to become a quagmire of indebtedness.

Even in the face of poor economic performance, however, investments in these last kinds of projects appear to have both a social and political attraction. To begin with, the selling of the city as a location for activity depends heavily upon the creation of an attractive urban imagery. City leaders can look upon the spectacular development as a 'loss leader' to pull in other forms of development. Part of what we have seen these past two decades is the attempt to build a physical and social imagery of cities suited for that competitive purpose. The production of an urban image of this sort also has internal political and social consequences. It helps counteract the sense of alienation and anomie that Simmel long ago identified as such a problematic feature of modern city life. It particularly does so when an urban terrain is opened for display, fashion and the 'presentation of self' in a surrounding of spectacle and play. If everyone, from punks and rap artists to the 'yuppies' and the haute bourgeoisie, can participate in the production of an urban image through their production of social space, then all can at least feel some sense of belonging to that place. The orchestrated production of an urban image can, if successful, also help create a sense of social solidarity, civic pride and loyalty to place and even allow the urban image to provide a mental refuge in a world that capital treats as more and more place-less. Urban entrepreneurialism (as opposed to the much more faceless bureaucratic managerialism) here meshes with a search for local identity and, as such, opens up a range of mechanisms for social control. Bread and circuses was the famous Roman formula that now stands to be reinvented and revived, while the ideology of locality, place and community becomes central to the political rhetoric of urban governance which concentrates on the idea of togetherness in defense against a hostile and threatening world of international trade and heightened competition.

The radical reconstruction of the image of Baltimore through the new waterfront and inner-harbor development is a good case in point. The redevelopment put Baltimore on the map in new way, earned the city the title of 'renaissance city' and put it on the front cover of *Time* magazine, shedding its image of dreariness and impoverishment. It appeared as a dynamic go-getting city, ready to accommodate outside capital and to encourage the movement in of capital and of the 'right' people. No matter that the reality is one of increased impoverishment and overall urban deterioration, that a thorough local enquiry based on interviews with community, civic and business leaders identified plenty of 'rot beneath the glitter' (Szanton 1986), that a Congressional Report of 1984 described the city as one of the 'neediest' in the US, and that a thorough study of the renaissance by Levine (1987) showed again and again how partial and limited the benefits were and how the city as a whole was accelerating rather than reversing its decline. The image of prosperity conceals all that, masks the underlying difficulties and projects an imagery of success that spreads internationally so that the British newspaper the *Sunday Times* (29 November 1987) can report, without a hint of criticism, that:

> Baltimore, despite soaring unemployment, boldly turned its derelict harbour into a playground. Tourists meant shopping, catering and transport, this in turn meant construction, distribution, manufacturing – leading to more jobs more residents, more activity. The decay of old Baltimore slowed, halted, then turned back. The harbour area is now among America's top tourist draws and urban unemployment is falling fast.

Yet it is also apparent that putting Baltimore on the map in this way, giving it a stronger sense of place and of local identity, has been successful politically in consolidating the power of influence of the local public-private partnership that brought the project into being. It has brought development money into Baltimore (though it is hard to tell if it has brought more in than it has taken out, given the absorption of risk by the public sector). It also has given the population at large some sense of placebound identity. The circus succeeds even if the bread is lacking. The triumph of image over substance is complete.

Critical perspectives on the entrepreneurial turn in urban governance under conditions of inter-urban competition

There has been a good deal of debate in recent years over the 'relative autonomy' of the local state in relation to the dynamics of capital accumulation. The turn to entrepreneurialism in urban governance seems to suggest considerable autonomy of local action. The notion of urban entrepreneurialism, as I have here presented it, does not in any way presume that the local state or the broader class alliance that constitutes urban governance is automatically (or even in the famous 'last instance') captive of solely capitalist class interests or that its decisions are pre-figured directly in terms reflective of the requirements of capital accumulation. On the surface, at least, this seems to render my account inconsistent with that Marxist version of local state theory put forward by, say, Cockburn (1977), and strongly dissented from by a range of other non-Marxist or neo-Marxist writers such as Mollenkopf (1983), Logan and Molotch (1987), Gurr and King (1987) and Smith (1988). Consideration of inter-urban competition, however, indicates a way in which a seemingly autonomous urban entrepreneurialism can be reconciled with the albeit contradictory requirements of continuous capital accumulation while guaranteeing the reproduction of capitalist social relations on ever wider scales and at deeper levels.

Marx advanced the powerful proposition that competition is inevitably the 'bearer' of capitalist social relations in any society where the circulation of capital is a hegemonic force. The coercive laws of competition force individual or collective agents (capitalist firms, financial institutions, states, cities) into certain configurations of activities which are themselves constitutive of the capitalist dynamic. But the 'forcing' occurs after the action rather than before. Capitalist development is always speculative – indeed, the whole history of capitalism can best be read as a whole series

of minuscule and sometimes grandiose speculative thrusts piled historically and geographically one upon another. There is, for example, no exact
prefiguring of how firms will adapt and behave in the face of market
competition. Each will seek its own path to survival without any prior
understanding of what will or will not succeed. Only after the event does
the 'hidden hand' (Adam Smith's phrase) of the market assert itself as
'an *a posteriori*, nature-imposed necessity, controlling the lawless caprice
of the producers' (Marx 1967: 336).

Urban governance is similarly and liable to be equally, if not even more,
lawless and capricious. But there is also every reason to expect that such
'lawless caprice' will be regulated after the fact by inter-urban competition. Competition for investments and jobs, particularly under conditions
of generalized unemployment, industrial restructuring and in a phase of
rapid shifts towards more flexible and geographically mobile patterns
of capital accumulation, will presumably generate all kinds of ferments
concerning how best to capture and stimulate development under particular local conditions. Each coalition will seek out its distinctive version
of what Jessop (1983) calls 'accumulation strategies and hegemonic
projects'. From the standpoint of long-run capital accumulation, it is
essential that different paths and different packages of political, social, and
entrepreneurial endeavors get explored. Only in this way is it possible for
a dynamic and revolutionary social system, such as capitalism, to discover
new forms and modes of social and political regulation suited to new
forms and paths of capital accumulation. If this is what is meant by
the 'relative autonomy' of the local state, then there is nothing about it
which makes urban entrepreneurialism in principle in any way different
from the 'relative autonomy' which all capitalist firms, institutions and
enterprises possess in exploring different paths to capital accumulation.
Relative autonomy understood in this way is perfectly consistent with,
and indeed is constitutive of, the general theory of capital accumulation
to which I would subscribe (Harvey 1982). The theoretical difficulty arises,
however, as in so many issues of this type, because Marxian as well as
non-Marxian theory treats of the relative autonomy argument as if it can
be considered outside of the controlling power of space relations and as if
inter-urban and spatial competition are either non-existent or irrelevant.

In the light of this argument, it would seem that it is the managerial
stance under conditions of weak inter-urban competition that would
render urban governance less consistent with the rules of capital accumulation. Consideration of that argument requires, however, an extended
analysis of the relations of the welfare state and of national Keynesianism
(in which local state action was embedded) to capital accumulation during
the 1950s and 1960s. This is not the place to attempt such an analysis, but

it is important to recognize that it was in terms of the welfare state and Keynesian compromise that much of the argument over the relative autonomy of the local state emerged. Recognizing that as a particular interlude, however, helps understand why civic boosterism and urban entrepreneurialism are such old and well-tried traditions in the historical geography of capitalism (starting, of course, with the Hanseatic League and the Italian city states). The recovery and reinforcement of that tradition and the revival of inter-urban competition these past two decades, suggests that urban governance has moved more rather than less into line with the naked requirements of capital accumulation. Such a shift required a radical reconstruction of central to local state relations and the cutting-free of local state activities from the welfare state and the Keynesian compromise (both of which have been under strong attack these past two decades). And, needless to say, there is strong evidence of turmoil in this quarter in many of the advanced capitalist countries in recent years.

It is from this perspective that it becomes possible to construct a critical perspective on the contemporary version of urban entrepreneuri-alism. To begin with, enquiry should focus on the contrast between the surface vigor of many of the projects for regeneration of flagging urban economies and the underlying trends in the urban condition. It should recognize that behind the mask of many successful projects, there lie some serious social and economic problems and that in many cities these are taking geographical shape in the form of a dual city of inner-city regeneration and a surrounding sea of increasing impoverishment. A critical perspective should also focus on some of the dangerous macro-economic consequences, many of which seem inescapable given the coercion exercised through inter-urban competition. The latter include regressive impacts on the distribution of income, volatility within the urban network and the ephemerality of the benefits which many projects bring. Concentration on spectacle and image rather than on the substance of economic and social problems can also prove deleterious in the long run, even though political benefits can all too easily be had.

Yet there is something positive also going on here that deserves close attention. The idea of the city as a collective corporation, within which democratic decision-making can operate has a long history in the pantheon of progressive doctrines and practices (the Paris Commune being, of course, the paradigm case in socialist history). There have been some recent attempts to revive such a corporatist vision both in theory (see Frug 1980) as well as in practice (see Blunkett and Jackson 1987). While it is possible, therefore, to characterize certain kinds of urban entrepre-neurialism as purely capitalistic in both method, intent and result, it is also useful to recognize that many of the problems of collective corporatist action originate not with the fact of some kind of civic boosterism, or even by virtue of who, in particular, dominates the urban class alliances that form or what projects they devise. For it is the generality of inter-urban competition within an overall framework of uneven capitalist geographical development which seems so to constrain the options that 'bad' projects drive out 'good' and well-intended and benevolent coalitions of class forces find themselves obliged to be 'realistic' and 'pragmatic' to a degree which has them playing to the rules of capitalist accumulation rather than to the goals of meeting local needs or maximizing social welfare. Yet even here, it is not clear that the mere fact of inter-urban competition is the primary contradiction to be addressed. It should be regarded, rather, as a condition which acts as a 'bearer' (to use Marx's phrase) of the more general social relations of any mode of production within which that competition is embedded. Socialism within one city is not, of course, a feasible project even under the best of circumstances. Yet cities are important power bases from which to work. The problem is to devise a geopolitical strategy of inter-urban linkage that mitigates inter-urban competition and shifts political horizons away from the locality and into a more generalizable challenge to capitalist uneven development. Working-class movements, for example, have proven historically to be quite capable of commanding the politics of place, but they have always remained vulnerable to the discipline of space relations and the more powerful command over space (militarily as well as economically) exercised by an increasingly internationalized bourgeoisie. Under such conditions, the trajectory taken through the rise of urban entrepreneurialism these past few years serves to sustain and deepen capitalist relations of uneven geographical development and thereby affects the overall path of capitalist development in intriguing ways. But a critical perspective on urban entrepreneurialism indicates not only its negative impacts but its potentiality for transformation into a progressive urban corporatism, armed with a keen geopolitical sense of how to build alliances and linkages across space in such a way as to mitigate if not challenge the hegemonic dynamic of capitalist accumulation to dominate the historical geography of social life.

CHAPTER 17

The geography of class power

First published in Socialist Register, *1998.*

It is imperative to reignite the political passions that suffuse the *Communist Manifesto*. It is an extraordinary document full of insights, rich in meanings and bursting with political possibilities. While we have not the right, as Marx and Engels wrote in their 1872 preface to the German edition, to alter what has become a key historical document, we have not only the right but the obligation to interpret it in the light of contemporary conditions and historical-geographical experience. '*The practical application of the principles*,' wrote Marx and Engels in that Preface, '*will depend*, as the *Manifesto* itself states *everywhere and at all times, on the historical conditions for the time being existing*.' This italicized phrase precisely delineates our present task.

The accumulation of capital has always been a profoundly geographical affair. Without the possibilities inherent in geographical expansion, spatial reorganization and uneven geographical development, capitalism would long ago have ceased to function as a political economic system. This perpetual turning to 'a spatial fix' to capitalism's internal contradictions (most notably registered as an overaccumulation of capital within a particular geographical area) coupled with the uneven insertion of different territories and social formations into the capitalist world market has created a global historical geography of capital accumulation whose character needs to be well understood. How Marx and Engels conceptualized the problem in the *Communist Manifesto* deserves some commentary for it is here that the communist movement – with representatives from many countries – came together to try to define a revolutionary agenda that would work in the midst of considerable geographical differentiation. This differentiation is just as important today as it ever was and the *Manifesto*'s weaknesses, as well as its strengths, in its approach to this problem need to be confronted and addressed.

The spatial fix in Hegel and Marx

In *The Philosophy of Right*, Hegel presented imperialism and colonialism as potential solutions to the internal contradictions of what he considered

369

to be a 'mature' civil society (Hegel 1967: 150–2). The increasing accumulation of wealth at one pole and the formation of a 'penurious rabble' trapped in the depths of misery and despair at the other, sets the stage for social instability and class war that cannot be cured by any internal transformation (such as a redistribution of wealth from rich to poor). Civil society is thereby driven by its 'inner dialectic' to 'push beyond its own limits and seek markets, and so its necessary means of subsistence, in other lands that are either deficient in the goods it has overproduced, or else generally backward in industry.' It must also found colonies and thereby permit a part of its population 'a return to life on the family basis in a new land' at the same time as it also 'supplies itself with a new demand and field for its industry'. All of this is fuelled by a 'passion for gain' that inevitably involves risk, so that industry, 'instead of remaining rooted to the soil and the limited circle of civil life with its pleasures and desires . . . embraces the element of flux, danger, and destruction.'

Having, in a few brief startling paragraphs, sketched the possibilities of an imperialist and colonial solution to the ever-intensifying internal contradictions of civil society, Hegel just as suddenly dropped the matter. He leaves us in the dark as to whether capitalism could be stabilized by appeal to some sort of 'spatial fix' in either the short or long run. Instead, he turns his attention to the concept of the state as the actuality of the ethical idea. This could be taken to imply that transcendence of civil society's internal contradictions by the modern state – an inner transformation – is both possible and desirable. Yet Hegel nowhere explains how the problems of poverty and of the increasing polarization in the distribution of wealth are actually to be overcome. Are we supposed to believe, then, that these particular problems can be dealt with by imperialism? The text is ambivalent. This is, as Avineri points out, 'the only time in his system, where Hegel raises a problem – and leaves it open' (Avineri 1972: 132).

How far Hegel influenced Marx's later concerns can be endlessly debated. Engels certainly believed that Marx was 'the only one who could undertake the work of extracting from Hegelian logic the kernel containing Hegel's real discoveries.' The language Marx uses to describe the general law of capitalist accumulation, for example, bears an eerie resemblance to

[1] Compare, for example, Hegel's argument in *The Philosophy of Right* that: 'When the standard of living of a large mass of people falls below a certain subsistence level – a level regulated automatically as the one necessary for a member of the society . . . the result is the creation of a rabble of paupers. At the same time this brings with it, at the other end of the social scale, conditions which greatly facilitate the concentration of disproportionate wealth in a few hands,' and Marx's conclusion in *Capital*, Volume I, that: 'as capital accumulates, the situation of the worker, be his payment high or low, must grow

that of Hegel.[1] It is even possible to interpret Volume 1 of *Capital* as a tightly orchestrated argument, buttressed by good deal of historical and material evidence, to prove that the propositions Hegel had so casually advanced, without any logical or evidentiary backing, were indubitably correct.[2] The internal contradictions that Hegel depicted were, in Marx's view, not only inevitable but also incapable of any internal resolution short of proletarian revolution. And this was, of course, the conclusion that Marx wanted to force not only upon the Hegelians but upon everyone else. But in order to make the argument stick, he also has to bear in mind the question that Hegel had raised but left open.

In this light, one other feature in the structure of argument in *Capital* makes sense. The last chapter of the book deals with the question of colonization. It seems, at first sight, an odd afterthought to a work which, in the preceeding chapter, announced expropriation of the expropriators and the death-knell of the bourgeoisie with a rhetoric reminiscent of the *Manifesto*. But in the light of Hegel's argument, the chapter acquires a particular significance.

Marx first seeks to show how the bourgeoisie contradicted its own myths as to the origin and nature of capital by the policies it advocated in the colonies. In bourgeois accounts (the paradigmatic case being that of Locke), capital (a thing) originated in the fruitful exercise of the producer's own capacity to labor, while labor power as a commodity arose through a social contract, freely entered into, between those who produced surplus capital through frugality and diligence, and those who chose not to do so. 'This pretty fancy', Marx thunders, is 'torn asunder' in the colonies. As long as the laborer can 'accumulate for himself – and this he can do as long as he remains possessor of his means of production – capitalist accumulation and the capitalist mode of production are impossible.' Capital is not a physical thing but a social relation. It rests on the 'annihilation of self-earned private property, in other words, the expropriation of the labourer.' Historically, this expropriation was 'written in the annals of mankind in letters of blood and fire', and Marx cites chapter, verse and the Duchess of Sutherland to prove his point. The same truth, however, is expressed in colonial land policies, such as those of Wakefield in Australia, in which the powers of private property and the

---

worse . . . It makes an accumulation of misery a necessary condition, corresponding to the accumulation of wealth. Accumulation of wealth at one pole is, therefore, at the same time accumulation of misery, the torment of labour, slavery, ignorance, brutalization and moral degradation at the opposite pole, i.e. on the side of the class that produces its own product as capital.' The parallel between the two texts is striking.

[2] See Harvey (1982) ch. 13, Harvey (1981) for further details of this argument.

state were to be used to exclude laborers from easy access to free land in order to preserve a pool of wage laborers for capitalist exploitation. Thus was the bourgeoisie forced to acknowledge in its programme of colonization what it sought to conceal at home: that wage labor and capital are both based on the forcible separation of the laborer from control over the means of production (Marx 1967). This is the secret of 'primitive' or 'original' capital accumulation.

The relation of all this to the question Hegel left open needs explication. If laborers can return to a genuinely unalienated existence through migration overseas or to some frontier region, then capitalist control over labor supply is undermined. Such a form of expansion may be advantageous to labor but it could provide no solution to the inner contradictions of capitalism. The new markets and new fields for industry which Hegel saw as vital could be achieved only through the re-creation of capitalist relations of private property and the associated power to appropriate the labor of others. The fundamental conditions which gave rise to the problem in the first place – alienation of labor – are thereby replicated. Marx's chapter on colonization appears to close off the possibility of any external 'spatial fix' to the internal contradictions of capitalism. Marx evidently felt obliged in *Capital* to close the door that Hegel had left partially ajar and consolidate his call for total revolution by denying that colonization could, in the long run, be a viable solution to the inner contradictions of capital.

But the door will not stay shut. Hegel's 'inner dialectic' undergoes successive representations in Marx's work and at each point the question of the spatial resolution to capitalism's contradictions can legitimately be posed anew. The chapter on colonization may suffice for the first volume of *Capital* where Marx concentrates solely on questions of production. But what of the third volume where Marx shows that the requirements of production conflict with those of circulation to produce crises of overaccumulation? Polarization then takes the form of 'unemployed capital at one pole and unemployed worker population at the other' and the consequent devaluation of both. Can the formation of such crises be contained through geographical expansions and restructurings? Marx does not rule out the possibility that foreign trade and growth of external markets, the export of capital for production, and the expansion of the proletariat through primitive accumulation in other lands, can counteract the falling rate of profit in the short run. But how long is the short run? And if it extends over many generations (as Rosa Luxemburg in her theory of imperialism implied), then what does this do to Marx's theory and its associated political practice of seeking for revolutionary transformations in the heart of civil society in the here and now?

The spatial dimension to the
Communist Manifesto

Many of these problems arise in the *Communist Manifesto*.[3] The manner of approach that Marx and Engels took to the problem of uneven geographical development and the spatial fix is in some respects deeply ambivalent. On the one hand, questions of urbanization, geographical transformation and 'globalization' are given a prominent place in the argument, but on the other hand the potential ramifications of geographical restructurings tend to get lost in a rhetorical mode that in the last instance privileges time and history over space and geography.

The opening sentence of the *Manifesto* situates the argument in Europe and it is to that transnational entity and its working classes that its theses are addressed. This reflects the fact that 'Communists of various nationalities' (French, German, Italian, Flemish and Danish as well as English are the languages envisaged for publication of the document) were assembled in London to formulate a working-class program. The document is, therefore, Eurocentric rather than international. But the importance of the global setting is not ignored. The revolutionary changes that brought the bourgeoisie to power were connected to 'the discovery of America, the rounding of the Cape' and the opening-up of trade with the colonies and with the East Indian and Chinese markets. The rise of the bourgeoisie is, from the very outset of the argument, intimately connected to its geographical activities and strategies:

> Modern industry has established the world market, for which the discovery of America paved the way. This market has given an immense development to commerce to navigation, to communication by land. This development has in turn, reacted on the extension of industry; in proportion as industry, commerce, navigation, railways extended, in the same proportion the bourgeoisie developed, increased its capital, and pushed into the background every class handed down from the Middle Ages.

By these geographical means, the bourgeoisie bypassed and suppressed place-bound feudal powers. By these means also the bourgeoisie converted the state (with its military, organizational and fiscal powers) into the executive of its own ambitions. And, once in power, the bourgeoisie continued to pursue its revolutionary mission in part via geographical transformations which are both internal and external. Internally, the creation of great

---

[3] All citations are from Marx and Engels, *Manifesto of the Communist party* Progress Publishers edition, Moscow 1952.

cities and rapid urbanization bring the towns to rule over the country (simultaneously rescuing the latter from the 'idiocy' of rural life and reducing the peasantry to a subaltern class). Urbanization concentrates productive forces as well as labor power in space, transforming scattered populations and decentralized systems of property rights into massive concentrations of political and economic power. 'Nature's forces' are subjected to human control: 'machinery, application of chemistry to industry and agriculture, steam navigation, railways, electric telegraphs, clearing of whole continents for cultivation, canalisation of rivers, whole populations conjured out of the ground . . .'

But this concentration of the proletariat in factories and towns makes them aware of their common interests. On this basis, they begin to build institutions, such as unions, to articulate their claims. Furthermore, the modern systems of communications put 'the workers of different localities in contact with each other', thus allowing 'the numerous local struggles, all of the same character' to be centralized into 'one national struggle between the classes'. This process, as it spreads across frontiers, strips the workers of 'every trace of national character', for each and everyone of them is subject to the unified rule of capital. The organization of working-class struggle concentrates and diffuses across space in a way that mirrors the actions of capital.

Marx expands on this idea in a passage that is so famous that we are apt to skim over it rather than read and reflect upon it with the care it deserves:

> The need for a constantly expanding market chases the bourgeoisie over the whole surface of the globe. It must settle everywhere, establish connexions everywhere . . . The bourgeoisie has through its exploitation of the world market given a cosmopolitan character to production and consumption in every country . . . All old established national industries have been destroyed or are daily being destroyed. They are dislodged by new industries whose introduction becomes a life and death question for all civilized nations, by industries that no longer work up indigenous raw material, but raw material drawn from the remotest zones; industries whose products are consumed, not only at home, but in every quarter of the globe. In place of the old wants, satisfied by the production of the country, we find new wants, requiring for their satisfaction the products of distant lands and climes. In place of the old local and national seclusion and self-sufficiency, we have intercourse in every direction, universal interdependence of nations. And as in material, so also in intellectual production. The intellectual creations of individual nations become common property. National one-sidedness and narrow-mindedness become more and more impossible, and from the numerous national and local literatures, there arises a world literature . . .

If this is not a compelling description of 'globalization' as we now know it, then it is hard to imagine what would be. The traces of Hegel's 'spatial fix' argument are everywhere apparent. But Marx and Engels add something:

> The bourgeoisie . . . draws all, even the most barbarian nations into civilization, the cheap prices of its commodities are the heavy artillery with which it batters down all Chinese walls, with which it forces the barbarians' intensely obstinate hatred of foreigners to capitulate. It compels all nations on pain of extinction, to adopt the bourgeois mode of production; it compels them to introduce what it calls civilization into their midst, i.e. to become bourgeois themselves. In one word, it creates a world after its own image.

The theme of the 'civilizing mission' of the bourgeoisie is here enunciated (albeit with a touch of irony). But a certain limit to the power of the spatial fix to work indefinitely and in perpetuity is implied. If the geographical mission of the bourgeoisie is the reproduction of class and productive relations on a progressively expanding geographical scale, then the bases for both the internal contradictions of capitalist and for socialist revolution likewise expand geographically. The conquest of new markets paves the way 'for more extensive and more destructive crises,' while 'diminishing the means whereby crises are prevented'. Class struggle becomes global. Marx and Engels therefore enunciate the imperative 'working men of all countries unite' as a necessary condition for an anti-capitalist and pro-socialist revolution.

Problematizing the *Manifesto*'s geography

The geographical element in the *Manifesto* has, to a large degree, been ignored in subsequent commentaries. When it has been the focus of attention, it has often been treated as unproblematic in relation to political action. This suggests a twofold response as we look back upon the argument. First, it is vital to recognize (as the *Manifesto* so clearly does) the ways in which geographical reorderings and restructurings spatial strategies and geopolitical elements, uneven geographical developments, and the like, are vital aspects to the accumulation of capital, both historically and today. It is likewise vital to recognize (in ways the *Manifesto* tends to underplay) that class struggle unfolds differentially across this highly variegated terrain and that the drive for socialism must take these geographical realities into account. But, secondly, it is equally important to problematize the actual account ('sketch' might be a more appropriate word) given in the *Manifesto* in order to develop a more sophisticated, accurate and politically useful understanding as to how the geographical dimensions to capital accumulation and class struggle play such a fundamental role in the perpetuation of bourgeois power and the suppression of worker rights and aspirations not only in particular places but also globally.

In what follows, I shall largely take the first response as a 'given' even though I am only too aware that it needs again and again to be reasserted within a movement that has not by any means taken on board some, let alone all, of its very basic implications. While Lefebvre perhaps exaggerates a touch, I think it worth recalling his remark that capitalism has survived in the twentieth century by one and only one means: 'by occupying space, by producing space' (Lefebvre 1976). How ironic if the same were to be said at the end of the twenty-first century!

My main concern here, then, is to problematize the account given in the *Manifesto*. This requires, tacitly or explicitly, a non-Hegelian counter-theory of the spatio-temporal development of capital accumulation and class struggle (Meszaros 1995; Harvey 1996). From such a perspective, I shall isolate six aspects of the *Manifesto* for critical commentary.

First, the division of the world into 'civilized' and 'barbarian' nations is, to say the least, anachronistic if not downright objectionable even if it can be excused as typical of the times. Furthermore, the centre-periphery model of capital accumulation which accompanies it is at best a gross oversimplification and at worst misleading. It makes it appear as if capital originated in one place (England or Europe) and then diffused outwards to encompass the rest of the world. Adoption of this stance seems to derive from uncritical acceptance of Hegels' teleology – if space is to be considered at all, it is as a passive recipient of a teleological process that starts from the centre and flows outwards to fill up the entire globe. Leaving aside the whole problem of where, exactly, capitalism was born and whether it arose in one and only one place or was simultaneously emerging in geographically distinctive environments (an arena of scholarly dispute that shows no sign of coming to a consensus) the subsequent development of a capitalism that had, by the end of the eighteenth century at least, come to concentrate its freest forms of development in Europe in general and Britain in particular, cannot be encompassed by such a diffusionist way of thinking. While there are some instances in which capital diffused outwards from a centre to a periphery (for example the export of surplus capital from Europe to Argentina or Australia in the late nineteenth century), such an account is inconsistent with what happened in Japan after the Meiji restoration or what is happening today as first South Korea and then China engages in some form of internalized primitive accumulation and inserts its labor power and its products into global markets.

The geography of capital accumulation deserves a far more principled treatment than the diffusionist sketch provided in the *Manifesto*. The problem does not lie in the sketchiness of the account per se, but in the failure to delineate a theory of uneven geographical development (often entailing uneven primitive accumulation) that would be helpful for charting the dynamics of working–class formation and class struggle across even the European, let alone the global, space. I would also argue for a more fully theorized understanding of the space/place dialectic in capitalist development (Harvey 1996). How do places, regions, territories evolve given changing space relations? We have observed how geopolitical games of power, for example, become interconnected with market position in a changing structure of space-relations which, in turn, privileges certain locations and territories for capitalist accumulation. It is also interesting to note how those national bourgeoisies that could not easily use spatial powers to circumvent feudalism ended up with fascism (Germany, Italy, Spain are cases in point). Since these are rather abstract arguments, I shall try to put some flesh and bones on them in what follows.

To begin with, the globe never has been a level playing-field upon which capital accumulation could play out its destiny. It was, and continues to be, an intensely variegated surface, ecologically, politically, socially and culturally differentiated. Flows of capital found some terrains easier to occupy than others in different phases of development. And in the encounter with the capitalist world market, some social formations adapted to aggressively insert themselves into capitalistic forms of market exchange while others did not, for a wide range of reasons and with consummately important effects. Primitive or 'original' accumulation can, and has occurred, in different places and times, albeit facilitated by contact with the market network that increasingly pins the globe together into an economic unity. But how and where that primitive accumulation occurs depends upon local conditions even if the effects are global. It is now a widely held belief in Japan, for example, that the commercial success of that country after 1960 was in part due to the non-competitive and withdrawn stance of China after the revolution and that the con-temporary insertion of Chinese power into the capitalist world market spells doom for Japan as a producer as opposed to a rentier economy. Contingency of this sort rather than teleology has a lot of play within capitalist world history. Furthermore, the globality of capital accumulation poses the problem of a dispersed bourgeois power that can become much harder to handle geopolitically precisely because of its multiple sites. Marx himself later worried about this political possibility. In 1858 he wrote (in a passage that Meszaros rightly makes much of (1996: xii)):

> For us the difficult question is this: the revolution on the Continent is imminent and its character will be at once socialist; will it not be *necessarily crushed* in this *little corner of the world*, since on a much larger terrain the development of bourgeois society is still *in the ascendant*.

It is chastening to reflect upon the number of socialist revolutions around the world that have been successfully encircled and crushed by the geopolitical strategies of an ascendant bourgeois power.

Second, the *Manifesto* quite correctly highlights the importance of reducing spatial barriers through innovations and investments in transport and communications as critical to the growth and sustenance of bourgeois power. Moreover, the argument indicates that this is an ongoing rather than already-accomplished process. In this respect, the *Manifesto* is prescient in the extreme. 'The annihilation of space through time' as Marx later dubbed it (adopting an expression that was quite common in the early nineteenth century as people adjusted to the revolutionary implications of the railroad and the telegraph) is deeply embedded in the logic of capital accumulation, entailing as it does the continuous, though often jerky, transformations in space relations that have characterized the historical-geography of the bourgeois era (from turnpikes to cyberspace). These transformations undercut the absolute qualities of space (often associated with feudalism) and emphasize the relativity of space relations and locational advantages, thus making the Ricardian doctrine of comparative advantage in trade a highly dynamic rather than stable affair. Furthermore, spatial tracks of commodity flows have to be mapped in relation to flows of capital, labor power, military advantage, technology transfers, information flows, and the like. In this regard, at least, the *Manifesto* was not wrong as much as underelaborated upon and underappreciated for its prescient statements.

Third, perhaps one of the biggest absences in the *Manifesto* is its lack of attention to the territorial organization of the world in general and of capitalism in particular. If, for example, the state was necessary as an 'executive arm of the bourgeoisie', then the state had to be territorially defined, organized and administered. While the right of sovereign independent states to coexistence was established at the Treaty of Westphalia in 1648 as a (distinctively shaky) European norm, the general extension of that principle across the globe took several centuries to take shape and is even now arguably not accomplished. The nineteenth century was the great period of territorial definitions (with most of the world's boundaries being established between 1870 and 1925 and most of those being drawn by the British and the French alone, the carve-up of Africa in 1885 being the most spectacular example). But state formation and consolidation is quite another step beyond territorial definition and it has proven a long-drawn-out and often unstable affair (particularly, for example, in Africa). It could well be argued that it was only after 1945 that decolonization pushed state formation worldwide a bit closer to the highly simplified model that the *Manifesto* envisages. Furthermore, the relativism introduced by revolutions in transport and communications coupled with the uneven dynamics of class struggle and uneven resource endowments means that territorial configurations cannot remain stable for long. Flows of commodities, capital, labour and information always render boundaries porous. There is plenty of play for contingency (including phases of territorial reorganization and redefinition) here, thus undermining the rather simplistic teleology that derives from Hegel but which can still be found in some versions of both capitalistic and communist ideas about what the future necessarily holds.

Fourth, the state is, of course, only one of many mediating institutions that influence the dynamics of accumulation and of class struggle worldwide. Money and finance must also be given pride of place. In this respect there are some intriguing questions about which the *Manifesto* remains silent, in part, I suspect, because its authors had yet to discover their fundamental insights about the dialectical relations between money, production, commodity exchange, distribution and production (as these are conceptualized, for example, in the Introduction to the *Grundrisse*). There are two ways to look at this (and I here take the question of money as both emblematic and fundamental). On the one hand, we can interpret world money as some universal representation of value to which territories relate (through their own currencies) and to which capitalist producers conform as they seek some measure of their performance and profitability. This is a very functionalist and undialectical view. It makes it seem as if value hovers as some ethereal abstraction over the activities of individuals as of nations (this is, incidentally, the dominant conception at work in the contemporary neoclassical ideology of globalization). In *Capital*, Marx looks upon world money differently, as a representation of value that arises out of a dialectical relation between the particularity of material activities (concrete labor) undertaken in particular places and times and the universality of values (abstract labor) achieved as commodity exchange becomes so widespread and generalized as to be a normal social act. But institutions mediate between particularity and universality so as to give some semblance of order and permanence to what is otherwise shifting sand. Central banks, financial institutions, exchange systems, state-backed local currencies and so on then become powerful mediators between the universality of money on the world market and the particularities of concrete labors conducted here and now around us. Such mediating institutions are also subject to change as, for example, powers shift from yen to deutschmarks to dollars and back again or as new institutions (like the IMF and the World Bank after 1945) spring up to take on new mediating roles. The point here is that there is always a problematic relation between local and particular conditions on the one hand and the universality of values achieved on the world market on the other, and that this internal relation is mediated by institutional structures which themselves acquire a certain kind of independent power. These mediating institutions are often territorially based and biased in important ways. They play a key role in determining what kinds of concrete labors and what kinds of class relations shall arise where and can sometimes even dictate patterns of uneven geographical development through their command over capital assembly and capital flows. Given the importance of European-wide banking and finance in the 1840s (the Rothschilds being prominent players in the events of 1848, for example) and the political-economic theories of the Saint-Simonians with respect to the power of associated capitals to change the world, the absence of any analysis of the mediating institutions of money and finance is surprising. Subsequent formulations (not only by Marx but also by Lenin, Hilferding and many others) may have helped to rectify matters, but the rather episodic and contingent treatment of the role of finance and money capital in organizing the geographical dynamics of capital accumulation may have been one of the *Manifesto*'s unwitting legacies (hardly anything was written on the topic between Hilferding and the early 1970s).

Fifth, the argument that the bourgeois revolution subjugated the countryside to the city as it similarly subjugated territories in a lesser state of development to those in a more advanced state, that processes of industrialization and rapid urbanization laid the seedbed for a more united working-class politics, is again prescient in the extreme at least in one sense. Reduced to its simplest formulation, it says that the production of spatial organization is not neutral with respect to class struggle. And that is a vital principle no matter how critical we might be with respect to the sketch of these dynamics as laid out in the *Manifesto*. The account offered runs like this:

> The proletariat goes through various stages of development. With its birth begins its struggle with the bourgeoisie. At first the contest is carried on by individual labourers, then by the workpeople of a factory, then by the operatives of one trade, in one locality, against the individual bourgeois who directly exploits them. At this stage the labourers still form an incoherent mass scattered over the country, and broken up by their mutual competition. If anywhere they unite to form more compact bodies this is

not yet the consequence of their own active union but of the union of the bourgeoisie... But with the development of industry the proletariat not only increases in number; it becomes concentrated in greater masses, its strength grows, and it feels that strength more... the collisions between individual workmen and individual bourgeois take more and more the character of collisions between two classes. Thereupon the workers begin to form combinations (trades unions)... This union (of the workers) is helped on by the improved means of communication that are created by modern industry and that place the workers of different localities in contact with one another. It was just this contact that was needed to centralise the numerous local struggles, all of the same character into one national struggle between classes...

For much of the nineteenth century, this account captures a common enough path to the development of class struggle. And there are plenty of of twentieth-century examples where similar trajectories can be discerned (the industrialization of South Korea being paradigmatic). But it is one thing to say that this is a useful descriptive sketch and quite another to argue that these are necessary stages through which class struggle must evolve en route to the construction of socialism. But if it is interpreted, as I have suggested, as a compelling statement of the non-neutrality of spatial organization in the dynamics of class struggle, then it follows that the bourgeoisie may also evolve its own spatial strategies of dispersal, of divide and rule, of geographical disruptions to the rise of class forces that so clearly threaten its existence. To the passages already cited, we find added the cautionary statement that: 'this organization of the proletarians into a class, and consequently into a political party, is continually being upset again by the competition between the worker's themselves.' And there are plenty of examples of bourgeois strategies to achieve that effect. From the dispersal of manufacturing from centres to suburbs in late nineteenth-century US cities to avoid concentrated proletarian power to the current attack on union power by dispersal and fragmentation of production processes across space (much of it, of course, to so-called developing countries where working-class organization is weakest) has proven a powerful weapon in the bourgeois struggle to enhance its power. The active stimulation of inter-worker competition across space has likewise worked to capitalist advantage, to say nothing of the problem of localism and nationalism within working-class movements (the position of the Second International in the First World War being the most spectacular case). In general, I think it fair to say that workers' movements have been better at commanding power in places and territories rather than in controlling spatialities, with the result that the capitalist class has used its superior powers of spatial manoeuvre to defeat place-bound proletarian/socialist revolutions (see Marx's 1858 worry cited above). The recent geographical and ideological assault on working-class forms of power through 'globalization' gives strong support to this thesis. While none of this is inconsistent with the basic underpinning of the argument in the *Manifesto*, it is, of course, quite different from the actual sketch of class-struggle dynamics set out as a stage model for the development of socialism in the European context.

Sixth, this leads us to one of the most problematic elements in the *Manifesto*'s legacy. This concerns the homogenization of the 'working man' and of 'labor powers' across a highly variegated geographical terrain as the proper basis for struggles against the powers of capital. While the slogan 'working men of all countries unite' may still stand (suitably modified to rid it of its gendered presupposition) as the only appropriate response to the globalizing strategies of capital accumulation, the manner of arriving at and conceptualizing that response deserves critical scrutiny. Central to the argument lies the belief that modern industry and wage labor, imposed by the capitalists ('the same in England as in France, in America as in Germany'), have stripped the workers 'of every trace of national character'. As a result:

> The working men have no country. We cannot take from them what they have not got. Since the proletariat must first of all acquire political supremacy, must rise to be the leading class of the nation, must constitute itself the nation, it is, so far, itself national, though not in the bourgeois sense of the word.
> National differences and antagonisms between peoples are daily more and more vanishing, owing to the development of the bourgeoisie, to freedom of commerce, to the world market, to uniformity in the mode of production and in the conditions of life corresponding thereto.
> The supremacy of the proletariat will cause them to vanish still faster. United action, of the leading civilised countries at least, is one of the first conditions for the emancipation of the proletariat.
> In proportion as the exploitation of one individual by another is put an end to, the exploitation of one nation by another will also be put an end to. In proportion as the antagonism between classes within the nation vanishes, the hostility of one nation to another will come to an end.

The guiding vision is noble enough but there is unquestionably a lot of wishful thinking here. At best, the *Manifesto* mildly concedes that the initial measures to be taken as socialists come to power will 'of course be different in different countries'. It also notes how problems arise in the translation of political ideas from one context to another – the Germans took on French ideas and adapted them to their own circumstances which were not so well-developed, creating a German kind of socialism of which Marx was highly critical in Part III of the *Manifesto*. In the practical world of politics, then, there is a certain sensitivity to uneven material conditions and local circumstances. And in the final section of the *Manifesto*, attention is paid to the different political conditions in France, Switzerland, Poland and Germany. From this Marx and Engels divine that the task of communists is to bring unity to these causes, to define the commonalities within the differences and to make a movement in which workers of the world can unite. But in so doing, the force of capital that uproots and destroys local place-bound loyalties and bonds is heavily relied upon to prepare the way.

There are I think two ways in which we can read this. On the one hand, the *Manifesto* insists, quite correctly in my view, that the only way to resist capitalism and transform towards socialism is through a global struggle in which global working-class formation, perhaps achieved in a step-wise fashion from local to national to global concerns acquires sufficient power and presence to fulfill its own historical potentialities.[4] In this case, the task of the communist movement is to find ways, against all odds, to properly bring together all the various highly differentiated and often local movements into some kind of commonality of purpose. The second reading is rather more mechanistic. It sees the automatic sweeping-away of national differences and differentiations through bourgeois advancement, the delocalization and denationalization of working-class populations and therefore of their political aspirations and movements. The task of the communist movement is to prepare for and hasten on the endpoint of this bourgeois revolution, to educate the working class as to the true nature of their situation and to organize, on that basis, their revolutionary potential to construct an alternative. Such a mechanistic reading is, in my view, incorrect even though substantial grounding for it can be found within the *Manifesto* itself.

The central difficulty lies in the presumption that capitalist industry and commodification will lead to homogenization of the working population. There is, of course, an undeniable sense in which this is true, but what it fails to appreciate is the way in which capitalism simultaneously differentiates, sometimes feeding off ancient cultural distinctions, gender relations, ethnic predilections and religious beliefs. It does this not only through the development of explicit bourgeois strategies of divide and control, but also by converting the principle of market choice into a mechanism for group differentiation. The result is the implantation of all

---

[4] I have elsewhere tried to adapt Raymond Williams concept of 'militant particularism' to capture this process and its inevitable contradictions: see Harvey 1996: ch. 1.

manner of class, gender and other social divisions into the geographical landscape of capitalism. Divisions such as those between cities and suburbs, between regions as well as between nations cannot be understood as residuals from some ancient order. They are not automatically swept away. They are actively produced through the differentiating powers of capital accumulation and market structures. Place-bound loyalties proliferate and in some respects strengthen rather than disintegrate through the mechanisms of class struggle as well as through the agency of both capital and labor working for themselves. Class struggle all too easily dissolves into a whole series of geographically fragmented communitarian interests, easily co-opted by bourgeois powers or exploited by the mechanisms of neo-liberal market penetration.

There is a potentially dangerous underestimation within the *Manifesto* of the powers of capital to fragment, divide and differentiate, to absorb, transform and even exacerbate ancient cultural divisions, to produce spatial differentiations, to mobilize geopolitically, within the overall homogenization achieved through wage labor and market exchange. And there is likewise an underestimation of the ways in which labor mobilizes through territorial forms of organization, building place-bound loyalties en route. The dialectic of commonality and difference has not worked out (if it ever could) in the way that the sketch supplied in the *Manifesto* implied, even if its underlying logic and its injunction to unite is correct.

'Working men of all countries, unite!'

The World Bank estimates that the global labor force doubled in size between 1966 and 1995 (it now stands at an estimated 2.5 billion men and women). But:

> the more than a billion individuals living on a dollar or less a day depend ... on pitifully low returns to hard work. In many countries workers lack representation and work in unhealthy, dangerous, or demeaning conditions. Meanwhile 120 millions or so are unemployed worldwide and millions more have given up hope of finding work.
>
> (World Bank 1995: 9)

This condition exists at a time of rapid growth in average levels of productivity per worker (reported also to have doubled since 1965 worldwide) and a rapid growth in world trade fuelled in part by reductions in costs of movement but also by a wave of trade liberalization and sharp increases in the international flows of direct investments. The latter helped construct transnationally integrated production systems largely organized through intra-firm trade. As a result:

the number of workers employed in export- and import-competing industries has grown significantly. In this sense, therefore, it could be said that
labour markets across the world are becoming more interlinked . . . Some
observers see in these developments the emergence of a global labour
market wherein 'the world has become a huge bazaar with nations peddling
their workforces in competition against one another, offering the lowest
prices for doing business' . . . The core apprehension is that intensifying
global competition will generate pressures to lower wages and labour
standards across the world.

(International Labour Office 1996: 2)

This process of ever-stronger interlinkage has been intensified by 'the
increasing participation in the world economy of populous developing
countries such as China, India and Indonesia.' With respect to China, for
example, the United Nations Development Programme reports:

> The share of labour-intensive manufactures in total exports rose from 36%
> in 1975 to 74% in 1990 . . . Between 1985 and 1993 employment in textiles
> increased by 20%, in clothing and fibre products by 43%, in plastic prod-
> ucts by 51%. China is now a major exporter of labour-intensive products
> to many industrial countries . . . For all its dynamic job creation, China still
> faces a formidable employment challenge. Economic reforms have released
> a 'floating population' of around 80 million most of whom are seeking
> work. The State Planning Commission estimates that some 20 million
> workers will be shed from state enterprises over the next five years and that
> 120 million more will leave rural areas hoping for work in the cities. Labour
> intensive economic growth will need to continue at a rapid pace if all these
> people are to find work.

(United Nations Development Program 1996: 94)

I quote this instance to illustrate the massive movements into the global
labor force that have been and are underway. And China is not alone in
this. The export-oriented garment industry of Bangladesh hardly existed
twenty years ago, but it now employs more than a million workers (80 per
cent of them women and half of them crowded into Dhaka). Cities like
Jakarta, Bangkok and Bombay, as Seabrook reports, have become meccas
for formation of a transnational working class, heavily dependent upon
women, under conditions of poverty, violence, pollution and fierce
repression (Seabrook 1996: ch. 6).

It is hardly surprising that the insertion of this proletarianized mass
into global trading networks has been associated with wide-ranging social
convulsions and upheavals as well as changing structural conditions, such
as the spiralling inequalities between regions (that left sub-Saharan Africa
far behind as East and Southeast Asia surged ahead) as well as between

classes. As regards the latter, 'between 1960 and 1991 the share of the richest 20 per cent rose from 70 per cent of global income to 85 per cent – while that of the poorest declined from 2.3 per cent to 1.4 per cent'. By 1991, 'more than 85 per cent of the world's population received only 15 per cent of its income' and 'the net worth of the 358 richest people, the dollar billionaires, is equal to the combined income of the poorest 45 per cent of the world population – 2.3 billion people' (UN Development Program 1996: 13). This polarization is simply astounding, rendering hollow the World Bank's extraordinary claim that international integration coupled with free-market liberalism and low levels of government interference (conditions oddly and quite erroneously attributed to repressive political regimes in Taiwan, South Korea and Singapore) is the best way to deliver growth and rising living standards for workers (World Bank 1996: 3).

It is against this background that it becomes easier to assess the power of the tales assembled by Seabrook:

> Indonesia, in the name of the free market system, promotes the grossest violations of human rights, and undermines the right to subsist of those on whose labour its competitive advantage rests. The small and medium-sized units which subcontract to the multinationals are the precise localities where the sound of the hammering, tapping, beating of metal comes from the forges where the chains are made for industrial bondage . . .
>
> Many transnationals are subcontracting here: Levi Strauss, Nike, Reebok. A lot of the subcontractors are Korean-owned. They all tend to low wages and brutal management. Nike and Levis issue a code of conduct as to criteria for investment; but in reality, under the tender system they always go for the lowest cost of production . . . Some subcontractors move out of Jakarta to smaller towns, where workers are even less capable of combining to improve their conditions.
>
> (Seabrook 1996: 103–5)

Or, at a more personal level there is the account given by a woman worker and her sister:

> We are regularly insulted as a matter of course. When the boss gets angry he calls the women dogs, pigs, sluts, all of which we have to endure patiently without reacting . . . We work officially from seven in the morning until three (salary less than $2 per day), but there is often compulsory overtime, sometimes – especially if there is an urgent order to be delivered – until nine. However tired we are, we are not allowed to go home. We may get an extra 200 rupiah (10 US cents) . . . We go on foot to the factory from where we live. Inside it is very hot. The building has a metal roof, and there is not much space for all the workers. It is very cramped. There are over 200 people working there, mostly women, but there is only one toilet for the

whole factory... when we come home from work, we have no energy left to do anything but eat and sleep.

(Seabrook 1996)

Home is a single room, two metres by three, costing $16 a month; it costs nearly 10 cents to get two cans of water and at least a $1.50 a day to eat.

In *Capital* Marx recounts the story of the milliner, Mary Anne Walkley, twenty years of age, who often worked thirty hours without a break (though revived by occasional supplies of sherry, port and coffee) until, after a particularly hard spell necessitated by preparing 'magnificent dresses for the noble ladies invited to the ball in honour of the newly imported Princess of Wales,' died, according to the doctor's testimony, 'from long hours of work in an over-crowded work-room, and a too small and badly ventilated bedroom.' Compare that with a contemporary account of conditions of labour in Nike plants in Vietnam:

(Mr Nguyen) found that the treatment of workers by the factory managers in Vietnam (usually Korean or Taiwanese nationals) is a 'constant source of humiliation,' that verbal abuse and sexual harassment occur frequently, and that 'corporal punishment' is often used. He found that extreme amounts of forced overtime are imposed on Vietnamese workers. 'It is a common occurrence,' Mr Nguyen wrote in his report, 'to have several workers faint from exhaustion, heat and poor nutrition during their shifts.' We were told that several workers even coughed up blood before fainting. Rather than crack down on the abusive conditions in the factories, Nike has resorted to an elaborate international public relations campaign to give the appearance that it cares about its workers. But no amount of public relations will change the fact that a full-time worker who makes $1.60 a day is likely to spend a fair amount of time hungry if three very simple meals cost $2.10.

(Herbert 1997)

The material conditions that sparked the moral outrage that suffuses the *Manifesto* have not gone away. They are embodied in everything from Nike shoes, Disney products, Gap clothing to Liz Claiborne products. And, as in the nineteenth century, part of the response has been reformist middle-class outrage backed by the power of working-class movements to regulate 'sweatshop labor' worldwide and develop a code of 'fair labor practices' perhaps certified by a 'fair labor label' on the products we buy (Goodman 1996; Greenhouse 1997a; 1997b).

The setting for the *Manifesto* has not, then, radically changed at its basis. The global proletariat is far larger than ever and the imperative for workers of the world to unite is greater than ever. But the barriers to that unity are far more formidable than they were in the already complicated

European context of 1848. The workforce is now far more geographically dispersed, culturally heterogeneous, ethnically and religiously diverse, racially stratified, and linguistically fragmented. The effect is to radically differentiate both the modes of resistance to capitalism and the definitions of alternatives. And while it is true that means of communication and opportunities for translation have greatly improved, this has little meaning for the billion or so workers living on less than a dollar a day possessed of quite different cultural histories, literatures and understandings (compared to international financiers and transnationals who use them all the time). Differentials (both geographical and social) in wages and social provision within the global working class are likewise greater than they have ever been. The political and economic gap between the most affluent workers in, say Germany and the US, and the poorest wage workers in Indonesia and Mali, is far greater than between the so-called aristocracy of European labour and their unskilled counterparts in the nineteenth century. This means that a certain segment of the working class (mostly but not exclusively in the advanced capitalist countries and often possessing by far the most powerful political voice) has a great deal to lose besides its chains. And while women were always an important component of the workforce in the early years of capitalist development, their participation has become much more general at the same time as it has become concentrated in certain occupational categories (usually dubbed 'unskilled') in ways that pose acute questions of gender in working-class politics that have too often been pushed under the rug in the past.

Ecological variations and their associated impacts (resource wars, environmental injustice, differential effects of environmental degradation) have also become far more salient in the quest for an adequate quality of life as well as for rudimentary healthcare. In this regard, too there is no level playing-field upon which class struggle can be evenly played out because the relation to nature is itself a cultural determination that can have implications for how any alternative to capitalism can be constructed at the same time as it provides a basis for a radical critique of the purely utilitarian and instrumental attitudes embedded in capitalist accumulation and exploitation of the natural world. How to configure the environmental with the economic, the political with the cultural, becomes much harder at the global level, where the presumption of homogeneity of values and aspirations across the earth simply does not hold.

Global populations have also been on the move. The flood of migratory movements seems impossible to stop. State boundaries are less porous for people and for labor than they are for capital, but they are still porous enough. Immigration is a very significant issue worldwide (including within the labor movement itself). Organizing labor in the face of the considerable ethnic, racial, religious and cultural diversity generated out of migratory movements poses particular problems that the socialist movement has never found easy to address let alone solve. Europe, for example, now has to face all of those difficulties that have been wrestled with for so many years in the US.

Urbanization has also accelerated to create a major ecological, political, economic and social revolution in the spatial organization of the world's population. The proportion of an increasing global population living in cities has doubled in thirty years, making for massive spatial concentrations of population on a scale hitherto regarded as inconceivable. It has proven far easier to organize class struggle in, say, the small-scale mining villages of the South Wales coalfield, or even in relatively homogeneous industrial cities like nineteenth-century Manchester (with a population of less than a million, albeit problematically divided between English and Irish laborers), than organizing class struggle (or even developing the institutions of a representative democracy) in contemporary Sao Paulo, Cairo, Lagos, Los Angeles, Shanghai, Bombay, and the like, with their teeming, sprawling and often disjointed populations reaching close to or over the twenty-million mark.

The socialist movement has to come to terms with these extraordinary geographical transformations and develop tactics to deal with them. This does not dilute the importance of the final rallying cry of the *Manifesto* to unite. The conditions that we now face make that call more imperative than ever. But we cannot make either our history or our geography under historical-geographical conditions of our own choosing. A geographical reading of the *Manifesto* emphasizes the non-neutrality of spatial structures and powers in the intricate spatial dynamics of class struggle. It reveals how the bourgeoisie acquired its powers *vis-à-vis* all preceding modes of production by mobilizing command over space as a productive force peculiar to itself. It shows how the bourgeoisie has continuously enhanced and protected its power by that same mechanism. It therefore follows that until the working-class movement learns how to confront that bourgeois power to command and produce space, it will always play from a position of weakness rather than of strength. Likewise, until that movement comes to terms with the geographical conditions and diversities of its own existence, it will be unable to define, articulate and struggle for a realistic socialist alternative to capitalist domination.

The implications of such an argument are legion and some clues as to strategies are already embedded in the *Manifesto*. Properly embellished, they can take us onto richer terrains of struggle. It is important to accept, for example, that the beginning point of class struggle lies with the particularity of the laboring body, with figures like Mary Anne Walkley and the billions of others whose daily existence is shaped through an often traumatic and conflictual relation to the dynamics of capital accumulation. The laboring body is, therefore, a site of resistance that achieves a political dimension through the political capacity of individuals to act as moral agents. To treat of matters this way is not to revert to some rampant individualism but to insist, as the *Manifesto* does, that the universality of class struggle originates with the particularity of the person and that class politics must translate back to that person in meaningful ways. The alienation of the individual is, therefore, an important beginning point for politics and it is that alienation that must be overcome.

But, and this is of course the crucial message of the *Manifesto*, that alienation cannot be addressed except through collective struggle and that means building a movement that reaches out across space and time in such a way as to confront the universal and transnational qualities of capital accumulation. Ways have to be found to connect the microspace of the body with the macrospace of what is now called 'globalization'. The *Manifesto* suggests this can be done by linking the personal to the local to the regional, the national, and ultimately the international. A hierarchy of spatial scales exists at which class politics must be constructed. But the 'theory of the production of geographical scale,' as Smith observes, 'is grossly underdeveloped' and we have yet to learn, particularly with respect to global working-class formation and body politics, how to 'arbitrate and translate' between the different spatial scales (Smith 1992). This is an acute problem that must be confronted and resolved if working-class politics is to be revived. I give just three examples.

The traditional beginning point for class struggle has been a particular space – the factory – and it is from there that class organization has been built up through union movements, political parties, and the like. But what happens when factories disappear or become so mobile as to make permanent organizing difficult if not impossible? And what happens when much of the workforce becomes temporary or casualized? Under such conditions, labor-organizing in the traditional manner loses its geographical basis and its power is correspondingly diminished. Alternative models of organizing must then be constructed. In Baltimore, for example, the campaign for a living wage (put together under the aegis of an organization called Baltimoreans United in Leadership Development – BUILD) appeals to an alternative possible strategy that works at the metropolitan scale – the movement is city-wide – and has as its objective directly affecting the base wage-level for the whole metropolitan area: everyone (temporary as well as permanent workers) should receive a living wage of at least $7.70 an hour plus benefits. To accomplish this goal, institutions of community (particularly the churches), activist organizations, student groups, as well as whatever union support can be procured, combine together with the aim of unionizing temporary workers and those on workfare, targeting the immoveable institutions in the metropolitan space (government, including sub-contracting, universities, hospitals, and the like). A movement is created in the metropolitan space that operates outside of traditional labor-organizing models but in a way that addresses new conditions.[5] The BUILD strategy of inserting a metropolitan-scale politics into the equations of class struggle is an interesting example of shifting a sense of spatial scale to counteract the spatial tactics which capital uses.

Consider a second example. Governmentality for contemporary capitalism has entailed the construction of important supranational authorities such as NAFTA and the European Union. Unquestionably, such constructions – the Maastricht Agreement being the paradigmatic case – are pro-capitalist. How should the left respond? The divisions here are important to analyze (in Europe the debate within the left is intense), but too frequently the response is an overly simplistic argument that runs along the following lines: 'because NAFTA and Maastricht are pro-capitalist we fight them by defending the nation state against supranational governance.' The argument here outlined suggests an entirely different response. The left must learn to fight capital at *both* spatial scales simultaneously. But, in so doing, it must also learn to coordinate potentially contradictory politics within itself at the different spatial scales, for it is often the case in hierarchical spatial systems (and ecological problems frequently pose this dilemma) that what makes good political sense at one scale does not make such good politics at another (the rationalization of, say, automobile production in Europe may mean plant closures in Oxford or Turin). Withdrawing to the nation-state as the exclusive strategic site of class organization and struggle is to court failure (as well as to flirt with nationalism and all that that entails). This does not mean the nation-state has become irrelevant; indeed it has become more relevant than ever. But the choice of spatial scale is not 'either/or' but 'both/and' even though the latter entails confronting serious contradictions. This means that the union movement in the US ought to put just as much effort into cross-border organizing (particularly with respect to Mexico) as it puts into fighting NAFTA, and that the European union movement must pay as much attention to procuring power and influence in Brussels and Strasbourg as each does in its own national capital.

Moving to the international level poses similar dilemmas and problems. It is interesting to note that the internationalism of labor struggle,

---

[5] For accounts of the work of BUILD, see Cooper (1997) and Harvey (1998).

while it hovers as an obvious and latent necessity over much of the labor movement, faces serious difficulties organizationally. I again, in part, attribute this to a failure to confront the dilemmas of integrating struggles at different spatial scales. Examples exist of such integrations in other realms. Movements around human rights, the environment and the condition of women illustrate the possible ways in which politics can get constructed (as well as some of the pitfalls to such politics) to bridge the micro-scale of the body and the personal on the one hand and the macro-scale of the global and the political-economic on the other. Nothing analogous to the Rio Conference on the environment or the Beijing Conference on women has occurred to confront global conditions of labor. We have scarcely begun to think of concepts such as 'global working-class formation' or even to analyse what that might mean. Much of the defence of human dignity in the face of the degradation and violence of labor worldwide has been articulated through the churches rather than through labor organization directly (the churches' ability to work at different spatial scales provides a number of models for political organization from which the socialist movement could well draw some important lessons). As in the case of BUILD at the local level, alliances between labor organizations and many other institutions in civil society appear now to be crucial to the articulation of socialist politics at the international scale. Many of the campaigns orchestrated in the US, for example, against global sweatshops in general or particular versions (such as Disney operations in Haiti and Nike in Southeast Asia) are organized quite effectively through such alliances. The argument here is not that nothing is being done or that institutions do not exist (the revitalization of the ILO might be an interesting place to start). But the reconstruction of some sort of socialist internationalism after 1989 has not been an easy matter, even if the collapse of the wall opened up new opportunities to explore that internationalism free of the need to defend the rump-end of the Bolshevik Revolution against the predatory politics of capitalist powers.[6]

How to build a political movement at a variety of spatial scales as an answer to the geographical and geopolitical strategies of capital is a problem that in outline at least the *Manifesto* clearly articulates. How to do it for our times is an imperative issue for us to resolve for our time and place. One thing, however, is clear: we cannot set about that task without recognizing the geographical complexities that confront us. The clarifications that a study of the *Manifesto*'s geography offer provide a

[6] *The Socialist Register* for 1994 examines many of these problems at length and the different contributions collectively reflect much of the complexity – both theoretical and practical – of constructing a new internationalist politics.

marvellous opportunity to wrestle with that task in such a way as to reignite the flame of socialism from Jakarta to Los Angeles, from Shanghai to New York City, from Porto Allegre to Liverpool, from Cairo to Warsaw, from Beijing to Turin. There is no magic answer. But there is at least a strategic way of thinking available to us that can illuminate the way. And that is what the *Manifesto* can still provide.

CHAPTER 18

The art of rent: globalization and the commodification of culture

*Prepared for the Conference on Global and Local, held at the
Tate Modern in London, February, 2001.*

That culture has become a commodity of some sort is undeniable. Yet there is also a widespread belief that there is something so special about cultural products and events (be they in the arts, theater, music, cinema, architecture or more broadly in localized ways of life, heritage, collective memories and affective communities) as to set them apart from ordinary commodities like shirts and shoes. It may be, of course, that we set them apart simply because we cannot bear to think of them as anything other than different, existing on some higher plane of human creativity and meaning than that located in the factories of mass production and consumption. Yet even when we strip away all residues of wishful thinking (often backed by powerful ideologies) we are still left with something very special about those products designated as 'cultural'. How, then, can the commodity status of so many of these phenomena be reconciled with their special character? The relation between culture and capital evidently calls for careful probing and nuanced scrutiny.

Monopoly rent and competition

I begin with some reflections on the significance of monopoly rents to understanding how contemporary processes of economic globalization relate to localities and cultural forms.

The category of 'monopoly rent' is an abstraction drawn from the language of political economy. To those more interested in affairs of culture, of aesthetics, of affective values, of social life and of the heart, such a term might appear far too technical and arid to bear much weight in human affairs beyond the possible calculi of the financier, the developer, the real-estate speculator and the landlord. But I hope to show that it has a much grander purchase: that, properly constructed, it can generate rich interpretations of the many practical and personal dilemmas arising in the nexus between capitalist globalization, local political-economic developments and the evolution of cultural meanings and aesthetic values.

All rent is based on the monopoly power of private owners of certain portions of the globe. Monopoly rent arises because social actors can realize an enhanced income-stream over an extended time by virtue of their exclusive control over some directly or indirectly tradeable item which is in some crucial respects unique and non-replicable. There are two situations in which the category of monopoly rent comes to the fore. The first arises because social actors control some special quality resource, commodity or location which, in relation to a certain kind of activity, enables them to extract monopoly rents from those desiring to use it. In the realm of production, Marx (1967, vol. 3: 775) argues, the most obvious example is the vineyard producing wine of extraordinary quality that can be sold at a monopoly price. In this circumstance, 'the monopoly price creates the rent.'

The locational version would be centrality (for the commercial capitalist) relative to, say, the transport and communications network or proximity (for the hotel chain) to some highly concentrated activity (such as a financial center). The commercial capitalist and the hotelier are willing to pay a premium for the land because of accessibility. These are the indirect cases of monopoly rent. It is not the land, resource or location of unique qualities which is traded but the commodity or service produced through their use. In the second case, the land or resource is directly traded upon (as when vineyards or prime real-estate sites are sold to multinational capitalists and financiers for speculative purposes). Scarcity can be created by withholding the land or resource from current uses and speculating on future values. Monopoly rent of this sort can be extended to ownership of works of art (such as a Rodin or a Picasso) which can be, and increasingly are, bought and sold as investments. It is the uniqueness of the Picasso or the site which here forms the basis for the monopoly price.

The two forms of monopoly rent often intersect. A vineyard (with its unique chateau and physical setting) renowned for its wines can be traded at a monopoly price directly, as can the uniquely flavored wines produced on that land. A Picasso can be purchased for capital gains and then leased to someone else who puts it on view for a monopoly price. The proximity to a financial center can be traded directly as well as indirectly to, say, the hotel chain that uses it for its own purposes. But the difference between the two rental forms is important. It is unlikely (though not impossible), for example, that Westminster Abbey and Buckingham Palace will be traded directly (even the most ardent privatizers might baulk at that). But they can be, and plainly are, traded upon through the marketing practices of the tourism industry (or in the case of Buckingham Palace, by the Queen).

Two contradictions attach to the category of monopoly rent. Both of them are important to the argument that follows.

First, while uniqueness and particularity are crucial to the definition of 'special qualities', the requirement of tradeability means that no item can be so unique or so special as to be entirely outside of the monetary calculus. The Picasso has to have a money value as does the Monet, the Manet, the aboriginal art, the archaeological artefacts, the historic buildings, the ancient monuments, the buddhist temples, and the experience of rafting down the Colorado, being in Istanbul or on top of Mount Everest. There is, as is evident from such a list, a certain difficulty of 'market formation' here. For while markets have formed around works of art and, to some degree, around archaeological artefacts (there are some well documented cases, as with Australian aboriginal art, of what happens when some art form gets drawn into the market sphere), there are plainly several items on this list that are hard to incorporate directly (this is the problem with Westminster Abbey). Many items may not even be easy to trade upon indirectly. The contradiction here is that the more easily marketable such items become, the less unique and special they appear. In some instances, the marketing itself tends to destroy the unique qualities (particularly if these depend on qualities such as wilderness, remoteness, the purity of some aesthetic experience, and the like). More generally, to the degree that such items or events are easily marketable (and subject to replication by forgeries, fakes, imitations or simulacra), the less they provide a basis for monopoly rent. I am put in mind here of the student who complained about how inferior her experience of Europe was compared to Disney World:

> At Disney World all the countries are much closer together, and they show you the best of each country. Europe is boring. People talk strange languages and things are dirty. Sometimes you don't see anything interesting in Europe for days, but at Disney World something different happens all the time and people are happy. It's much more fun. It's well designed.
>
> (cited in Kelbaugh 1997: 51)

While this sounds a laughable judgement, it is sobering to reflect on how much Europe is attempting to redesign itself to Disney standards (and not only for the benefit of American tourists). But, and here is the heart of the contradiction, the more Europe becomes disneyfied, the less unique and special it becomes. The bland homogeneity that goes with pure commodification erases monopoly advantages. If monopoly rents are to be realized, then some way has to be found to keep commodities or places unique and particular enough (and I will later reflect on what this might mean) to maintain a monopolistic edge in an otherwise commodified and often fiercely competitive economy. But why, in a neo-liberal world where competitive markets are supposedly dominant, would monopoly of any sort be tolerated, let alone be seen as desirable? We here encounter the second contradiction which, at root, turns out to be a mirror image of the first. Competition, as Marx long ago observed, always tends towards monopoly (or oligopoly) simply because the survival of the fittest in the war of all against all eliminates the weaker firms. The fiercer the competition, the faster the trend towards oligopoly, if not monopoly. It is therefore no accident that the liberalization of markets and the celebration of market competition in recent years has produced incredible centralization of capital (Microsoft, Rupert Murdoch, Bertelsmann, financial services, and a wave of takeovers, mergers and consolidations in airlines, retailing and even in old-line industries like automobiles, petroleum, and the like). This tendency has long been recognized as a troublesome feature of capitalist dynamics, hence the anti-trust legislation in the United States and the work of the monopolies and mergers commissions in Europe. But these are weak defenses against an overwhelming force.

This structural dynamic would not have the importance it does were it not for the fact that capitalists actively cultivate monopoly powers. They thereby realize far-reaching control over production and marketing and hence stabilize their business environment to allow rational calculation and long-term planning, the reduction of risk and uncertainty, and more generally guarantee themselves a relatively peaceful and untroubled existence. The visible hand of the corporation, as Chandler terms it, has consequently been of far greater importance to capitalist historical geography than the invisible hand of the market made so much of by Adam Smith and paraded *ad nauseam* before us in recent years as the guiding power in the neo-liberal ideology of contemporary globalization. But it is here that the mirror image of the first contradiction comes most clearly into view: market processes crucially depend upon the individual monopoly of capitalists (of all sorts) over the means of production of surplus value including finance and land (all rent, recall, is a return to the monopoly power of private ownership of any portion of the globe). The monopoly power of private property is, therefore, both the beginning and the end-point of all capitalist activity. A non-tradeable juridical right exists at the very foundation of all capitalist trade, making the option of non-trading (hoarding, withholding, miserly behavior) an important problem in capitalist markets. Pure market competition, free commodity exchange and perfect market rationality are, therefore, rather rare and chronically unstable devices for coordinating production and consumption decisions. The problem is to keep economic relations competitive enough while sustaining the individual and class monopoly privileges of private property that are the foundation of capitalism as a political-economic system.

This last point demands one further elaboration to bring us closer to the topic at hand. It is widely but erroneously assumed that monopoly power of the grand and culminating sort is most clearly signalled by the centralization and concentration of capital in mega-corporations. Conversely, small firm size is widely assumed, again erroneously, to be a sign of a competitive market situation. By this measure, a once competitive capitalism has become increasingly monopolized over time. The error arises because the economic theory of the firm ignores entirely its spatial context even though it does accept (on those rare occasions where it does deign to consider the matter) that locational advantage involves 'monopolistic competition'. In the nineteenth century, for example, the brewer, the baker, the candlestick maker were all protected to considerable degree from competition in local markets by the high cost of transportation. Local monopoly powers were omnipresent and very hard to break in everything from energy to food supply. By this measure, nineteenth century capitalism was far less competitive than now. It is at this point that the changing conditions of transport and communications enter in as crucial determining variables. As spatial barriers diminished through the capitalist penchant for 'the annihilation of space through time', so many local industries and services lost their local protections and monopoly privileges. They were forced into competition with producers in other locations, at first relatively close by but then with producers much further away. The historical geography of the brewing trade is very instructive in this regard. In the nineteenth century, most people drank local brew because they had no choice. By the end of the nineteenth century, beer production and consumption in Britain had regionalized to a considerable degree and remained so until the 1960s (foreign imports, with the exception of Guinness were unheard of). But then the market became national (Newcastle Brown and Scottish Youngers appeared in London and the south) before becoming international (imports suddenly became all the rage). If one drinks local brew now it is by choice, usually out of some mix of principled attachment to locality or because of some special quality of the beer (based on the technique, the water, or whatever) that supposedly differentiates it from others. Plainly, the economic space of competition has changed in both form and scale over time. Other barriers to spatial movement also exist. Protective tarriffs, for example, typically protect monopoly privileges within the space of the nation-state.

You can now, perhaps, more clearly divine the drift of my argument. The recent bout of globalization has significantly diminished the monopoly protections given historically by high transport and communications costs while the removal of institutional barriers to trade (protectionism) has likewise diminished the monopoly rents to be procured by that means. But capitalism cannot do without monopoly powers and craves means to assemble them. So the question upon the agenda is how to assemble monopoly powers in a situation where the protections afforded by the so-called 'natural monopolies' of space and location and the political protections of national boundaries have been seriously diminished if not eliminated.

The obvious answer is to centralize capital in mega-corporations or to set up looser alliances (as in airlines and automobiles) that dominate markets. And we have seen plenty of that. The second path is to secure ever more firmly the monopoly rights of private property through international commercial laws that regulate all global trade. Patents and so-called 'intellectual property rights' have consequently become a major field of struggle through which monopoly powers more generally get asserted. The pharmaceutical industry, to take a paradigmatic example, has acquired extraordinary monopoly powers in part through massive centralizations of capital and in part through the protections of patents and licensing agreements. And it is hungrily pursuing even more monopoly powers as it seeks to establish property rights over genetic materials of all sorts (including those of rare plants in tropical rainforests traditionally collected by indigenous inhabitants). As monopoly privileges from one source diminish, so we witness a desperate attempt to preserve and assemble them by other means. Innumerable books and articles are being written on this process. I cannot possibly review them here. I do want, however, to look more closely at those aspects of this process that impinge most directly upon the problems of local development and cultural activities. I wish to show first, that there are continuing struggles over the definition of the monopoly powers that might be accorded to location and localities and that the idea of 'culture' is more and more entangled with attempts to reassert such monopoly powers precisely because claims to uniqueness and authenticity can best be articulated as distinctive and non-replicable cultural claims. I begin with the most obvious example of monopoly rent given by 'the vineyard producing wine of extraordinary quality that can be sold at a monopoly price'.

Adventures in the wine trade

The wine trade, like brewing, has become more and more international over the past thirty years and the stresses of international competition have produced some curious effects. Under pressure from the European Union, for example, international wine producers have agreed (after long legal battles and intense negotiations) to phase out the use of 'traditional expressions' on wine labels, which could eventually include terms like 'chateau' and 'domaine' as well as generic terms like 'champagne', 'burgundy', 'chablis' or 'sauternes'. In this way the European wine industry, led by the French, seeks to preserve monopoly rents by insisting upon the unique virtues of land, climate and tradition (lumped together under the French term 'terroir') and the distinctiveness of its product certified by a name. Reinforced by institutional controls like 'appellation contrôlée', the French wine trade insists upon the authenticity and originality of its product which grounds the uniqueness upon which monopoly rent can be based. Australia is one of the countries that agreed to this move. Chateau Tahbilk in Victoria obliged by dropping the 'Chateau' from its label, airily pronouncing that 'we are proudly Australian with no need to use terms inherited from other countries and cultures of bygone days'.[1] To compensate, they identified two factors which, when combined, 'give us a unique position in the world of wine'. Theirs is one of only six worldwide wine regions where the meso-climate is dramatically influenced by inland water mass (the numerous lakes and local lagoons moderate and cool the climate). Their soil is of a unique type (found in only one other location in Victoria) described as red/sandy loam colored by a very high ferric-oxide content, which 'has a positive effect on grape quality and adds a certain distinctive regional character to our wines.' These two factors are brought together to define 'Nagambie Lakes' as a unique viticultural region (to be authenticated, presumably, by the Australian Wine and Brandy Corporation's Geographical Indications Committee, set up to identify viticultural regions throughout Australia). Tahbilk thereby establishes a counterclaim to monopoly rents on the grounds of the unique mix of environmental conditions in the region where it is situated. It does so in a way that parallels and competes with the uniqueness claims of 'terroir' and domaine pressed by French wine producers.

But we then encounter the first contradiction. All wine is tradeable and therefore in some sense comparable no matter where it is from. Enter Robert Parker and the *Wine Advocate* which he publishes regularly. Parker evaluates wines for their taste and pays no particular mind to 'terroir' or any other cultural-historical claims. He is notoriously independent (most other guides are supported by influential sectors of the wine industry). He ranks wines on a scale according to his own distinctive taste. He has an extensive following in the US, a major market. If he gives a wine from Bordeaux sixty-five points and an Australian wine ninety-five points, then

[1] Tahblik Wine Club (2000), *Wine Club Circular*, issue 15, June 2000, Tahblik Winery and Vineyard, Tahblik, Victoria, Australia.

prices are affected. The Bordeaux wine producers are terrified of him. They have sued him, denigrated him, abused him and even physically assaulted him. He challenges the bases of their monopoly rents. Monopoly claims, we can conclude, are as much 'an effect of discourse' and an outcome of struggle as they are a reflection of the qualities of the product. But if the language of 'terroir' and tradition is to be abandoned, then what kind of discourse can be put in its place? Parker and many others in the wine trade have in recent years invented a language in which wines are described in terms such as 'flavor of peach and plum, with a hint of thyme and gooseberry'. The language sounds bizarre but this discursive shift, which corresponds to rising international competition and globalization in the wine trade, takes on a distinctive role reflecting the commodification of wine consumption along standardized lines.

But wine consumption has many dimensions that open paths to profitable exploitation. For many it is an aesthetic experience. Beyond the sheer pleasure (for some) of a fine wine with the right food, there lie all sorts of other referents within the western tradition that track back to mythology (Dionysus and Bacchus), religion (the blood of Jesus and communion rituals) and traditions celebrated in festivals, poetry, song and literature. Knowledge of wines and 'proper' appreciation is often a sign of class and is analyzable as a form of 'cultural' capital (as Bourdieu would put it). Getting the wine right may have helped to seal more than a few major business deals (would you trust someone who did not know how to select a wine?). Style of wine is related to regional cuisines and thereby embedded in those practices that turn regionality into a way of life marked by distinctive structures of feeling (it is hard to imagine Zorba the Greek drinking Mondavi Californian jug wine, even though the latter is sold in Athens airport).

The wine trade is about money and profit but it is also about culture in all of its senses (from the culture of the product to the cultural practices that surround its consumption and the cultural capital that can evolve alongside among both producers and consumers). The perpetual search for monopoly rents entails seeking out criteria of speciality, uniqueness, originality and authenticity in each of these realms. If uniqueness cannot be established by appeal to 'terroir' and tradition, or by straight description of flavor, then other modes of distinction must be invoked to establish monopoly claims and discourses devised to guarantee the truth of those claims (the wine that guarantees seduction or the wine that goes with nostalgia and the log fire are current advertising tropes in the US). In practice, what we find within the wine trade is a host of competing discourses, all with different truth claims about the uniqueness of the product. But, and here I go back to my starting-point, all of these discursive shifts and swayings, as well as many of the shifts and turns that have occurred in the strategies for commanding the international market in wine, have at their root not only the search for profit but also the search for monopoly rents. In this, the language of authenticity, originality, uniqueness, and special unreplicable qualities looms large. The generality of a globalized market produces, in a manner consistent with the second contradiction I earlier identified, a powerful force seeking to guarantee not only the continuing monopoly privileges of private property but the monopoly rents that derive from depicting commodities as incomparable.

Urban entrepreneurialism, monopoly rent and global forms

Recent struggles within the wine trade provide a useful model for understanding a wide range of phenomena within the contemporary phase of globalization. They have particular relevance to understanding how local cultural developments and traditions get absorbed within the calculi of political economy through attempts to garner monopoly rents. It also poses the question of how much the current interest in local cultural innovation and the resurrection and invention of local traditions attaches to the desire to extract and appropriate such rents. Since capitalists of all sorts (including the most exuberant of international financiers) are easily seduced by the lucrative prospects of monopoly powers, we immediately discern a third contradiction: that the most avid globalizers will support local developments that have the potential to yield monopoly rents (even if the effect of such support is to produce a local political climate antagonistic to globalization!). Emphasizing the uniqueness and purity of local Balinese culture may be vital to the hotel, airline and tourist industry but what happens when this encourages a Balinese movement that violently resists the 'impurity' of commercialization? The Basque country may appear a potentially valuable cultural configuration precisely because of its uniqueness but ETA is not amenable to commercialization. Let us probe a little more deeply into this contradiction as it impinges upon urban development politics. To do so requires, however, briefly situating that politics in relation to globalization. Urban entrepreneurialism has become important both nationally and internationally in recent decades. By this I mean that pattern of behavior within urban governance that mixes together state powers (local, metropolitan, regional, national or supranational) and a wide array of organizational forms in civil society (chambers of commerce, unions, churches, educational and research institutions, community groups, NGOs, and the like) and private interests (corporate and individual) to form coalitions to promote or manage urban/regional development of some sort or other. There is now an extensive literature on this topic which shows that the forms, activities and goals of these governance systems (variously know as 'urban regimes', 'growth machines' or 'regional growth coalitions') vary widely depending upon local conditions and the mix of forces at work within them. The role of this urban entrepreneurialism in relation to the neo-liberal form of globalization has also been scrutinized at length, most usually under the rubric of local-global relations and the so-called 'space-place dialectic'. Most geographers who have looked into the problem have rightly concluded that it is a categorical error to view globalization as a causal force in relation to local development. What is at stake here, they rightly argue, is a rather more complicated relationship across scales in which local initiatives can percolate upwards to a global scale, and vice versa, at the same time as processes within a particular definition of scale – inter-urban and inter-regional competition being the most obvious examples – can rework the local/regional configurations of what globalization is about. Globalization should not be seen, therefore, as an undifferentiated unity but as a geographically articulated patterning of global capitalist activities and relations.

But what, exactly, does it mean to speak of 'a geographically articulated patterning'? There is, of course, plenty of evidence of uneven geographical development (at a variety of scales) and at least some cogent theorizing to understand its capitalistic logic. Some of it can be understood in conventional terms as a search on the part of mobile capitals (with financial, commercial and production capital having different capacities in this regard) to gain advantages in the production and appropriation of surplus values by moving around. Trends can indeed be identified which fit with simple models of 'a race to the bottom' in which the cheapest and most easily exploited labor power becomes the guiding beacon for capital mobility and investment decisions. But there is plenty of countervailing evidence to suggest that this is a gross oversimplification when projected as a monocausal explanation of the dynamics of uneven geographical development. Capital in general just as easily flows into high wage regions as into low and often seems to be geographically guided by quite different criteria to those conventionally set out in both bourgeois and Marxist political economy.

The problem in part (but not wholly) derives from the habit of ignoring the category of landed capital and the considerable importance of long-term investments in the built environment which are by definition geographically immobile (except in the relative accessibility sense). Such investments, particularly when they are of a speculative sort, invariably call for even further waves of investments if the first wave is to prove profitable (to fill the convention center, we need hotels which require better transport and communications, which calls for an expansion of the convention center . . .). So there is an element of circular and cumulative causation at work in the dynamics of metropolitan area investments (look, for example, at the whole Docklands redevelopment in London and the financial viability of Canary Wharf which pivots on further investments both public and private). This is what urban growth machines are often all about: the orchestration of investment process dynamics and the provision of key public investments at the right place and time to promote success in inter-urban and inter-regional competition.

But this would not be as attractive as it is were it not for the ways in which monopoly rents might also be captured. A well-known strategy of developers, for example, is to reserve the choicest and most rentable piece of land in some development in order to extract monopoly rent from it after the rest of the project is realized. Savvy governments with requisite powers can engage in the same practices. The government of Hong Kong, as I understand it, is largely financed by controlled sales of public domain land for development at very high monopoly prices. This converts, in turn, into monopoly rents on properties which makes Hong Kong very attractive to international financial investment capital working through property markets. Of course, Hong Kong has other uniqueness claims given its location upon which it can also trade very vigorously in offering monopoly advantages. Singapore, incidentally, set out to capture monopoly rents, and was highly successful in so doing, in somewhat similar fashion, though by very different political-economic means.

Urban governance of this sort is mostly oriented to constructing patterns of local investments not only in physical infrastructures such as transport and communications, port facilities, sewage and water, but also in the social infrastructures of education, technology and science, social control, culture and living qualities. The aim is to create sufficient synergy within the urbanization process for monopoly rents to be created and realized by both private interests and state powers. Not all such efforts are successful, of course, but even the unsuccessful examples can partly or largely be understood in terms of their failure to realize monopoly rents. But the search for monopoly rents is not confined to the practices of real-estate development, economic initiatives and government finance. It has a far wider application.

Collective symbolic capital, marks of distinction and monopoly rents

If claims to uniqueness, authenticity, particularity and speciality underlie the ability to capture monopoly rents, then on what better terrain is it possible to make such claims than in the field of historically constituted cultural artefacts and practices and special environmental characteristics (including, of course, the built, social and cultural environments)? All such claims are, as in the wine trade, as much an outcome of discursive constructions and struggles as they are grounded in material fact. Many rest upon historical narratives, interpretations and meanings of collective memories, significations of cultural practices, and the like: there is always a strong social and discursive element at work in the construction of such claims. Once established, however, such claims can be pressed home hard in the cause of extracting monopoly rents since there will be, in many people's minds at least, no other place than London, Cairo, Barcelona, Milan, Istanbul, San Francisco or wherever, to gain access to whatever it is that is supposedly unique to such places.

The most obvious point of reference where this works is in contemporary tourism, but I think it would be a mistake to let the matter rest there. For what is at stake is the power of collective symbolic capital, of special marks of distinction that attach to some place, which have a significant drawing power upon the flows of capital more generally. Bourdieu, to whom we owe the general usage of these terms, unfortunately restricts them to individuals (rather like atoms floating in a sea of structured aesthetic judgements) when it seems to me that the collective forms (and the relation of individuals to those collective forms) might be of even greater interest. The collective symbolic capital which attaches to names and places like Paris, Athens, New York, Rio de Janeiro, Berlin and Rome is of great import and gives such places great economic advantages relative to, say, Baltimore, Liverpool, Essen, Lille and Glasgow. The problem for these latter places is to raise their quotient of symbolic capital and to increase their marks of distinction to better ground their claims to the uniqueness that yields monopoly rent. Given the general loss of other monopoly powers through easier transport and communications and the reduction of other barriers to trade, the struggle for collective symbolic capital becomes even more important as a basis for monopoly rents. How else can we explain the splash made by the Guggenheim Museum in Bilbao with its signature Gehry architecture? And how else can we explain the willingness of major financial institutions, with considerable international interests, to finance such a signature project?

The rise of Barcelona to prominence within the European system of cities, to take another example, has in part been based on its steady amassing of symbolic capital and its accumulating marks of distinction. In this the excavation of a distinctively Catalan history and tradition, the marketing of its strong artistic accomplishments and architectural heritage (Gaudi of course) and its distinctive marks of lifestyle and literary traditions, have loomed large, backed by a deluge of books, exhibitions, and cultural events that celebrate distinctiveness. This has all been show-cased with new signature architectural embellishments (Norman Foster's radio communications tower and Meier's gleaming white Museum of Modern Art in the midst of the somewhat degraded fabric of the old city) and a whole host of investments to open up the harbor and the beach, reclaim derelict lands for the Olympic Village (with cute reference to the utopianism of the Icarians) and turn what was once a rather murky and even dangerous nightlife into an open panorama of urban spectacle. All of this was helped on by the Olympic Games which opened up huge oppor-tunities to garner monopoly rents (Juan Samaranch, president of the International Olympic Committee, just happened to have large real-estate interests in Barcelona).

But Barcelona's initial success appears headed deep into the first contradiction. As opportunities to pocket monopoly rents galore present themselves on the basis of the collective symbolic capital of Barcelona as a city (property prices have skyrocketed and the Royal Institute of British Architects awards the whole city its medal for architectural accomplish-ments), so their irresistible lure draws more and more homogenizing multinational commodification in its wake. The later phases of waterfront development look exactly like every other in the western world, the stupe-fying congestion of the traffic leads to pressures to put boulevards through parts of the old city, multinational stores replace local shops, gentrification removes long-term residential populations and destroys older urban fabric, and Barcelona loses some of its marks of distinction. There are even unsubtle signs of disneyfication. This contradiction is marked by questions and resistance. Whose collective memory is to be celebrated here (the anarchists like the Icarians who played such an impor-tant role in Barcelona's history, the republicans who fought so fiercely against Franco, the Catalan nationalists, immigrants from Andalusia, or a long-time Franco ally like Samaranch)? Whose aesthetics really count (the famously powerful architects of Barcelona like Bohigas)? Why accept disneyfication of any sort? Debates of this sort cannot easily be stilled precisely because it is clear to all that the collective symbolic capital that Barcelona has accumulated depends upon values of authenticity, uniqueness and particular non-replicable qualities. Such marks of local distinction are hard to accumulate without raising the issue of local empowerment, even of popular and oppositional movements. At that point, of course, the guardians of collective symbolic and cultural capital (the museums, the universities, the class of benefactors, and the state apparatus) typically close their doors and insist upon keeping the riff-raff out (though in Barcelona the Museum of Modern Art, unlike most institutions of its kind, has remained amazingly and constructively open to popular sensibilities). The stakes here are significant. It is a matter of determining which segments of the population are to benefit most from the symbolic capital to which everyone has, in their own distinctive ways, contributed. Why let the monopoly rent attached to that symbolic capital be captured only by the multinationals or by a small powerful segment of the local bourgeoisie? Even Singapore, which created and appropriated monopoly rents so ruthlessly and so successfully (mainly out of its locational and political advantage) over the years, saw to it that the benefits were widely distributed through housing, healthcare and education.

For the sorts of reasons that the recent history of Barcelona exemplifies, the knowledge and heritage industries, cultural production, signature architecture and the cultivation of distinctive aesthetic judgements have become powerful constitutive elements in the politics of urban entrepreneurialism in many places (though most particularly in Europe). The struggle to accumulate marks of distinction and collective symbolic capital in a highly competitive world is on. But this entrains in its wake all of the localized questions about whose collective memory, whose aesthetics and who benefits. The initial erasure of all mention of the slave trade in the reconstruction of Albert Dock in Liverpool, for example, generated protests on the part of the excluded population of Caribbean background, and the holocaust memorial in Berlin has sparked long-drawn-out controversies. Even ancient monuments such as the Acropolis, whose meaning one would have thought by now would be well settled, are subject to contestation. Such contestations can have widespread, even if indirect, political implications. Consider, for example, the arguments that have swirled around the reconstruction of Berlin after German reunification. All manner of divergent forces are colliding there as the struggle to define Berlin's symbolic capital unfolds. Berlin, rather obviously, can stake a claim to uniqueness on the basis of its potentiality to mediate between east and west. Its strategic position in relation to the uneven geographical development of contemporary capitalism (with the opening up of the ex-Soviet Union) confers obvious advantages. But there is also another kind of battle for identity being waged which invokes collective memories, mythologies, history, culture, aesthetics and tradition. I take up just one particularly troubling dimension of this struggle, one that is not necessarily dominant and whose capacity to ground claims to monopoly rent under global competition is not at all clear or certain.

A faction of local architects and planners (with the support of certain parts of the local state apparatus) seeks to revalidate the architectural forms of eighteenth- and nineteenth-century Berlin and in particular to highlight the architectural tradition of Schinkel to the exclusion of much else. This might be seen as a simple matter of elitist aesthetic preference, but it is freighted with a whole range of meanings that have to do with collective memories, monumentality, the power of history and political identity in the city. It is also associated with that climate of opinion (articulated in a variety of discourses) which defines who is or is not a Berliner and who has a right to the city in narrowly defined terms of pedigree or adhesions to particular values and beliefs. It excavates a local history and an architectural heritage that is charged with nationalist and romanticist connotations. In a context where the ill-treatment of and violence against immigrants is widespread, it may even offer tacit legitimation to such actions. The Turkish population (many of whom are now Berlin-born) has suffered many indignities and has in any case largely been forced out from the city center. Its contribution to Berlin as a city is totally ignored. Furthermore, this romanticist/nationalist style fits with a traditional approach to monumentality that broadly replicates in contemporary plans (though without specific reference and maybe even without knowing it) Albert Speer's plans (drawn up for Hitler in the 1930s) for a monumental foreground to the Reichstag. This is not, fortunately, all that is going on in the search for collective symbolic capital in Berlin. Norman Foster's reconstruction of the Reichstag, for example, or the collection of international modernist architects brought in by the multinationals (largely in opposition to local architects) to dominate the Potsdamer Platz, are hardly consistent with it. And the local romanticist response to the threat of multinational domination could, of course, merely end up being an innocent element of interest in a complex achievement of diverse marks of distinction for the city (Schinkel, after all, has considerable architectural merit and a rebuilt eighteenth-century castle could easily lend itself to disneyfication). But the potential downside of the story is of interest because it highlights how the contradictions of monopoly rent can all too easily play out. Were these narrower plans and exclusionary aesthetics and discursive practices to become dominant, then the collective symbolic capital created would be hard to trade freely upon because its very special qualities would position it largely outside globalization. The collective monopoly powers that urban governance can potentially command can always be orchestrated in opposition to the banal cosmopolitanism of multinational globalization.

The two dilemmas – veering so close to pure commercialization as to lose the marks of distinction that underlie monopoly rents or constructing marks of distinction that are so special as to be very hard to trade upon – are perpetually present. But, as in the wine trade, there are always strong discursive effects at play in defining what is or is not so special about a product, a place, a cultural form, a tradition, an architectural heritage.

Discursive battles become part of the game, and advocates (in the media and academia, for example) gain their audience as well as their financial support in relation to these processes. There is much to achieve, for example, by appeals to fashion (interestingly, being a center of fashion is one way for cities to accumulate considerable collective symbolic capital). Capitalists are well aware of this and must therefore wade into the culture wars, as well as into the thickets of multiculturalism, fashion and aesthetics, because it is precisely through such means that monopoly rents stand to be gained, if only for a while. And if, as I claim, monopoly rent is always an object of capitalist desire, then the means of gaining it through interventions in the field of culture, history, heritage, aesthetics and meanings must necessarily be of great import for capitalists of any sort.

Monopoly rent and spaces of hope

By now, critics will complain at the seeming economic reductionism of the argument. I make it seem, they will say, as if capitalism produces local cultures, shapes aesthetic meanings and so dominates local initiatives as to preclude the development of any kind of difference that is not directly subsumed within the circulation of capital. I cannot prevent such a reading, but this would be a perversion of my message. For what I hope to have shown, by invoking the concept of monopoly rent within the logic of capital accumulation, is that capital has ways to appropriate and extract surpluses from local differences, local cultural variations and aesthetic meanings of no matter what origin. The music industry in the US, for example, succeeds brilliantly in appropriating the incredible grassroots and localized creativity of musicians of all stripes (almost invariably to the benefit of the industry rather than to the benefit of the musicians). The shameless commodification and commercialization of everything is, after all, one of the hallmarks of our times.

But monopoly rent is a contradictory form. The search for it leads global capital to value distinctive local initiatives (and, in certain respects, the more distinctive the initiative, the better). It also leads to the valuation of uniqueness, authenticity, particularity, originality and all manner of other dimensions to social life that are inconsistent with the homogeneity presupposed by commodity production. And if capital is not to totally destroy the uniqueness that is the basis for the appropriation of monopoly rents (and there are many circumstances where it has done just that), then it must support a form of differentiation and allow of divergent and to some degree uncontrollable local cultural developments that can be antagonistic to its own smooth functioning. It is within such spaces that all manner of oppositional movements can form, even presupposing, as is often the case, that oppositional movements are not already firmly entrenched there. The problem for capital is to find ways to co–opt, subsume, commodify and monetize such differences just enough to be able to appropriate monopoly rents therefrom. The problem for oppositional movements is to use the validation of particularity, uniqueness, authenticity, culture and aesthetic meanings in ways that open up new possibilities and alternatives rather than to allow them to be used to create a more fertile terrain from which monopoly rents can be extracted by those who have both the power and the compulsive inclination to do so. The widespread, though usually fragmented, struggles that ensue between capitalistic appropriation and artistic creativity can lead a segment of the community concerned with cultural matters to side with a politics opposed to multinational capitalism.

It is by no means clear, however, that the conservativism and even reactionary exclusionism that often attaches to 'pure' values of authenticity, originality and an aesthetic of particularity of culture is an adequate foundation for a progressive oppositional politics either. It can all too easily veer into local, regional or nationalist identity politics of the neo-fascist sort of which there are already far too many troubling signs throughout much of Europe. This is a central contradiction with which the left must in turn wrestle. The spaces for transformational politics are there because capital can never afford to close them down and the left opposition is gradually learning how better to use them. The fragmented oppositional movements to neo–liberal globalization as manifest in Seattle, Prague, Melbourne, Bangkok and Nice and now, more constructively, at the World Social Forum in Porto Alegre (in opposition to the annual meetings of the business elites and government leaders in Davos), indicate such an alternative politics. It is not wholly antagonistic to globalization but wants it on very different terms. It is no accident, of course, that it is Porto Alegre rather than Barcelona, Berlin, San Francisco or Milan that has opened itself to this initiative. For in that city, the forces of culture and of history are being mobilized by a political movement (led by the Brazilian Worker's Party) in a quite different way, seeking a different kind of collective symbolic capital to that flaunted in the Guggenheim Museum in Bilbao or the extension to the Tate Gallery in London. The marks of distinction being accumulated in Porto Alegre derive from its struggle to fashion an alternative to globalization that does not trade on monopoly rents in particular or cave in to multinational capitalism in general. In focusing on popular mobilization, it is actively constructing new cultural forms and new definitions of authenticity, originality and tradition. That is a hard path to follow, as previous examples such as the remarkable experiments in Red Bologna in the 1960s and 1970s showed.

Socialism in one city is not a viable concept. But then it is quite clear that no alternative to the contemporary form of globalization will be delivered to us from on high either. It will have to come from within multiple local spaces conjoining into a broader movement.

It is here that the contradictions faced by the capitalists as they search for monopoly rent assume a certain structural significance. By seeking to trade on values of authenticity, locality, history, culture, collective memories and tradition, they open a space for political thought and action within which alternatives can be both devised and pursued. That space deserves intense exploration and cultivation by oppositional movements. It is one of the key spaces of hope for the construction of an alternative kind of globalization. One in which the progressive forces of culture appropriate those of capital rather than the other way round.

Bibliography

Abers, R. (1998), 'Learning Democratic Practice: Distributing Government Resources through Popular Participation in Porto Alegre, Brazil', in M. Douglass and J. Friedmann (eds), *Cities for Citizens: Planning and the Rise of Civil Society in a Global Age*, New York.

Agnew, J. and Duncan, J. (eds) (1989), *The Power of Place: Bringing Together the Geographical and Sociological Imaginations*, Boston.

Althusser, L. (1969), *For Marx*, Harmondsworth, Middlesex.

Altvater, E. (1973), 'Notes on Some Problems of State Interventionism', *Kapitalistate*, no. 1, pp. 96–108 and no. 2, pp. 76–83.

Amin, S. (1973), *Accumulation on a World Scale*, New York.

Anderson, B. (1998), *The Specter of Comparisons*, London.

Avineri, S. (1972), *Hegel's Theory of the Modern State*, London.

Aydalot, P. (1976), *Dynamique Spatiale et Développement Inégal*, Paris.

Ball, M. (1983), *Housing Policy and Economic Power: the Political Economy of Owner Occupation*, London.

Baran, P. (1957), *The Political Economy of Growth*, New York.

Barnbrock, J. (1976), 'Ideology and Location Theory: a Critical Enquiry into the Work of J. H. Von Thünen', *Doctoral Dissertation*, Department of Geography and Environmental Engineering, The Johns Hopkins University, Baltimore.

Barratt Brown, M. (1974), *The Economics of Imperialism*, Harmondsworth, Middlesex.

Becker, C. (1975), *Human Capital: a Theoretical and Empirical Analysis with Special Reference to Education*, New York.

Berkowitz, B. (1984), 'Economic Development Really Works: Baltimore, MD', in R. Bingham and J. Blair (eds), *Urban Economic Development*, Beverly Hills.

Berman, M. (1982), *All That is Solid Melts into Air*, New York.

Bianchini, F. (1991), 'The Arts and the Inner Cities', in B. Pimlott and S. Macgregor (eds), *Tackling the Inner Cities*, Oxford.

Bird, J., Curtis, B., Putnam, T., Robertson, G., and Tickner, L. (eds) (1993), *Mapping Futures: Local Cultures Global Change*, London.

Blaut, J. (2000), *Eight Eurocentric Historians*, New York.

Bleaney, P. (1976), *Underconsumption Theories*, London.

Blunkett, D. and Jackson, K. (1987), *Democracy in Crisis: the Town Halls Respond*, London.

412

Boas, G. and Wheeler, H. (1953), *Lattimore the Scholar*, Baltimore.

Boddy, M. (1984), 'Local Economic and Employment Strategies', in M. Boddy and C. Fudge (eds), *Local Socialism*, London.

Bouinot, J. (ed.) (1987), *L'action Economiques des Grandes Villes en France et à l'Etranger*, Paris.

Boyd-White, J. (1990), *Justice as Translation*, Chicago.

Breitbart, M. (ed.) (1979), 'Anarchism and Environment', *Antipode*, 11, 33–41.

Breitbart, M. (1981), 'Peter Kropotkin, the Anarchist Geographer', in D. Stoddart (ed.), *Geography, Ideology and Social Concern*, Oxford.

Brown, L. (1949), *The Story of Maps*, Boston.

Buchanan, K. (1970), *The Transformation of the Chinese Earth*, New York.

Buchanan, K. (1974), 'Reflections on a "dirty word"', *Dissent*, 31, 25–31.

Bukharin, N. (1972 edn), *Imperialism*, London.

Bunge, W. (1977), 'The First Years of the Detroit Geographical Expedition', in R. Peet (ed.), *Radical Geography*, Chicago.

Buttimer, A. (1974), *Values in Geography*, Resource paper no. 24, Association of American Geographers, Washington, DC.

Capel, H. (1981), 'Institutionalization of Geography and Strategies of Change', in D. Stoddart (ed.), *Geography, Ideology and Social Concern*, Oxford.

Carter, E., Donald, J. and Squires, J. (eds) (1993), *Space and Place: Theories of Identity and Location*, London.

Castells, M. (1983), *The City and the Grassroots*, Berkeley.

Chalmers, T. (1900 edn), *The Christian and Civic Economy of Large Towns*, 3 vols, Clifton, NJ.

Chang, S. (1931), *The Marxian Theory of the State*, Philadelphia.

Chisholm, M. (1962), *Rural Settlement and Land Use*, London.

Chorley, R. and Haggett, P. (eds) (1967), *Models in Geography*, London.

Cochrane, A. (ed.) (1987), *Developing Local Economic Strategies*, Milton Keynes.

Cockburn, C. (1977), *The Local State: Management of Cities and People*, London.

Cohen, G. (1978), *Karl Marx's Theory of History: a Defence*, Oxford.

Cole, H., Freeman, C., Jahoda, M., and Pavitt, K. (1973), *Thinking about the Future: a Critique of the Limits to Growth*, London.

Collett, L. (1977), *Marxism and Hegel*, London.

Cooke, P. (1989), *Localities*, London.

Cooke, P. (1990), 'Locality, Structure and Agency: a Theoretical Analysis', *Cultural Anthropology*, 5, 3–15.

Cooper, S. (1997), 'When Push Comes to Shove: Who is Welfare Reform Really Helping?', *The Nation*, 2 June 1997, 11–15.

Corey, K. (1972), 'Advocacy in Planning: a Reflective Analysis', *Antipode*, 4, 46–63.

Cox, K. and Mair, A. (1989), 'Levels of Abstraction in Locality Studies', *Antipode*, 21, 121–32.

Cronon, W. (1983), *Changes in the Land*, New York.

Davies, H. (1980), 'The Relevance of Development Control', *Town Planning Review*, 51, 7–24.

Davis, H. (1978), *Toward a Marxist Theory of Nationalism*, New York.

Davis, M. (1990), *City of Quartz: Excavating the Future in Los Angeles*, London.

Dear, M. and Scott, A. (eds) (1981), *Urbanization and Urban Planning in Capitalist Society*, New York.

Dempsey, B. (1960), *The Frontier Wage*, Chicago.

Diamond, J. (1997), *Guns, Germs, and Steel: the Fates of Human Societies*, New York.

Dorpalen, A. (1942), *The World of General Haushofer*, New York.

Douglass, M. and Friedmann, J. (eds) (1998), *Cities for Citizens: Planning and the Rise of Civil Society in a Global Age*, New York.

Duncan, S. and Savage, M. (1989), 'Space, Scale and Locality', *Antipode*, 21, 179–206.

Dworkin, D. and Roman, L. (eds) (1993), *Views beyond the Border Country: Raymond Williams and Cultural Politics*, London.

Eagleton, T. (ed.) (1989), *Raymond Williams: Critical Perspectives*, Cambridge.

Elkin, S. (1987), *City and Regime in the American Republic*, Chicago.

Emmanuel, A. (1972), *Unequal Exchange*, London.

Engels, F. (1940 edn), *The Dialectics of Nature*, New York.

Engels, F. (1941 edn), *Origin of the Family, Private Property and the State*, New York.

Etzioni, A. (1997), 'Community watch', *The Guardian*, 28 June 1997, 9.

Fanon, F. (1967), *The Wretched of the Earth*, Harmondsworth, Middlesex.

Firey, W. (1960), *Man, Mind and the Land*, Glencoe, IL.

Fisher, S. (ed.) (1993), *Fighting Back in Appalachia*, Philadelphia.

Foucault, M. (1984), 'Preface' to G. Deleuze and F. Guattari, *Anti-Oedipus: Capitalism and Schizophrenia*, London.

Frank, A. (1969), *Capitalism and Underdevelopment in Latin America*, New York.

Frug, G. (1980), 'The City as a Legal Concept', *Harvard Law Review*, 93 (6) 1059–153.

Gardner, L. (1971), *Economic Aspects of New Deal Diplomacy*, Boston.

Gertler, M. (1988), 'The Limits to Flexibility: Comments on the Post-Fordist Vision of Production and its Geography', *Transactions, Institute of British Geographers*, New Series, 13, 419–32.

Giddens, A. (1981), *A Contemporary Critique of Historical Materialism*, London.

Gilroy, P. (1987), *There ain't no Black in the Union Jack*, London.

Glacken, C. (1967), *Traces on the Rhodian Shore*, Berkeley.

Godelier, M. (1972), *Rationality and Irrationality in Economics*, London.

Gold, D., Lo, C. and Wright, E. (1975), 'Recent Developments in Marxist Theories of the Capitalist State', *Monthly Review*, no. 5, 30–43 and no. 6, 36–51.

Goodman, E. (1996), 'Why not a Labor Label?' *Baltimore Sun*, 19 July 1996, 25A.

Goodman, R. (1979), *The Last Entrepreneurs*, Boston, MA.

Gorz, A. (1973), *Socialism and Revolution*, New York.

Gottlieb, M. (1976), *Long Swings in Urban Development*, New York.

Gramsci, A. (1971), *Selections from the Prison Notebooks*, London.

Granovetter, M. (1985), 'Economic Action and Social Structure: the Problem of Embeddedness', *American Journal of Sociology*, 91, 481–510.

Green, L. (1988), 'Retailing in the New Economic Era', in G. Sternlieb and J. Hughes (eds), *America's New Market Geography*, New Brunswick, NJ.

Greenhouse, S. (1997a), 'Voluntary Rules on Apparel Labor Proving Elusive', *New York Times*, 1 February 1997, 1.

Greenhouse, S. (1997b), 'Accord to Combat Sweatshop Labor Faces Obstacles', *New York Times*, 13 April 1997, 1.

Gregory, D. and Urry, J. (eds) (1985), *Social Relations and Spatial Structures*, London.

Gundle, S. (1986), 'Urban Dreams and Metropolitan Nightmares: Models and Crises of Metropolitan Local Governments in Italy', in B. Szajkowski (ed.), *Marxist Local Governments in Western Europe and Japan*, London.

Gurr, T. and King, D. (1987), *The State and the City*, London.

Hall, P. (1966), *Von Thünen's Isolated State*, London.

Hall, S. (1989), 'Politics and Letters', in T. Eagleton (ed.), *Raymond Williams: Critical Perspectives*, Cambridge.

Hallowell, A. (1955), *Culture and Experience*, Philadelphia.

Hartshorne, R. (1939), *The Nature of Geography: a Critical Survey of Current Thought in the Light of the Past*, Lancaster, PA.

Harvey, D. (1973), *Social Justice and the City*, London.

Harvey, D. (1974), 'Class-monopoly Rent, Finance Capital and the Urban Revolution', *Regional Studies*, 8, 239–55.

Harvey, D. (1975a), 'The Geography of Capitalist Accumulation: a Reconstruction of the Marxian Theory', *Antipode*, 7 (no. 2) 9–21.

Harvey, D. (1975b), 'The Political Economy of Urbanization: the Case of the United States', in G. Gappert and H. Rose (eds), *The Social Economy of Cities*, Beverley Hills.

Harvey, D. (1977a), 'Labor, Capital and Class Struggle around the Built Environment in Advance Capitalist Societies', *Politics and Society*, 6, 265–95.

Harvey, D. (1977b), 'Population, Resources and the Ideology of Science', in R. Peet (ed.), *Radical Geography*, Chicago.

Harvey, D. (1981), 'The Spatial Fix: Hegel, Von Thünen, and Marx, *Antipode*, 13, no. 2, 1–12.

Harvey, D. (1982), *The Limits to Capital*, Oxford.

Harvey, D. (1985a), *The Urbanization of Capital*, Oxford.

Harvey, D. (1985b), *Consciousness and the Urban Experience*, Oxford.

Harvey, D. (1989a), *The Urban Experience*, Oxford.

Harvey, D. (1989b), *The Condition of Postmodernity*, Oxford.

Harvey, D. (1996), *Justice, Nature and the Geography of Difference*, Oxford.

Harvey, D. (1998), 'The Body as an Accumulation Strategy', *Society and Space*, 40, 16, 401–21.

Harvey, D. (2000), 'Cosmopolitanism and the Banality of Geographical Evils', *Public Culture*, 12, no. 2, 529–64.

Harvey, D. (forthcoming), 'The Spaces of Utopia', in L. Bower, D. Goldberg and M. Musheno (eds), *Justice and Social Identities*, Oxford.

Harvey, D. and Smith, N. (1976), 'From Capitals to Capital', in B. Ollman and E. Vernoff (eds), *The Left Academy*, vol. 2, New York.

Hays, S. (1959), *The Conservation Movement and the Gospel of Efficiency*, Cambridge, MA.

Hayter, T. and Harvey, D. (eds) (1993), *The Factory and the City: the Story of the Cowley Automobile Workers in Oxford*, Brighton.

Hegel, G. (1967 edn), *Philosophy of Right*, New York.

Held, D. (1995), *Democracy and the Global Order: from the Modern State to Cosmopolitan Governance*, Stanford, CA.

Herbert, B. (1997), 'Brutality in Vietnam', *New York Times*, 28 March 1997, A29.

Hérodote (1975), *Stratégies, géographies, idéologies*, Paris.

Herodotus (1954 edn), *The Histories*, Harmondsworth, Middlesex.

Hirschman, A. (1976), 'On Hegel, Imperialism and Structural Stagnation', *Journal of Development Economics*, 3, 1–8.

Hobson, J. (1938), *Imperialism*, London.

Hofstadter, R. (1967), *The Paranoid Style in American Politics and other Essays*, New York.

Horvath, R. and Gibson, K. (1984), 'Abstraction in Marx's Method', *Antipode*, 16, 12–25.

Hudson, W. (1970), *Modern Moral Philosophy*, London.

Humboldt, A. von (1811), *Essai Politique sur le Royaume de la Nouvelle Espagne*, Paris.

Humboldt, A. von (1849–52), *Cosmos*, London.

Ibn Khaldûn (1958 edn), *The Muqadimma*, London.

Ingold, T. (1993), 'Globes and Spheres: The Topology of Environmentalism', in K. Milton (ed.), *Environmentalism: the View from Anthropology*, London.

International Labour Office (1996), *World Employment 1996/97: National Policies in a Global Context*, Geneva.

Isard, W. (1956), *Location and Space-economy*, Cambridge, MA.

Jacks, G. and Whyte, R. (1939), *Vanishing Lands*, New York.

Jacobs, J. (1984), *Cities and the Wealth of Nations*, New York.

Jessop, B. (1983), 'Accumulation Strategies, State Forms, and Hegemonic Projects', *Kapitalistate*, 10/11, 89–112.

Judd, D. and Ready, R. (1986), 'Entrepreneurial Cities and the New Politics of Economic Development', in G. Peterson and C. Lewis (eds), *Reagan and the Cities*, Washington, DC.

Judis, J. (1981), 'Setting the Stage for Repression', *The Progressive*, 45, April 1981, 22–30.

Julien, C-A., Bruhat, J., Bourgin, C., Crouzet, M. and Renouvin, P. (1949), *Le Politiques d'Expansion Impérialiste*, Paris.

Kaldor, M. (1978), *The Disintegrating West*, Harmondsworth, Middlesex.

Kant, I. (1999 edn), *Géographie (Physique Géographie)*, Paris.

Kapp, K. (1950), *The Social Costs of Private Enterprise*, Cambridge, MA.

Keith, M. and Pile, S. (1993), *Place and the Politics of Identity*, London.

Kelbaugh, D. (1997), *Common Place*, Seattle, Washington.

Keynes, J. (1951), *Essays in Biography*, New York.

Keynes, J. (1936), *The General Theory of Employment, Interest and Money*, New York.

Kneese, A., Ayres, R. and D'Arge, R. (1970), *Economics and the Environment*, Washington, DC.

Knox, P. (1994), 'The Stealthy Tyranny of Community Spaces', *Environment and Planning A*, 26, 170–3.

Kropotkin, P. (1898), *Fields, Factories and Workshops*, London.

Kuhn, T. (1962), *The Structure of Scientific Revolutions*, Chicago.

Kuznets, S. (1961), *Capital in the American Economy: its Formation and Financing*, Princeton, NJ.

Laclau, E. (1975), 'The Specificity of the Political: Around the Poulantzas-Miliband Debate', *Economy and Society*, 5, no. 1, 87–110.

Landes, D. (1998), *The Wealth and Poverty of Nations: Why Some are So Rich and Some So Poor*, New York.

Lattimore, E. (1934), *Turkestan Review*, New York.

Lattimore, O. (1949), *The Situation in Asia*, New York.

Lattimore, O. (1962), *Studies in Frontier History: Collected Papers, 1928–58*, London.

Lefebvre, H. (1976), *The Survival of Capitalism*, New York.

Leitner, H. (1989), 'Cities in Pursuit of Economic Growth: the Local State as Entrepreneur', MS, *Department of Geography, University of Minnesota*, Minneapolis.

Lenin, V. (1949 edn), *The State and Revolution*, New York.

Lenin, V. (1963 edn), 'Imperialism, the Highest Stage of Capitalism', in *Selected Works, Volume 1*, Moscow.

Lenin, V. I. (1963 edn), 'The Right of Nations to Self-determination', in *Selected Works, Volume 1*, 595–648, Moscow.

Levi-Strauss, C. (1966), *The Savage Mind*, Chicago.

Levi-Strauss, C. (1973), *Tristes tropiques*, New York.

Levine, M. (1987), 'Downtown Redevelopment as an Urban Growth Strategy: a Critical Appraisal of the Baltimore Renaissance', *Journal of Urban Affairs*, 9 (2) 103–23.

Lewontin, R. (1982), 'Organism and Environment', in H. Plotkin (ed.), *Learning, Development and Culture*, Chichester.

Ley, D. and Samuels, M. (eds) (1978), *Humanistic Geography: Prospects and Problems*, Chicago.

Livingstone, D. (1992), *The Geographical Tradition*, Oxford.

Logan, J. and Molotch, H. (1987), *Urban Fortunes: the Political Economy of Place*, Berkeley.

Lomnitz-Adler, C. (1991), 'Concepts for the Study of Regional Culture', *American Ethnologist*, 18, 195–214.

Lösch, A. (1954), *The Economics of Location*, New Haven.

Luxemburg, R. (1976 edn), *The National Question: Selected Writings*, New York.

Luxemburg, R. (1968 edn), *The Accumulation of Capital*, London.

Lyall, K. (1982), 'A Bicycle Built for Two: Public-Private Partnership in Baltimore', in S. Fosler and R. Berger (eds), *Public-Private Partnerships in American Cities*, Lexington, MA.

Mackinder, H. (1962), *Democratic Ideals and Reality*, New York.

MacPherson, C. (1962), *The Political Theory of Possessive Individualism: from Hobbes to Locke*, New York.

Malthus, T. (1968 edn), *Principles of Political Economy*, New York.

Malthus, T. (1970 edn), *An Essay on the Principle of Population and a Summary View of the Principle of Population*, Harmondsworth, Middlesex.

Markusen, A. (1986), 'Defense spending: a successful industrial policy', *International Journal of Urban and Regional Research*, 10, 105–22.

Marshall, A. (1949), *Principles of Economics*, London.

Marx, K. (1938 edn), *Critique of the Gotha Program*, New York.

Marx, K. (1963 edn), *The Poverty of Philosophy*, New York.

Marx, K. (1964 edn), *The Economic and Philosophic Manuscripts of 1844*, New York.

Marx, K. (1967), *Capital*, 3 volumes, New York.

Marx, K. (1969a edn), *Theories of Surplus Value, Volume 1*, London.

Marx, K. (1969b edn), *Theories of Surplus Value, Volume 2*, London.

Marx, K. (1970 edn), *A Contribution to the Critique of Political Economy*, New York.

Marx, K. (1972 edn), *Theories of Surplus Value, Volume 3*, London.

Marx, K. (1973 edn), *Grundrisse*, Harmondsworth, Middlesex.

Marx, K. and Engels, F. (1952 edn), *Manifesto of the Communist Party*, Moscow.

Marx, K. and Engels, F. (1955 edn), *Selected Correspondence*, Moscow.

Marx, K. and Engels, F. (1970 edn), *The German Ideology*, New York.

Marx, K. and Engels, F. (1972 edn), *On Colonialism*, New York.

Marx, K. and Engels, F. (1974 onwards), *Collected Works*, 24 volumes to date, New York.

Massey, D. (1991), 'The Political Place of Locality Studies', *Environment and Planning A*, 23, 267–81.

Meadows, D., Meadows, D., Randers, J. and Behrens, W. (1972), *The Limits to Growth*, New York.

Merrifield, A. (1993), 'Place and Space: a Lefebvrian Reconciliation', *Transactions of the Institute of British Geographers*, new series, 18, 516–31.

Merrington, J. (1975), 'Town and Country in the Transition to Capitalism', *New Left Review*, 93, 71–92.

Meszaros, I. (1972), 'Ideology and Social Science', *Socialist Register*, 1972.

Meszaros, I. (1995), *Beyond Capital*, New York.

Miliband, R. (1969), *The State in Capitalist Society*, London.

Mill, J. (1965 edn), *Principles of Political Economy*, Toronto.

Mollenkopf, J. (1983), *The Contested City*, Princeton, NJ.

Molotch, H. (1976), 'The City as a Growth Machine: Towards a Political Economy of Place', *American Journal of Sociology*, 82, 309–32.

Morishima, M. and Catephores, G. (1978), *Value, Exploitation and Growth*, Maidenhead, Berkshire.

Murray, F. (1983), 'Pension Funds and Local Authority Investments, *Capital and Class*, 230, 89–103.

Navarro, V. (1974), 'What does Chile mean? an Analysis of Events in the Health Sector before and after Allende's Administration', *mimeo*, Department of Medical Care and Hospitals, The Johns Hopkins University, Baltimore, MD.

Needham, J. (1954), *Science and Civilization in China*, Cambridge.

Newman, R. (1983), 'Owen Lattimore and his Enemies', *Antipode*, 15 no. 3, 12–26.

Newman, R. (1992), *Owen Lattimore and the 'Loss' of China*, California.

Noyelle, T. and Stanback, T. (1984), *The Economic Transformation of American Cities*, Totawa, NJ.

Nussbaum, M., with respondents (1996), *For Love of Country: Debating the Limits of Patriotism*, Boston.

O'Connor, J. (1973), *The Fiscal Crisis of the State*, New York.

O'Malley, J. (1970), 'Introduction', to K. Marx, *Critique of Hegel's 'Philosophy of Right'*, Cambridge.

Offe, K. (1973), 'The Abolition of Market Control and the Problem of Legitimacy', *Kapitalistate*, no. 1, 109–16 and no. 2, 73–83.

Ollman, B. (1971), *Alienation: Marx's Conception of Man in Capitalist Society*, Cambridge.

Ollman, B. (1973), 'Marxism and Political Science: Prologomenon to a Debate on Marx's Method', *Politics and Society*, 3, 491–510.

Orans, M. (1966), 'Surplus', *Human Organization*, 24, 24–32.

Park, R. (1967), *On Collective Control and Social Behavior*, Chicago.

Pearson, H. (1957), 'The Economy has no Surplus: a Critique of a Theory of Development', in K. Polanyi, C. Arensberg and H. Pearson (eds), *Trade and Market in Early Empires*, Glencoe, IL.

Peet, R. (ed.) (1977), *Radical Geography*, Chicago.

Pelczynski, Z. (1962), 'Introductory Essay', in Knox, T. (ed.), *Hegel's Political Writings*, New York.

Peterson, P. (1981), *City Limits*, Chicago.

Piaget, J. (1970), *Structuralism*, New York.

Piaget, J. (1972), *The Principles of Genetic Epistemology*, London.

Plant, R. (1977), 'Hegel and Political Economy', *New Left Review*, 103, 79–92; 104, 103–13.

Pleskovic, B. and Stiglitz, J. (eds) (1999), *Annual World Bank Conference on Economic Development*, Washington, DC.

Polanyi, K. (1968), *Primitive, Archaic and Modern Economies: Essays of Karl Polanyi*, ed. G. Dalton, Boston.

Poulantzas, N. (1973), *Political Power and Social Classes*, London.

Poulantzas, N. (1975), *Classes in Contemporary Capitalism*, London.

Poulantzas, N. (1976), 'The Capitalist State: a Reply to Miliband and Laclau', *New Left Review*, 95, 63–83.

Pred, A. (1984), 'Place as Historically Contingent Process: Structuration and the Time-Geography of Becoming Places', *Annals of the Association of American Geographers*, 74, 279–97.

Putnam, R. (1993), *Making Democracy Work: Civic Traditions in Modern Italy*, Princeton, NJ.

Quaini, M. (1982), *Geography and Marxism*, Totawa, NJ.

Ratzel, F. (1923), *Politische Geographie*, Munich.

Reclus, E. (1982), *L'Homme et la Terre*, Paris.

Rees, G. and Lambert, J. (1985), *Cities in Crisis: the Political Economy of Post-war Development in Britain*, London.

Ricardo, D. (1951a edn), *Principles of Political Economy*, Cambridge.

Ricardo, D. (1951b edn), *The Works and Correspondence of David Ricardo, Volume 2*, Cambridge.

Ritter, C. (1822–59), *Die Erdkunde*, Berlin.

Roman, L. (1993), '"On the Ground" with Antiracist Pedagogy and Raymond Williams's Unfinished Project to Articulate a Socially Transformative Critical Realism', in D. Dworkin and L. Roman (eds), *Raymond Williams: Critical Perspectives*, Cambridge.

Said, E. (1979), *Orientalism*, New York.

Said, E. (with Raymond Williams) (1989), 'Appendix: Media, margins and modernity', in R. Williams, *The Politics of Modernism*, London.

Sandercock, L. (1998), *Towards Cosmopolis*, New York.

Sassen-Koob, S. (1988), *Global Cities*, Princeton, NJ.

Sauer, C. (1952), *Agricultural Origins and Dispersals*, New York.

Sayer, A. (1981), 'Defensible Values in Geography: Can Values be Science Free?', in D. Herbert and R. Johnston (eds), *Geography and the Urban Environment, Volume 4*, New York.

Sayer, A. (1989), 'Post-Fordism in Question', *International Journal of Urban and Regional Research*, 13, 666–95.

Schmidt, A. (1971), *The Concept of Nature in Marx*, London.

Schoenberger, E. (1988), 'From Fordism to Flexible Accumulation: Technology, Competitive Strategies and International Location', *Environment and Planning, Series D, Society and Space*, 6, 245–62.

Scott, A. (1988), *New Industrial Spaces: Flexible Production Organisation and Regional Development in North America and Western Europe*, London.

Seabrook, J. (1996), *In the Cities of the South: Scenes from a Developing World*, London.

Skelton, R. (1958), *Explorers' Maps: Chapters in the Cartographic Record of Geographical Discovery*, New York.

Smith, M. (1988), *City, State and Market*, Oxford.

Smith, M. and Keller, M. (1983), 'Managed Growth and the Politics of Uneven Development in New Orleans', in Fainstain et al. (eds), *Restructuring the City: the Political Economy of Urban Redevelopment*, New York.

Smith, N. (1987), 'Dangers of the Empirical Turn', *Antipode*, 19, 59–68.

Smith, N. (1990), *Uneven Development: Nature, Capital and the Production of Space*, Oxford.

Smith, N. (1992), 'Geography, Difference and the Politics of Scale', in J. Doherty, E. Graham and M. Malek (eds), *Postmodernism and the Social Sciences*, London.

Snedeker, G. (1993), 'Between Humanism and Social Theory: the Cultural Criticism of Raymond Williams', *Rethinking Marxism*, 6, 104–13.

Spoehr, A. (1956), 'Cultural Differences in the Interpretation of Natural Resources', in W. Thomas (ed.), *Man's role in Changing the Face of the Earth*, Chicago.

Stoddart, D. (ed.) (1981), *Geography, Ideology and Social Concern*, Oxford.

Stoker, R. (1986), 'Baltimore: the Self-evaluating City?' in C. Stone and H. Sanders (eds), *The Politics of Urban Development*, Lawrence, Kansas.

Strabo (1903–6), *The Geography of Strabo*, London.

Sweezy, P. (1942), *The Theory of Capitalist Development*, New York.

Swyngedouw, E. (1986), 'The Socio-spatial Implications of Innovations in

Industrial Organisation', Working paper no. 20, *Johns Hopkins European Center for Regional Planning and Research*, Lille.

Swyngedouw, E. (1989), 'The Heart of the Place: the Resurrection of Locality in an Age of Hyperspace', *Geografiska Annaler*, 71, Series B, 31–42.

Swyngedouw, E. (1992a), 'Territorial Organization and the Space/Technology Nexus', *Transactions, Institute of British Geographers*, New series, 17, 417–33.

Swyngedouw, E. (1992b), 'That Mammon Quest; "Glocalization", Interspatial Competition and the Monetary Order: the Construction of New Scales', in M. Dunford and G. Kafkalas (eds), *Cities and Regions in the New Europe*, London.

Szanton, P. (1986), *Baltimore 2000*, Baltimore.

Tarascio, V. (1966), *Pareto's Methodological Approach to Economics*, Chapel Hill, NC.

Thomas, B. (1973), *Migration and Economic Growth*, London.

Ulmen, G. (1978), *The Science of Society*, The Hague.

United Nations Development Program (1996), *Human Development Report, 1996*, New York.

United States Court of Appeals, District of Columbia (1954), *Brief of Appellee (USA versus Lattimore)*, No. 12,609, Washington, DC.

United States Senate, Subcommittee of the Committee of the Judiciary to Investigate the Administration of the Internal Security Act and other Internal Security Laws (1951–2), *The Institute of Pacific Relations*, 82nd Congress, 1st and 2nd sessions, Washington, DC.

Vogt, W. (1948), *The Road to Survival*, New York.

Walker, R. (1976), 'The Suburban Solution', *Doctoral Dissertation*, Department of Geography and Environmental Engineering, The Johns Hopkins University.

Weinberg, A. (1963), *Manifest Destiny*, Chicago.

Whitaker, J. (1975), *The Early Economic Writings of Alfred Marshall, 1867–90*, New York.

Williams, R. (1961), 'The Achievement of Brecht', *Critical Quarterly*, 3, 153–62.

Williams, R. (1973), *The Country and the City*, London.

Williams, R. (1977), *Marxism and Literature*, Oxford.

Williams, R. (1980), *Problems in Materialism and Culture*, London.

Williams, R. (1983), *Beyond 2000*, London.

Williams, R. [1960] (1988a), *Border Country*, London.

Williams, R. [1979] (1988b), *The Fight for Manod*, London.

Williams, R. [1964] (1988c), *Second Generation*, London.

Williams, R. (1989a), *Resources of Hope*, London.

Williams, R. (1989b), *The Politics of Modernism*, London.

Williams, R. [1985] (1989c), *Loyalties*, London.

Williams, R. [1989] (1990), *People of the Black Mountains: the Beginning*, London.

Williams, R. [1990] (1992), *People of the Black Mountains: the Eggs of the Eagle*, London.

Wilson, W. (1987), *The Truly Disadvantaged*, Chicago.

Wittgenstein, L. (1958), *Philosophical Investigations*, Oxford.

Wolf, L. (1976), 'National Economic Planning, a New Economic Policy for America', *Antipode*, 8 (no. 2), 64–74.

World Bank (1995), *World Development Report: Workers in an Integrating World*, New York.

Zeldin, T. (1994), *An Intimate History of Humanity*, New York.

Zinke, G. (1967), 'The Problem of Malthus: Must Progress End in Overpopulation?', *University of Colorado Studies, Series in Economics, no. 5*, Boulder, Colorado.

Index

SPACES OF CAPITAL

309, 327, 379–80, 390
university, 6, 8, 16–17, 27–30, 33, 37,
184, 217, 297, 325
urban governance, 188, 194, 205
urbanization, 5, 9–10, 12, 30, 32,
71–3, 75, 80–3, 111, 128–57,
188–207, 221, 245–7, 251, 313,
325–6, 345–68, 374, 389, 402–9;
see also cities
utopianism, 40–1, 112, 189, 198,
202–3, 233

Vogt, W., 63–4
von Thünen, J.-H., 10, 284, 288–95,
297–9, 310

wages, 14, 43–6, 48–9, 55–7, 65, 80,
84–5, 110, 144, 289–91, 297–8,
304–5, 314–15, 354, 390–2; *see
also* labor
war, 95, 120, 310–11, 342
Williams, R., 16–17, 160, 163–87,
193, 383
Wilson, E. O., 21–2
Wittfogel, K., 6, 94, 100–2, 107
women, 144, 238, 317, 385, 392
World Bank, 196, 209, 214, 217–18,
228, 233, 342, 380, 384, 386
world market, 242, 250, 253–66, 296,
369, 373–7, 380; *see also*
globalization